the Animal and the Environment

the Animal and the Environment

F. John Vernberg

Belle W. Baruch Professor
of Marine Ecology
Director, Belle W. Baruch
Coastal Research Institute
University of South Carolina

Winona B. Vernberg

Research Professor of Biology
Department of Biology
University of South Carolina

Holt Rinehart and Winston, Inc.
New York Chicago San Francisco Atlanta Dallas
Montreal Toronto London Sydney

Library of Congress Catalog Card Number: 74-113767
SBN: 03-079645-8
Printed in the United States of America
0123 22 987654321

To Marcie, Eric, and Amy

Acknowledgments

We wish to express our appreciation for the use of the following figures and tables.

Figures

2.4 J. Hedgpeth. 1957. Classification of marine environments. In J. W. Hedgpeth (ed.) *Treatise on Marine Ecology and Paleoecology,* Vol. 1, pp. 17–28. By permission of the Geological Society of America. **2.5** E. Odum. 1959. *Fundamentals of Ecology,* second edition. By permission of the W. B. Saunders Company. Also permission of the Duke University Press. **2.6** R. Johnson. 1965. Temperature variation in the infaunal environment of a sand flat. *Limnol. Oceanog.,* 10:114–120. **2.7** R. Johnson. 1967. Salinity of interstitial water in a sandy beach. *Limnol. Oceanog.,* 12:1–7. **2.8** H. Sanders *et al.,* 1965. Salinity and faunal distribution in the Pocasset River, Massachusetts. *Limnol. Oceanog.,* 10:(Suppl., Nov.)R216–R229. **2.9** V. Dellow. 1950. Intertidal ecology at Narrow Neck Reef, New Zealand (studies in intertidal zonation 3). *Pacific Science,* 4:355–374. By permission of the University of Hawaii Press. **2.11** H. Kurz and K. Wagner. 1957. Tidal marshes of the Gulf and Atlantic coasts of Northern Florida and Charlestown, South Carolina. *Florida State University Studies,* No. 24. By permission of the Florida State University Press. **2.13** Courtesy of Dr. P. J. DeCoursey. **2.15** S. English. 1962. A limnological study of Sandy Lake, Pennsylvania, with special reference to algal stratification and productivity. Unpublished Ph.D. Thesis, Univ. of Pittsburgh. **2.17** Courtesy of Dr. P. J. DeCoursey. **2.20** Courtesy of Dr. P. J. DeCoursey. **2.25** Courtesy of Dr. P. J. DeCoursey. **2.26** Courtesy of Dr. P. J. DeCoursey. **3.1** J. Brett. 1952. Temperature tolerance in young Pacific salmon, genus *Oncorhynchus. J. Fish Res. Bd. Canada,*. 9:265–323. By permission of the Fisheries Research Board of Canada. **3.2** P. Doudoroff. 1945. The resistance and acclimatization of marine fishes to temperature changes. II. Experiments with *Fundulus* and *Atherinops. Biol. Bull.,* 88: 194–206. **3.6** R. Ballinger and C. McKinney. 1966. Developmental temperature tolerance of certain anuran species. *J. Exptl. Zool.,* 161:21–28. By permission of the Wistar Press. **3.7** J. Brett. 1970. Temperature and fish. In: *Marine Biology,* Vol. 1, O. Kinne ed., Interscience Publishers, London, England. By permission of John Wiley and Sons, Inc. **3.8** J. Brett. 1944. Some lethal temperature relations of Algonquin Park fishes. *Univ. Toronto Biol. Ser.,* 52:1–49. By permission of the University of Toronto Press. **3.9** *Ibid.* **3.10** F. Fry and M. Gibson. 1953. Lethal temperature experiments with speckled trout x lake trout hybrids. *J. Heredity,* 44:56–57. By permission of the American Genetic Association. **3.11** W. Vernberg and F. Vernberg. 1963. Influence of parasitism on thermal resistance of the mud-flat snail *Nassarius obsoleta* Say. *Exp. Parasitol.,* 14:330–332. By permission of the Academic Press, Inc. **3.12** L.

Hutchins. 1947. The bases for temperature zonation in geographical distribution. *Ecol. Monographs,* 17:325–335. By permission of the Duke University Press. **3.13** Reprinted with permission from F. Vernberg *et al.* 1963. The influence of temperature and salinity on ciliary activity of excised cell tissue of molluscs from North Carolina. *Comp. Biochem. Physiol.,* 8:271–285. Pergamon Press. C. Schlieper *et al.,* 1960. Temperature and salinity relationships in marine bottom invertebrates. *Experientia,* 16:1–8. **3.14** O. Kinne. 1958. Adaptation to salinity variations—some facts and problems. In C. L. Prosser (ed.), *Physiological Adaptation,* pp. 92–106. Am. Physiol. Soc., Washington, D. C. Also by permission of Veb Gustav Fischer Verlag Jena (Zool. Jb. Physiol. Bd. 67, 5. 407–486). **3.15** M. Gordon. 1960. Anaerobiosis in marine sandy beaches. *Science,* 132:616–617. Copyright 1960 by the American Society for the Advancement of Science. By permission of *Science.* **3.16** N. Mattox. 1949. Effects of drying on certain marine snails from Puerto Rico. *Ecology,* 30:242–244. By permission of the Duke University Press. **3.17** D. McLeese. 1956. Effects of temperature, salinity and oxygen on the survival of the American Lobster. *J. Fish. Res. Bd. Canada,* 13:247–272. By permission of the Fisheries Research Board of Canada. **4.2** Weymouth *et al.* 1944. Total and tissue respiration in relation to body weight: a comparison of the kelp crab with other crustaceans and with mammals. *Physiol. Zool.,* 17:50–71. By permission of the University of Chicago Press. **4.3** *Ibid.* **4.4** F. J. Vernberg. 1956. Study of the oxygen consumption of excised tissues of certain marine decapod Crustacea in relation to habitat. *Physiol. Zool.,* 29:227–234. By permission of the University of Chicago Press. **4.5** F. J. Vernberg. 1959a. Studies on the physiological variation between tropical and temperate zone fiddler crabs of the genus *Uca.* Oxygen consumption of whole organisms. *Biol. Bull.,* 117:163–184. **4.8** T. H. Bullock. 1955. Compensation for temperature in the metabolism and activity of poikilotherms. *Biol. Rev.* 30:311–342. By permission of the Cambridge University Press. **4.9** H. Precht. 1958. Concepts of the temperature adaptation of unchanging reaction systems of cold-blooded animals. In C. L. Prosser (ed.), *Physiological Adaptation,* pp. 50–77. Am. Physiol. Soc., Washington, D. C. **4.10** C. L. Prosser. 1958. General summary: the nature of physiological adaptation. In C. L. Prosser (ed.), *Physiological Adaptation,* pp. 167–180. Am. Physiol. Soc., Washington, D. C. **4.11** F. J. Vernberg. 1959a. Studies on the physiological variation between tropical and temperate zone fiddler crabs of the genus *Uca.* Oxygen consumption of whole organisms. *Biol. Bull.,* 117:163–184. **4.12** F. J. Vernberg. 1959b. Studies on the physiological variation between tropical and temperate zone fiddler crabs of the genus *Uca.* IV. The influence of temperature acclimation on oxygen consumption of whole organisms. *Biol. Bull.,* 117:582–593. **4.13** Reprinted with permission from F. J. Vernberg and W. B. Vernberg. 1966. Studies on physiological variation between tropical and temperate zone fiddler crabs of the genus *Uca.* VII. Metabolic-temperature acclimation responses in southern hemisphere crabs. *Comp. Biochem. Physiol.,* 19:489–524. Pergamon Press. **4.14** F. J. Vernberg and J. Costlow. 1966. Studies on the physiological variation between tropical and temperate zone fiddler crabs of the genus *Uca.* IV. Oxygen consumption of larval and young crabs reared in the laboratory. *Physiol. Zool.,* 39:36–52. By permission of the University of Chicago Press. **4.15** *Ibid.* **4.16** W. B. Vernberg and F. J. Vernberg. 1968. Physiological diversity in metabolism in marine and terrestrial Crustacea. *Am. Zool.,* 8:449–458. **4.17** T. Haws and C. Goodnight. 1962. Some aspects of the hematology of two species of catfish in relation to their habitats. *Physiol. Zool.,* 35:8–17. By permission of the University of Chicago Press. **4.18** C. L. Prosser and F. Brown, Jr. 1961. *Comparative Animal Physiology.* Saunders, Philadelphia. **4.19** J. Redmond. 1955. The respiratory function of hemocyanin in Crustacea. J. Cell. Comp. Physiol., 46:209–247. By permission of the Wistar Press. **4.20** M. Straub. 1957. Weitere Untersuchungen zur Temperaturadaption der Sauerstoffbindung des Blutes von *Rana esculenta* L. In: Zeitschrift fur vergleichende Physiologie Bd. 39:S. 507–523. Berlin-Göttingen-Heidelberg: Springer, 1957. **4.21** J. Redmond. 1968. The respiratory function of hemocyanin. In F. Ghiretti (ed.), *Biochemistry and Physiology of Haemocyanins,* pp. 5–23. By permission of Academic Press, Inc. **5.2** A. P. M. Lockwood, 1964. *Animal Body Fluids and Their Regulation.* Harvard University Press, Cambridge, Mass. By permission of Heinemann Educational Books Ltd. **5.3** *Ibid.* **5.4** Reprinted with permission from W. Potts and G. Parry. 1964. *Osmotic and Ionic Regulation in Animals.* Pergamon Press. **5.5** Reproduced by permission of the National Research Council of Canada from the Canadian Journal of Zoology, *46,* pp. 427–437 (1968). **5.7** A. P. M. Lockwood, 1962. The osmoregulation of Crustacea. *Biol. Rev.,* 37:257–305. By permission of the Cambridge University Press. **5.8** P. Scholander *et al.* 1957. Super cooling and osmoregulation in Arctic fish. *J. Cell. Comp. Physiol.,* 49:5–24. By permission of the Wistar Press. **5.10** O. Kinne. 1958. Adaptation to salinity variations —some facts and problems. In C. L. Prosser (ed.), *Physiological Adaptation,* pp. 92–105. Am. Physiol. Soc., Washington, D. C. By permission of the American Physiological Society. Also by Veb Gustav Fischer Verlag Jena. **5.11** J. Needham. 1938. Contributions of chemical physiology to the problem of reversibility in evolution. *Biol. Rev.,* 13:225–251. By permission of the Cambridge University Press. **6.1** H. Cott. 1957. *Adaptive Coloration in Animals.* Methuen, London. By permission of Methuen and Co. Ltd. **6.2** *Ibid.* **6.3** *Ibid.* **6.4** *Ibid.* **6.6** R. Alexander. 1962. Evolutionary change in cricket acoustical communication. *Evolution,* 16:443–467. By permission of the Society for Study of Evolution. **6.7** M. Salmon. 1967. Coastal distribution, display and sound production by Florida fiddler crabs (genus *Uca*). *Anim. Behav.,* 15:449–459. **6.8** Reprinted with permission from H. Winn. 1964. The biological significance of fish sounds. In *Marine Bio-Acoustics, Proc. Symp., Bimini, Bahamas,* April 1963. Pergamon Press. **6.9** W. Ohsawa and H. Tsukudo. 1955. Extruding response of the periwinkle, *Nodilittorina granularis* Gray). *J. Inst. Polytech., Osaka City University, Ser.* D, 6:71–96. **6.10** P. Marler and W. Hamilton III. 1966. *Mechanisms of Animal Behavior.* Wiley, New York. By permission of John Wiley and Sons, Inc. **6.11** Schone. 1961. Complex behavior. *In* T. H. Watermann (ed.), *The Physiology of Crustacea.* Volume II. pp. 465–520. By permission of Academic Press. **6.12** B. Bayne. 1963. Responses of *Mytilus edulis* larvae to increases in hydrostatic pressure. *Nature,* 198:406–407. **7.1** R. Darnell. 1968. Animal nutrition in relation to secondary production. *Am. Zool.,* 8:83–93. By permission of the *American Zoologist.* **7.2** El-Wailly. 1966. Energy requirements for egg-laying and incubation in the zebra finch, *Taeniopygia castonotis. Condor,* 68:582–594. By permission of the Cooper Ornithological Society. **7.3** G. Comita. 1964. The energy budget of *Diaptomus siciloides Lilljeborg. Verh. Internat. Verein. Limnol.,* 15:646–653. By permission of E. Schweizerbartsche Verlagsbuchhandlung. **7.4** V. Dethier. 1961. Behavioral aspects of protein ingestion by the blowfly *Phormia regina* Meigen. *Biol. Bull.,* 121:456–470. **7.5** C. Holling. 1966. Response of invertebrate predators to prey density. *Memoirs Entomol. Soc. Canada,* 48:3–86. By permission of the Entomological Society of Canada. **7.6** K. Rao. 1953. Rate of water propulsion in *Mytilus californianus* as a function of latitude. *Biol. Bull.,* 104:171–181. **7.8** Reprinted with permission from J. Jennings. 1965. *Feeding, Digestion and Assimilation in Animals.* Pergamon, New York. Pergamon Press Ltd. **8.1** A. Farmanfarmaian *et al.* 1958. Annual reproductive cycles in four species of West Coast starfishes. *J. Exptl. Zool.,* 138:355–367. By permission of the Wistar Press. **8.2** A. Sastry. 1966a. Temperature effects in reproduction of the bay scallop, *Aequipecten irradians* Lamarck. *Biol. Bull.,* 130:118–134. **8.3** S. Marshall and A. Orr. 1964. Grazing by copepods in the sea. In D. J. Crisp (ed.), *Grazing in Terrestrial and Marine Environments,* pp. 227–238. By permission of Blackwell Scientific Publications, Ltd. **8.4** A. Sastry. 1966b. Variation in reproduction of latitudinally separated populations of two marine invertebrates. *Amer. Zool.*, 6:325–326. **8.5** *Ibid.* **8.6** P. Licht. 1966. Reproduction in lizards: influence of temperature on photoperiodism in testicular recrudescence. *Science,* 154:1168–1670. Copyright 1966 by the American Association for the Advancement of Science. By permission of *Science.* **8.7** A. Sastry. 1966a. Temperature effects in reproduction of the bay scallop, *Aequipecten irradians* Lamarck. *Biol. Bull.*, 130:118–134. **8.8** R. Ruibal. 1955. A study of altitudinal races in *Rana pipiens. Evolution,* 9:322–338. By permission of the Society for Study of Evolution. **8.9** R. Smith. 1964. On the early development of *Nereis diversicolor* in different salinities. *J. Morphol.*, 114:437–464. By permission of the Wistar Press. **9.1** H. Andrewartha and L. Birch. 1954. *The Distribution and Abundance of Animals.* By permission of the University of Chicago Press. **9.3** C. Istock. 1966. Distribution, coexistence, and competition of whirligig beetles. *Evolution,* 20:211–234. By permission of the Society for Study of Evolution. **9.4** M. Reeve. 1963. The filter-feeding of *Artemia.* I. In pure cultures of plant cells. *J. Exptl. Biol.*, 40: 195–205. By permission of the Company of Biologists Limited. **9.5** C. Holling. 1965. The functional response of predators to prey density and its role in mimicry and population regulation. *Mem. Entomol. Soc. Can.* No. 45. By permission of the Entomological Society of Canada. **9.6** *Ibid.* **9.7** C. Holling. 1959. The components of predation as revealed by a study of small mammal predation of the European sawfly. *Can.* Entomol., 91:293–320. By permission of the Entomological Society of Canada. **9.8** C. Holling. 1965. The functional response of predators to prey density and its role in mimicry and population regulation. *Mem. Entomol. Soc. Can.* No. 45. By permission of the Entomological Society of Canada. **10.2** A. Bartel and D. Davenport. 1956. A technique for the investigation of chemical responses in aquatic animals. *Brit. J. Anim. Behav.*, 4: 117–119. **10.3** G. Kearn. 1967. Experiments on host-finding and host-specificity in the monogenean skin parasite *Entobdella soleae. Parasitology,* 57:585–605. By permission of Cambridge University Press. **10.4** W. Rogers. 1966. Reversible inhibition of a receptor governing infection with some nematodes. *Exptl. Parasitol.,* 19:15–20. By permission of Academic Press, Inc. **10.5** Reprinted with permission from W. Vernberg and F. Vernberg. 1965. Interrelationships between parasites and their hosts. I. Comparative metabolic patterns of thermal acclimation of larval trematodes with that of their host. *Comp. Biochem. Physiol.*, 14:557–566. Pergamon Press. **10.6** A. Stiven. 1964. Experimental studies on the epidemiology of the host-parasite system, *Hydra* and *Hydramoeba hydroxena.* II. The components of a simple epidemic. *Ecol. Monographs,* 34: 119–142. By permission of the Duke University Press. **10.7** *Ibid.* **10.8** *Ibid.* **10.9** L. Roberts. 1961. The influence of population density on patterns and physiology of growth in *Hymenolepis deminuta* (Cestoda: Cyclophyllidea) in the definitive host. *Exptl. Parasitol.,* 11:332–371. By permission of Academic Press, Inc. **10.10** J. Holmes. 1961. Effects of concurrent infections of *Hymenolepis diminuta* (Cestoda) and *Moniliformis dubius* (Acanthocephala). I. General effects and comparison with crowding. *J. Parasitol.,* 47:209–216. By permission of the American Society of Parasitologists. **11.1** E. Kormondy. 1969. *Concepts of Ecology.* Prentice-Hall, Englewood Cliffs, N.J. Copyright 1969. By permission of Prentice-Hall, Inc. Also R. H. Whittaker. 1956. Vegetation of the Great Smoky Mountains. *Ecol. Monographs,* 26:1–80. By permission of the Duke University Press. **11.2** E. Odum. 1959. *Fundamentals of Ecology,* 2nd ed. Saunders, Philadelphia. By permission of W. B. Saunders Company. **11.6** G. Woodwell. 1967. Radiation and the patterns of nature. *Science,* 156:461–470. Copyright 1967 by the American Association for the Advancement of Science. By permission of *Science.* **11.7** E. Odum. 1968. Energy flow in ecosystems: a historical review *Am. Zool.*, 8:11–18. By permission of the *American Zoologist.* **11.8** J. Teal. 1957. Community metabolism in a temperate cold spring. *Ecol. Monographs,* 23:41–78. By permission of the Duke

University Press. **11.9** M. Engelmann. 1961. The role of soil arthropods in the energetics of an old field community. *Ecol. Monographs,* 31:221–238. By permission of the Duke University Press. **11.10** J. Phillipson. 1966. *Ecological Energetics.* Inst. Biol. Studies No. 1, St. Martin, New York. By permission of Edward Arnold (Publishers) Ltd. **11.11** E. Odum. 1959. *Fundamentals of Ecology,* 2nd ed. By permission of W. B. Saunders Company. **11.12** E. Kormondy. *Concepts of Ecology.* Prentice-Hall, Englewood Cliffs, N.J. Copyright 1969. By permission of Prentice-Hall, Inc. Also H. Odum. 1957. Trophic structure and productivity of Silver Springs, Florida. *Ecol. Monographs,* 27:55–112. By permission of the Duke University Press. **11.13** H. Sverdrup *et al.,* 1942. *The Oceans, Their Physics, Chemistry and General Biology.* Copyright 1942. By permission of Prentice-Hall, Inc. **12.2** From J. Mihursky. 1967. On possible constructive uses of thermal additions to estuaries. *BioScience,* 17:698–702. By permission of *BioScience.* **12.3** *Ibid.* **12.4** G. Woodwell *et al.* 1967. DDT residues in an east coast estuary: a case of biological concentration of a persistent insecticide. *Science,* 156: 821–824. Copyright 1967 by the American Association for the Advancement of Science. By permission of *Science.* **12.6** G. Woodwell. 1967. Radiation and the patterns of nature. *Science,* 156:461–470. Copyright 1967 by the American Association for the Advancement of Science. By permission of *Science.* **12.7** *Ibid.* **12.9** Reprinted with permission from T. Rice. 1965. The role of plants and animals in the cycling of radionuclides in the marine environment. *Health Physics,* 11:953–964. Pergamon Press. Quote on pp. 354 and 355 reprinted with permission from K. Mather, *The Earth Beneath Us.* 1964, Random House, Inc.

Tables

2.1 E. Glueckauf. 1951. The composition of atmospheric air. In T. F. Malone (ed.) *Compendium of Meterology,* pp. 3–9. Am. Meterol. Soc. Boston, Mass. **2.2** H. U. Sverdrup, M. W. Johnson and R. H. Fleming, THE OCEANS, Their Physics, Chemistry, and General Biology, copyright 1942, p. 191. By permission of Prentice-Hall, Inc. **2.3** G. Truesdale, *et al.* 1955. The solubility of oxygen in pure water and sea-water. *J. Appl. Chem.,* London, 5:53–63. By permission of the Society of Chemical Industry. **2.5** K. Bowden. Circulation and diffusion. In G. H. Lauff (ed.), *Estuaries,* pp. 15–36. A.A.A.S. Publ. No. 83, Washington, D.C. Copyright 1967 by American Association for the Advancement of Science. **3.1** H. Wells. 1961. The fauna of oyster beds, with special reference to the salinity factor. *Ecol. Monographs,* 31:239–266. By permission of the Duke University Press. **3.2** Reprinted with permission from F. Vernberg *et al.* 1963. The influence of temperature and salinity on ciliary activity of excised cell tissue of molluscs from North Carolina. *Comp. Biochem. Physiol.,* 8:271–285. Pergamon Press. **4.2** S. Basu. 1959. Active respiration of fish in relation to ambient concentrations of oxygen and carbon dioxide. *J. Fish. Res. Bd. Canada,* 16:175–212. By permission of the Fisheries Research Board of Canada. **4.4** F. J. Vernberg and W. B. Vernberg. 1964. Metabolic adaptation of animals from different latitudes. *Helg. wiss. Meeres.,* 9:476–487. **4.5** Reprinted with permission from F. J. Vernberg and W. B. Vernberg. 1966. Studies on physiological variation between tropical and temperate zone fiddler crabs of the genus *Uca.* VII. Metabolic-temperature acclimation responses in southern hemisphere crabs. *Comp. Biochem. Physiol.,* 19:489–524. Pergamon Press. **5.1** Reprinted with permission from W. Potts and G. Parry. 1964. *Osmotic and Ionic Regulation in Animals.* Pergamon Press. **5.2** R. Smith. 1963. A comparison of salt loss rate in three species of brackish-water nereid polychaetes. *Biol. Bull.,* 125:332–343. **7.1** G. Comita. 1964. The energy budget of *Diaptomus siciloides* Lilljeborg. *Verh. Internat. Verein. Limnol.,* 15:646–653. By permission of E. Schweizerbart'sche Verlagsbuchhandlung. **7.2** M.

Mullin and E. Brooks. 1967. Laboratory culture, growth rate, and feeding behavior of a planktonic marine copepod. *Limnol. Oceanog.*, 12:657–666. **7.3** C. Yonge. 1928. Feeding mechanisms in the invertebrates. *Biol. Rev.*, 3:21–76. By permission of the Cambridge University Press. **8.1** A. Giese *et al.* 1959. Organic productivity in the reproductive cycle of the purple sea urchin. *Biol. Bull.* 116:49–58. **11.2** L. Pomeroy. 1961. Isotopic and other techniques for measuring benthic primary production. In M. S. Doty (ed.), *Primary Productivity Measurement, Marine and Freshwater,* pp. 97–102 Proc. Conf. Primary Productivity Measurement. 10th Pacific Sci. Cong., Hawaii. **11.3** J. Ryther. 1959. Potential productivity of the sea. *Science,* 130:602–608. Copyright 1959 by the American Association for the Advancement of Science. By permission of *Science.* D. Westlake. 1965. In C. R. Goldman (ed.), *Primary Productivity in Aquatic Environments.* Reprinted by permission of the Regents of the University of California. **11.4** J. Ovington. 1962. Quantitative ecology and the woodland ecosystem concept. In J. B. Cragg, (ed.), *Advances in Ecological Research,* Vol. I, pp. 103–192. Academic Press, New York. By permission of Academic Press, Inc. **12.1** G. Woodwell *et al.* 1967. DDT residues in an east coast estuary: a case of biological concentration of a persistent insecticide. *Science,* 156:821–824. Copyright 1967 by the American Association for the Advancement of Science. By permission of *Science.* **12.2** C. F. Wurster and B. Wingate. 1968. DDT residues and declining reproduction in the Bermuda petrel. *Science,* 159:979–981. Copyright 1968 by the American Association for the Advancement of Science. By permission of *Science.*

Preface

Our goal in writing this book is to focus attention on the interaction between the animal and its environment. A number of compelling reasons stimulated our interests in this direction.

One of the important crises of present human society, and one that will become increasingly acute, is the problem of the haphazard modification of the environment. These modifications in some cases may have beneficial effects, but many are extremely damaging to the total ecology of the biosphere. Although some scientific evidence is available to permit predictions of what might result from man's unending desire to change the face of the earth, a broadly based body of information on the interaction of the environment and animals is not available. National scientific programs responsive to this fundamental need of society have begun, but they are in their infancy. Understanding of the complex interaction between organisms and their environment is not only an urgent practical necessity but also a prelude to planning that has fundamental biological significance. On the basis of our teaching and research experience, we felt that a textbook emphasizing the animal in its environment was needed to stimulate and encourage students to ask the important environmental questions and to seek the answers, even though it is necessary to cross traditional boundaries between scientific disciplines.

The book is for any biologist, but we had in mind the advanced undergraduate or beginning graduate student. Although the subject material is interdisciplinary, this textbook has relevance to many courses, including ecology, physiological ecology, environmental physiology, and comparative physiology. Obviously, we have not dealt comprehensively with the various current problems presented, but this was not our intention. Although we have included animals from terrestrial habitats, our selected field naturally leads us to draw many examples from the sea and other aquatic sources. But since examples simply illustrate a point, we suggest that nonmarine biologists broaden the background with examples typical of their specialties. Emphasis is directed toward the ability of the individual organism to meet the demands of its environment, but populations and communities of organisms also are discussed.

The many colleagues and friends who helped with constructive criticism are now formally and gratefully acknowledged. We are particularly indebted to two people: Dr. Sandra Woods who, among other things, checked reference source details and read and helpfully criticized the manuscript in its various stages, and Mrs. Vicky Macintyre, who not only typed the manuscript but also tried to keep our use of the English language within acceptable bounds.

A number of people critically read either all or part of the chapters, including Drs. Anna Ruth Brummett, Edward Cormondy, John Crites, Patricia De Coursey, C. P. Hickman, Jr., Conrad Istock, Dan Livingstone, Charles Lytle, Richard Miller, Michael Salmon, A. N. Sastry, Howard Schneiderman, Robert J. Shields, William Woods, Stanley Wecker, Dr. and Mrs. William Hamilton, Mr. Don Hoss, Mrs. Marilyn Barnes, Mr. William Kirby-Smith, and Mr. John Tonzetich.

Finally, we would like to thank our children, Marcia, Eric, and Amy, who have been quite understanding of our rather considerable preoccupation with the book.

Columbia, South Carolina F. J. V.
May 1970 W. B. V.

Contents

the Animal and the Environment

1

Introduction: The Organismic-Environmental Complex

If modern society is to manage and utilize existing natural resources intelligently, an understanding of the complex interaction between organisms and their environment is imperative. With the great advances in technology in recent years, man has developed unprecedented means for altering the environment. Because of his limited knowledge of the organismic-environmental complex, however, he has little scientific basis for predicting the results of his alterations of the earth's environment. As a consequence of man's potentially destructive capacity, coupled with the rapidly expanding human population, time and space for learning to manage the earth's resources are rapidly running out.

An organism lives in a state of dynamic equilibrium with the environment, and the environment is in a constant state of flux. The organism must be able to make internal adjustments to these external changes, not only to survive, but also to carry on the growth and reproduction that are necessary for continuity of both life of the individual and, perhaps more importantly, the species. Hence, if an organism is to succeed, an equilibrium between its total functional capabilities and the external environmental factors must be reached. If any one ambient factor or combination of factors changes significantly, the organism may survive by either establishing a new equilibrium or migrating to a more favorable environment. Otherwise the animal will tolerate the change only up to a point, and then die.

The vast diversity of habitats that animals and plants occupy is tribute to their ability to tolerate and compensate for environmental variation,

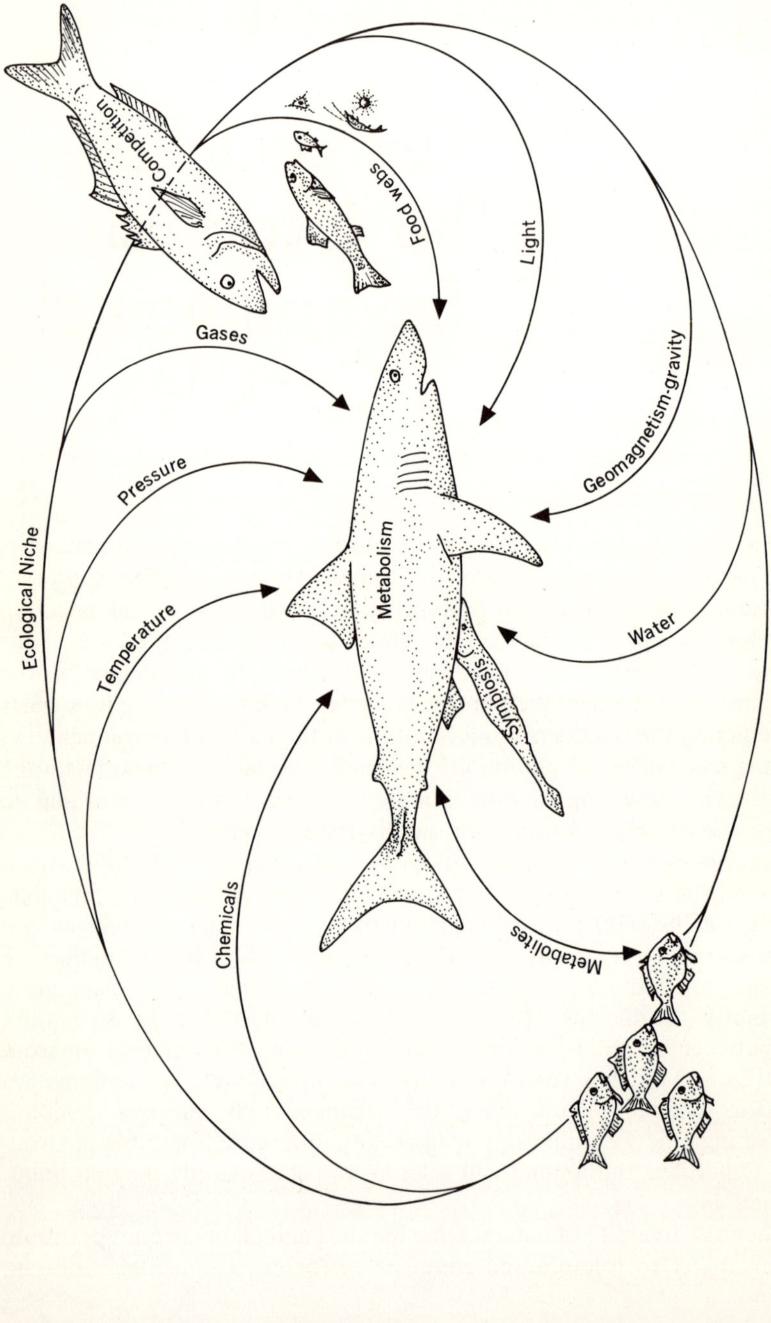

Figure 1-1. Interactions of an organism and its environment.

and one of the most exciting challenges to present-day biologists is to gain an understanding of the evolution of the various strategies that permit this success. The purpose of this book is to emphasize this complex interaction between organisms and their environment.

THE DYNAMIC EQUILIBRIUM BETWEEN AN ORGANISM AND ITS ENVIRONMENT

No organism can exist in the absence of an environment, but must live in equilibrium with an environment which consists of many variable factors. The environment commonly is subdivided into two major components: (1) abiotic factors, and (2) biotic factors. The abiotic component consists of various physical and chemical factors, such as oxygen, temperature, pressure, and pH, while the biotic component includes all interorganismic relationships, such as competition, predator-prey relationships, and symbiosis. In a few instances, a factor may not be easily classified as either biotic or abiotic; for example, a metabolite (chemical factor) produced by one organism may act as a prey attractant when released to the external environment (biotic interaction).

The interactions of an organism and its complex environment, represented in an oversimplified form in Fig. 1-1, can be analyzed by breaking down these interactions into the following levels of biological organization:

1. *Individual organism*—its functional and morphological adaptations to the external environment.
2. *Population*—characteristics and dynamics of interbreeding individuals.
3. *Animal association*—two or more species exhibiting various types of interdependence with each other.
4. *The ecosystem*—a number of populations considered as a naturally occurring aggregation and referred to as a community. An ecosystem in turn consists of the biotic community and its abiotic environment.

The arrangement of the subject material in this book follows a progression from the individual organism to the multiorganismic levels of biological organization: Chapters 3 to 8 are concerned with organismic responses; Chapters 8 and 9, with populations; Chapter 10, with animal associations; Chapter 11, with ecosystems; and Chapter 12, with the place of man in the environment.

The environment enveloping the intact organism is termed the *external environment,* in contrast to the combination of conditions existing within the organism, which is known as the *internal environment.* Although these two environments are separated from one another by a cellular membrane or a body covering, both show varying degrees of interdependence in different animals. One of the major evolutionary trends in the animal kingdom has been the development of mechanisms enabling an animal to maintain a relatively constant internal environment independent of fluctuations in the external environment. This attempt at constancy within narrow limits is known as *homeostasis* and any stratagem to achieve this steady state is termed a *homeostatic mechanism.* Not only do species differ widely in their homeostatic capabilities, but the stages in the life cycle of one species may also differ from one another. For example, the body temperature of fishes varies with that of the external environment, whereas most adult mammals are able to maintain a relatively constant body temperature. But the body temperature of many young mammals changes in response to ambient thermal change.

Within the organism the ultimate seat of control and regulation of life processes is found in the genetic material of the cell. Collectively, the sum total of all genes carried by an organism is its *genotype.* This genotype is constant throughout the diploid phases of the life cycle, except as modified by mutations. The expression of the genes in terms of function and/or structure at any one point in time is called the *phenotype.* Obviously one genotype will have a number of phenotypes, depending on environmental changes.

Great physiological diversity exists in the range of responses to environmental flux found in various animal groups. While some have metabolic machinery enabling them to cope with wide fluctuations in the intensity or concentration of a given factor, others are exceedingly sensitive to slight changes. For example, many temperate-zone fish can withstand temperatures from 0 to 30°C, but some Antarctic fish die if the temperature exceeds 5 or 6°C. The environment also influences the morphology of an organism, as exemplified by salamanders which develop larger gills when reared in oxygen-deficient water in comparison to other individuals of the same species grown in oxygen-rich waters.

Individuals that interbreed represent the population level of biotic organization. Various characteristics, such as the dynamics of population growth and the mechanism of control of population size, are basically different from those of the individual. These characteristics of populations are also influenced by environmental parameters.

Two or more species may interact with one another. The specific type of affinity between species may range from the intense metabolic dependence of a parasite upon a specific host to a more loosely dependent relationship, such as that between some marine fish and the shrimp that clean their gills. In other biotic relationships, animals may compete with other individuals of the same species or with other species for many factors, including environmental resources such as food or space, and for mates. Just as an individual is influenced by abiotic factors, so are biotic interactions.

Although not graphically represented in Fig. 1-1, the ecosystem, which also has an anatomy and physiology, is obviously influenced by environmental changes.

A biological system is affected not only by the abiotic environment but a degree of reciprocity exists. Under some circumstances the biotic system can alter the abiotic component, as was pointed out by Darwin, who noted that the burrowing of earthworms altered soil structure. This modification of the habitat could result in a change in the kinds of organisms found there, which in turn could lead to further environmental alteration.

Traditionally, the study of an organism and its environment has been relegated to that biological discipline called *ecology*. Various definitions of ecology have been proposed, but the one of Allee *et al.* (1949) clearly emphasizes the all-encompassing character of this field of study: "Ecology may be defined broadly as the science of the interrelation between living organisms and their environment, including both the physical and biotic environments, and emphasizing interspecies as well as intraspecies relations." Ecologists are in general agreement on the definition of ecology, but the scope and major research areas of the field have been and still are in a state of dynamic flux. Deevey (1964) maintains that ecology is not a field like geography, which has circumscribed boundaries, but rather is an "exceptionally multidisciplinary kind of biology." Because of the multifaceted nature of ecology, some biologists view it as a philosophical concept rather than as a functionally definable discipline.

Various terms have been used by ecologists to indicate more precisely their area of research. Some ecologists are concerned with the ecology of specific taxonomic groups—that is, molluscan ecologist, avian ecologist, or *Drosophila* ecologist. Others, such as a salt-marsh ecologist, plankton ecologist, or tree-hole ecologist, study the ecology of a specific habitat. Still others emphasize the theoretical approach to ecology, in which mathematical models are utilized for such topics as population dynamics, niche diversity, and ecosystem analysis. Another arbitrary method of subdividing the field has been to recognize various levels of biological organization, ranging from the molecular through the cellular, tissue-organ, organismal, population, to the community ecosystem level. At present, the terminology of systems analyses is used to denote the three most commonly studied levels of organization: (1) the individual, (2) the population, and (3) the community ecosystem.

Ecologic thought has not only influenced studies at various levels of biological organization, but it has also tended to permeate other disciplines of biology such as genetics, systematics, and physiology. Although these various interdisciplinary interrelationships will be discussed in this book, greater emphasis will be placed on the physiological response of organisms to environmental fluctuations. This phase of ecology is called *physiological ecology.*

In the detailed historical account of the development of ecology by Allee *et al.* (1949), the authors point out that from the time of the Renaissance, major contributions to ecological growth were made along four channels: development physiology, response physiology, relation of the species to the environment, and organic evolution. Thus, although ecology has been recognized as a distinct science only relatively recently, its roots are multidisciplinary and extend far into the scientific past, with natural history being one area that contributed much to present-day ecology.

In 1881 Semper summarized much of the ecological work prior to that time. He was well aware of the interrelationships between animals and the environment and stressed many problems which are significant today. In his book *Animal Life as Affected by the Natural Conditions of Existence,* he considered "the influence of inanimate surroundings," including such topics as drought, respiratory gases, forces of water (i.e., currents), and the effects of pressure of ocean depths. In addition, he developed a section on "the influence of living surroundings," in which he discussed coloring, mimicry, behavior, and various symbiotic relationships. Davenport published two volumes called *Experimental Morphology* in 1897 and 1899, in which he gave an excellent chronological history of ecological work up to the beginning of the twentieth century. He raised a number of questions which are still pertinent and which have not yet been fully resolved. For example, in discussing how some animals can survive in higher temperature environments while others cannot, he asked, "How has that acclimatization been effected?" (Davenport, Part 1, p. 252). In another section, again discussing differences in response to temperature, he stated: "The interpretation of the difference in sensitiveness to cold of the honey bee (*Apis mellifera*) and the red ant (*Formica rufa*) . . . must wait for further knowledge."

Since the Ecological Society of America was founded in 1916, there has been an ever-increasing number of ecological studies, with certain areas of investigation becoming prominent. These include: ecosystem analysis, with much emphasis on quantitative studies and mathematical models; population studies; behavioral ecology; and physiological ecology. Further, there has also been an increasing awareness of the application of ecological principles to the problems of human society.

This upsurge of research activity in the ecology of intact organisms has been accompanied by a great outpouring of research papers dealing with living and nonliving matter at the molecular level in relation to the environment. However, as Blair (1964) has pointed out, "insofar as the biologist is concerned they have not negated the fact that we live in a world in which protoplasm is organized into individuals, populations, and communities with challenging properties, mechanisms and principles to be explored at all levels." Although the methods and types of studies are constantly changing, the question at the center of all ecological problems is one which has long puzzled biologists: Why does an organism live where it does, and how does it adjust to the demands of its environment?

Consideration of all the parameters of the environmental complex in analyzing species-environment interaction is called the *holocoenotic* environmental approach. Although the goal of ecologists is to adhere to this approach, the practicality of observation and present-day experimentation allows the analysis of only one or a few factors at any one time. Admittedly, this limitation on the study of organisms is artificial and may lead to equivocal generalities, since variations in one factor may appreciably alter the expression of a second factor. However, a comparison of data from increasingly sophisticated laboratory and field studies will provide more meaningful bases for understanding why organisms are found where they are, and for predicting how they will respond in a modified environment.

Although the environment is complex, both in terms of the number of factors and in the interrelationships among the different factors, Mason and Langenbein (1957) suggested that a factor should meet the following criteria before being classified as an environmental phenomenon:

1. It must be operationally significant to an organism.
2. It must be directly effective at some time during the life of the organism.
3. It must be effective as to sequence, as ordered by the ontogeny (development) of the organism.

Thus they stress a functional concept of the environment which is "organism-directed, organism-timed, organism-ordered, and organism-spaced."

For each species a certain sector of the total range of expression of an abiotic environmental factor is compatible with life. At either end of this range or gradient there is a point beyond which an organism is unable to survive. The broad middle sector of a gradient has been given various names: the *zone* of *compatibility,* the *zone* of *tolerance,* the *biokinetic zone,* or the *zone* of *capacity adaption.* The region at either end of the zone of compatibility is called the *lethal zone* or the *zone* of *resistance*

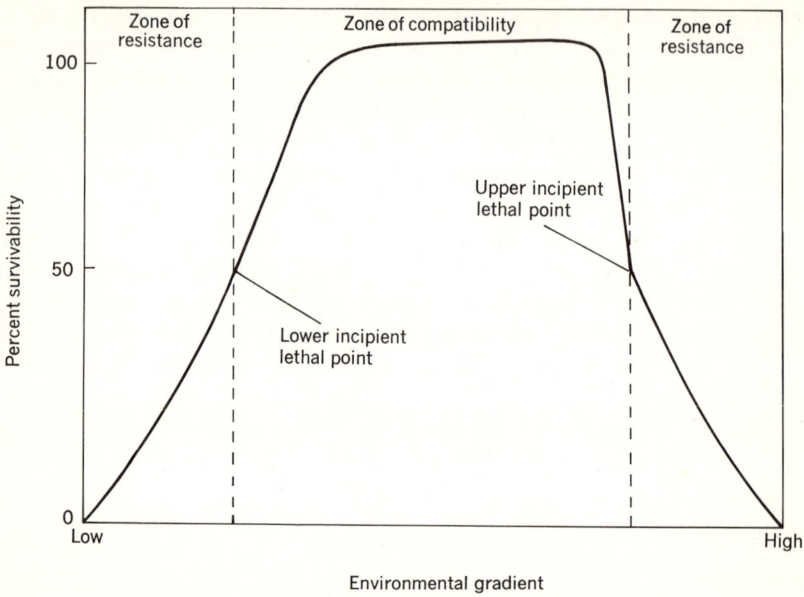

Figure 1-2. The zones of resistance and compatibility.

(Fig. 1-2). The point of transition between these two zones is referred to as the *upper* or *lower incipient lethal point.* These terms were stressed initially in studies of thermal lethal limits, but they can be generally applied when studying other environmental factors. Fry (1964) defines the upper and lower incipient thermal lethal point "as that temperature which kills a stated fraction of the population (usually 50 percent) within an indefinitely prolonged exposure." The definition of "indefinitely prolonged exposure" in terms of time varies with different research workers. This point of transition between the zones of resistance and compatibility is difficult to characterize, since factors such as sex, season, starvation, or interaction of environmental factors cause it to vary (see Chapter 3).

It is not, however, a simple matter quantitatively to circumscribe these limits for a species. Within a given interbreeding population, there is a considerable degree of genetic diversity. The sum total of all genetic material of a population is called a *gene pool,* whereas any one individual would typically have but one sample of genes from this pool. If a sufficient number of individuals of a population are studied, the physiological diversity of that population can be estimated.

Within the zone of compatibility, the organism survives exposure to a given external environmental complex, but the various physiological functions may be differentially influenced. A wide interspecific variation exists in the capacity adaptations of organisms, allowing them to function in a fluctuating environment that is not lethal. The description of these responses and the investigation of basic mechanisms involved are challenges that must be met in any concerted effort to understand the dynamic equilibrium between an organism and its environment. Chapters 4 to 8 deal with capacity adaptations.

ADAPTATION

Although the ecological dynamics of organisms are typically studied in terms of present environments, the ability of an organism to survive in a particular environmental complex is the result of evolutionary changes over an undetermined period of time. One of the basic characteristics of protoplasm is its ability to respond to environmental stimuli, for even the simplest unicellular animals respond to environmental stresses. Many responses of an organism to fluctuating ambient factors are adaptive in the sense that they enable the organism to survive. Obviously, those organisms that are unable to survive an environmental regime in a given region will perish and fail to perpetuate the species. Thus the environment acts as a selective force on a population so that there is a tendency for the evolution of mechanisms that are able to withstand environmental stresses. Organisms may invade new habitats, but in general they exist in and are adapted to environments where their ancestors lived.

Homeostatic mechanisms allow physiological maintenance of a more or less constant internal environment, even when the external environment changes. Some organisms, called *conformers,* have very limited homeostatic capabilities to withstand a particular abiotic factor, and their body functions appear to conform to changes in the environment; for example, if the temperature decreases, the rate of physiological processes decreases. Other organisms, known as *regulators,* maintain or regulate their body functions at a rate which tends to be independent of a widely fluctuating ambient environment (Fig. 1-3). Various intergrades exist between these two general groups of organisms. One species may be able to regulate one bodily function but not others, or the regulatory ability may depend on the stage of the life cycle. Therefore there is great diversity in the ability of various species to function under the different environmental conditions that characterize the various habitats available to organisms.

Adaptation, like so many terms in biology, has different meaning to different biologists. *Evolutionary adaptation* connotes a genetic basis for the adjustment among the parts of an organism and between the organism

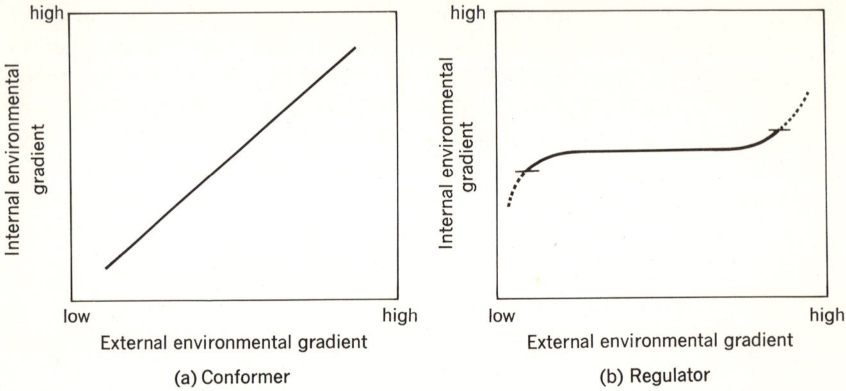

Figure 1-3. Two generalized response patterns of the organism's internal environment to a fluctuating external environment: (*a*) conformer: the internal environment conforms to that of the external environment; (*b*) regulator: the internal environment remains relatively unchanged when the external environment is varied.

and its environment (Allee *et al.,* 1949). Within the existing genome of an organism, adaptive morphological and/or physiological responses to environmental stress may occur and need not rely on mutations. Some biologists refer to these responses as *nongenetic* or *environmentally induced adaptation. Physiological adaptation,* as viewed by Prosser (1964), is "any property of an organism which favors survival in a specific environment, particularly a stressful one." He includes in this definition both environmentally and genetically determined responses. Two terms used in discussing differences in responses of organisms are environmentally induced (phenotypic) variation and genotypic variation. The responses of an individual organism to a certain environmental complex may be labile. In order to determine whether the observed responses of an animal are nongenetic or genetic, certain experimental approaches have been used: (1) the breeding and rearing of organisms under different environmental conditions that are controlled; and (2) acclimation studies in which responses of organisms exposed to a new environmental condition are observed. The response to an environmental complex may be altered if organisms experience different environmental preconditioning. For example, an organism kept in a water bath maintained at 25°C for three weeks has a relatively constant heart rate of 50 beats per minute when measured at 25°C. If, however, it is kept at 10°C for three weeks and then transferred to 25°C, its heart rate increases to a steady rate of 120 beats per minute. Each of these rates of heart beat is a phenotypic (nongenetic) expression dependent on previous environmental conditioning. However, one individual does not have unlimited capabilities of response, the limits of variation being imposed by the genetic make-up (genotype) of the organism. Only by mutations and selection can these limits possibly be extended. This phenomenon may be demonstrated as follows: If organisms having identical genetic make-up are placed in different environments, the result may be different bodily functions or forms. For example:

AA (genotype) in B_1 (environmental complex)
= AA (duplicated genes) + C_1 (bodily function or form)
AA (genotype) in B_2 (a second environment)
= AA + C_2 (new rate of bodily function or form)
AA (genotype) in B_3 (a third environment)
= AA + C_3(third rate of bodily function or form)
AA (genotype) in B_4 (a fourth environment) = the organism dies

Phenotypic variations may be observed at different points in the developmental history of an individual. Furthermore, transient environmental changes may influence the embryo or larval stages so that a phenotypic expression may be irreversible and become fixed, a process termed *canalization* (Waddington, 1957). Hence, within the genetic limitations of the species, the response of an organism to environmental change differs according to the time of exposure in the life cycle and to the magnitude of the change.

HABITAT AND NICHE

Every organism has special qualities that enable it to perform certain biological functions in a specific environment. Typically, an ecologist describes the place where an organism is found as its *habitat;* for example, certain fiddler crabs live in salt marshes. Elton (1949) described a habitat as an area that seems to possess a certain uniformity with respect to physiography, vegetation, or some other quality which the investigator decides is important. On the other hand, the term *niche* refers to the status or function of a species in a community; for example, phytoplankton are primary producers. The niche of an organism is dependent not only upon the organism's adaptive capabilities in response to a fluctuating abiotic environment, but also upon its dynamic encounter with individuals of other species. This concept has been expressed by Hutchinson (1965) in terms of a species occupying a multidimensional portion of the environment (hyperspace) which may vary with time. These boundaries are determined by the limits of "relevant variables in the life of a species of organism." In the absence of interspecific competition, the space occupied by the species would be called the *fundamental niche*. When other species are in competition, each has a *realized niche*, which is now smaller in volume. The niche for each species is distinctive and characteristic. It has been suggested that no two species can occupy the same niche indefinitely. One of the central and most difficult problems in this phase of ecology is the analysis of these "relevant variables" in terms of their functional significance. Unfortunately, detailed studies of all stages in the life history of an individual species and its interaction with closely related species are lacking, despite the vital importance of these studies to an understanding of niche diversity and the evolution of communities and ecosystems.

Within a general habitat that can be characterized on a regional basis by a range of environmental factors such as temperature, salinity, rainfall, and vegetation, different species may not occur uniformly throughout the habitat, but may be found only in a restricted area called a *microhabitat*, which has distinctive environmental conditions (*microenvironment*) unlike those of the general habitat. For example, the rock shore intertidal zone is a well-known generalized habitat. Certain sea anemones are found here, but only in a specific microhabitat, the undersurface or sides of rocks. Microhabitats may vary widely with respect to environmental conditions, although they may be separated from one another by only a short distance; for example, the surface of a rock in an intertidal zone may reach a temperature of 44°C while simultaneously the temperature may be only 25°C beneath the same rock. Typically, attached organisms living on the rock's surface have higher thermal-tolerance levels than those in the shaded area. The study of microenvironments and an interpretation of their biotic importance is a relatively new but important development in ecology. With recent technological developments such as miniaturization of equipment and automation, a greater understanding of the microenvironment can be expected.

SUMMARY

The organism is a dynamic metabolizing biological unit and it is greatly influenced by an enveloping and fluctuating environment. The successful organism is one that has evolved various adaptive strategies which enable it to meet the stresses of the external environment while maintaining a relatively stable internal environment. Although the adaptive capacity of each organism is determined by its genotype, organismic responses can be expressed in various ways when subjected to different environmental complexes. The study of this organismic-environment interrelationship is not only fundamental to many biological disciplines, especially ecology and physiology, but is also highly relevant to many of the man-induced problems facing modern society, such as pollution and effective use of natural resources.

References

Allee, W. C., A. E. Emerson, O. Park, T. Park, and K. P. Schmidt. 1949. *Principles of Animal Ecology*. Saunders, Philadelphia.

Blair, W. F. 1964. The case for ecology. *BioScience,* 14:17–19.

Davenport, C. B. 1897–1899. *Experimental Morphology,* Vols. I and II. Macmillan, London and New York.

Deevey, E. S., Jr. 1964. General and historical ecology. *BioScience,* 14:33–35.

Elton, C. 1949. Population interspersion: an essay on animal community patterns. *J. Ecol.,* 37:1–23.

Fry, F. E. J. 1964. Animals in aquatic environments: fishes. *In* D. B. Dill (ed.), *Handbook of Physiology, Section 4, Adaptation to the Environment,* pp. 715–728. Am. Physiol. Soc., Washington, D.C.

Hutchinson, G. E. 1965. *The Ecological Theater and the Evolutionary Play.* Yale University Press, New Haven, Conn.

Mason, H. L., and J. H. Langenbein. 1957. Language analysis and the concept environment. *Ecology,* 38:325–340.

Prosser, C. L. 1964. Perspectives of adaptation: theoretical aspects. *In* D. B. Dill (ed.), *Handbook of Physiology. Section 4: Adaptation to the Environment,* pp. 11–25. Am. Physiol. Soc., Washington, D.C.

Semper, K. G. 1881. *Animal Life as Affected by the Natural Conditions of Existence.* Appleton, New York.

Waddington, C. H. 1957. *The Strategy of the Genes.* G. Allen, London.

2

The Biosphere

DESCRIPTION OF MAJOR ABIOTIC FACTORS IN THE BIOSPHERE

Although the earth abounds with diverse forms of plants and animals inhabiting many distinctly different areas, life as we know it is restricted to a relatively thin surface zone around our planet. This restricted life-supporting zone is called the *biosphere.* It is not, however, entirely hospitable to life; in one region, the *parabiosphere,* severe environmental conditions permit inhabitation by some organisms only in the resistant stages of their life cycle. The parabiosphere includes broad areas such as the higher altitudes and a greater part of the cold polar regions, as well as numerous localized regions such as volcanos, geysers, and grossly polluted areas. The region of the biosphere in which active metabolic processes of organisms are possible is called the *eubiosphere.* Fear has been expressed that the parabiosphere is increasing in size at the expense of the eubiosphere, owing to the destructive activities of man.

Within the biosphere, many specific identifiable factors interact differentially but continuously to produce a number of distinctive habitats. The sum total of these complex interrelations is known as the *environment.* For convenience of discussion, the environment can be divided into two components: (1) biotic, that is, all interactions between biological entities; and (2) abiotic, that is, all physical and chemical factors in the environment.

Biological systems live and dynamically interact in this abiotic environmental complex. The influence of these factors on organisms will be discussed in detail in future chapters. The immediate goals of this chapter are to examine briefly the characteristics of the major abiotic factors that are important to animal life and to characterize the principal divisions of the biosphere. Not all abiotic factors have been studied with an equal degree of fervor. Principally because of the complexity of the environment and the difficulty of experimentally manipulating variables in the laboratory, research workers have emphasized those factors considered to be most important and most easily controlled.

Temperature

Temperature is a measure of the intensity of heat in terms of a standardized unit, and is commonly expressed as degrees on either the Fahrenheit or Celsius scale. Heat is a form of energy, and as such, is necessary for the very existence of life. Therefore it is important to measure not only the intensity of heat, but also the capacity for heat. The temperature at the earth's surface is governed by the brightness of the sun; the constancy of brightness has remained virtually unchanged for about 3 billion years. In fact, average terrestrial temperatures now probably do not differ radically from those at the earth's beginning (Schwarzschild, 1967). The total amount of heat entering the biosphere from the sun must balance the amount lost per unit time if temperatures are to remain unchanged, since the flow of geothermal heat from the interior of the earth is small by comparison and probably has been negligible for at least the past 500 million years. The estimate of this energy flow is referred to as the *heat budget.* Heat budgets have been projected for the total surface of the earth as well as for special environments.

Temperature also has a profound effect on other physical factors and thus influences the physiochemical properties of matter. As a body increases in temperature, it expands so that an enclosed mass of gas or vapor exerts a greater pressure. If heated enough, not only will a solid melt or a liquid change to vapor, but their elastic, electric, and magnetic properties will also be modified.

Water

Early in this century, Henderson (1913) listed in his stimulating book, *The Fitness of the Environment,* the following characteristics of water which are favorable to biological systems.

1. A tremendous quantity of water exists on earth in three forms: gaseous, liquid, and solid. Not only is 75 percent of the earth's surface covered by water, but the atmosphere also contains an abundance of aqueous vapor, and the polar region is ice-covered. Moreover, at least 60 percent of active protoplasm is water.
2. Water is an extremely inert body in relation to most other chemical substances.
3. Thermal properties

a. Heat capacity: Because of the high heat capacity of water, oceans and lakes tend to maintain a relatively constant temperature, and therefore the temperature of the biosphere is relatively stable. This property of water is functionally important to animal life. For example, a 165-lb man at rest produces sufficient heat to raise his body temperature more than 32°C, but if the heat capacity of his body were to correspond to that of many other substances, this amount of heat production would raise his temperature 100–150°C.

b. Latent heats of melting and evaporation: The latent heat of melting is the number of calories required to convert 1 gram of solid at the freezing point into 1 gram of liquid at the same temperature. This value is about 90 for water, and means that the amount of heat necessary to melt ice is the same as that required to raise the temperature of the resulting ice water to 80°C. The latent heat of evaporation is defined as the number of calories required to change 1 gram of liquid into vapor. For water, this value is 536; therefore as much heat is required to boil away 1 gram of water as to raise the temperature of 536 grams through 1°C. These properties of water are important not only because they moderate the temperature of the biosphere, but also because they play a basic role in the evaporation of water and its precipitation as rain and as dew in the hydrological (water) cycle (Fig. 2-1).

c. Thermal conductivity: Although water is a poor thermal conductor compared to metals, among the common liquids it is excellent: for example, the conductivity value for silver is 1.10; for water, 0.0125; for alcohol, 0.00048; and for benzene, 0.00033.

d. Expansion before freezing: Water is unique in that it expands when nearing the temperature of freezing. One example that illustrates the importance of this characteristic is the formation of ice in a lake or other body of water. If ice did not have buoyancy and were denser than water, the coldest water would sink to the bottom and freeze. This ice might never melt because the warmer water would stay at the surface. Instead, ice forms only at the surface and melts quickly with the onset of warm weather.

4. No other compound compares to water as a solvent. Many compounds are dissolved in sea water, and many minerals go into solution with water. In the internal environment, water carries compounds in solution in body fluids and protoplasm.

5. The extent and type of ionization in water far surpasses that in any other solvent. Ionization influences most electrical phenomena and many chemical phenomena of solutions.

6. Water has the greatest surface tension of all common liquids, except mercury. Surface tension not only helps to determine the height to

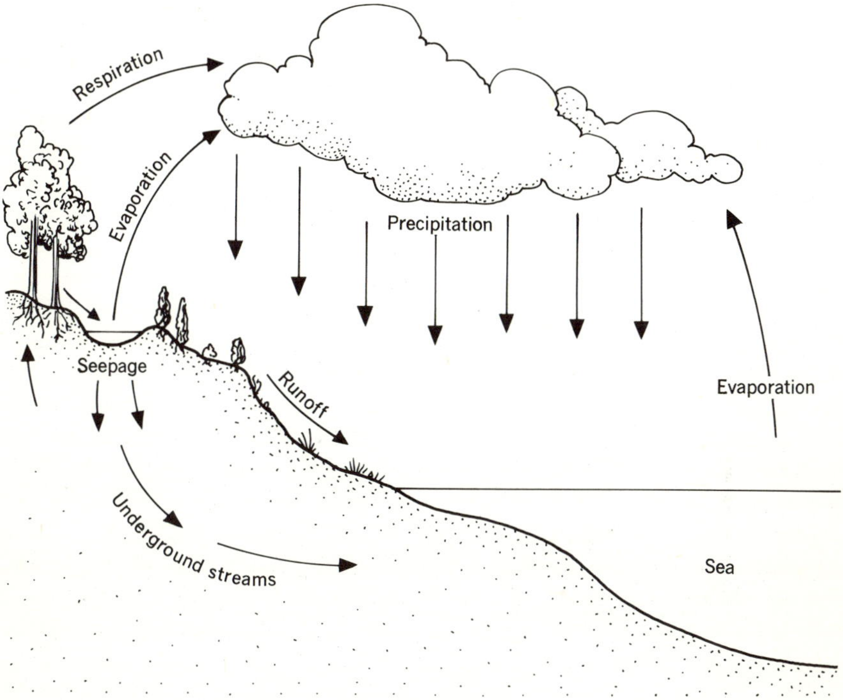

Figure 2-1. The hydrological cycle.

which a liquid will rise in a capillary system, but very importantly, it also influences absorptive phenomena.

Henderson (1913) concluded that no other known substance could be substituted for water as a basic abiotic environmental factor.

Gases

The earth is surrounded by a multilayered envelope of atmosphere (Fig. 2-2), and the layer in which man lives, the *troposphere,* has a relatively constant composition of gases (Table 2-1). In general, the composition of these gases in other media on earth, such as water or soil, is in equilibrium with the atmosphere. In special ecological habitats such as the anaerobic regions of deep lakes or sand-mud flats, however, the composition of gases is altered dramatically. Slight differences occur in the atmosphere at different latitudes. Regnault, in 1852, reported the highest value for oxygen at Paris and Berlin as 20.96 ± 0.001 percent, and the lowest value over the Arctic Ocean as 20.91 ± 0.001 percent. Reports of

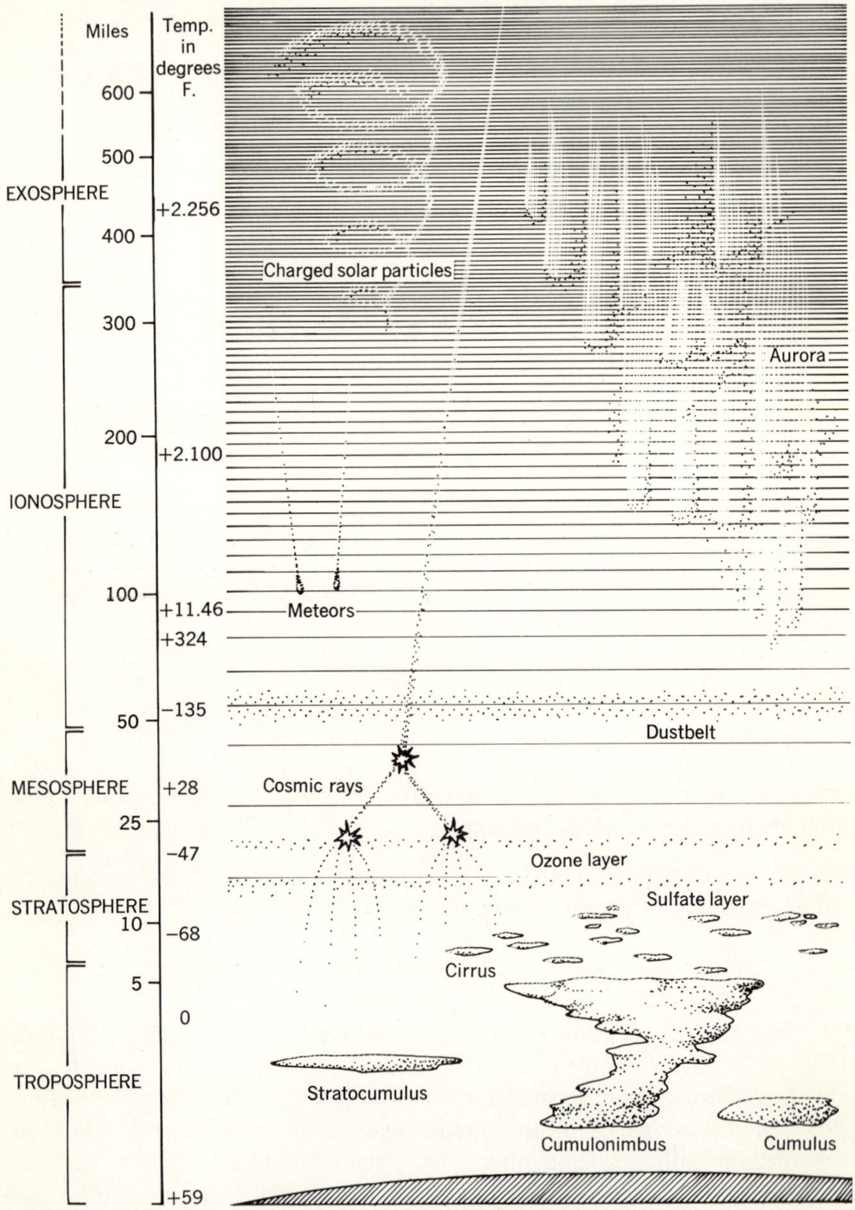

Figure 2-2. The earth's atmosphere.

Table 2-1. Composition of Atmospheric Air.

Component	*Dry Air*
Nitrogen	78.084 ± 0.004 % vol
Oxygen	20.946 ± 0.002 % vol
CO_2	0.033 ± 0.001 % vol
Argon	0.934 ± 0.001 % vol
Neon	18.18 ± 0.040 ppm
Helium	5.24 ± 0.004 ppm
Krypton	1.14 ± 0.010 ppm
CH_4	2.00 ppm

SOURCE: After Glueckauf, 1951.

variations in CO_2 in air over the sea apparently are correlated with temperature and with the affinity of sea water for CO_2 due to the presence of carbonate and bicarbonate. For example, polar air near Spitsbergen, Norway, sometimes contains as much as 20 parts per million (ppm) less CO_2 than does land and tropical air. As the result of increasing industrialization, the composition of the atmosphere may vary locally and be transitory, especially in certain metropolitan areas. For example, there has been an appreciable increase in atmospheric CO_2 during the past 50 years, and the increase is calculated to be equal to the amount produced by the combustion of fuels during this period (Glueckauf, 1951). Despite some variations, however, gases in the atmosphere are relatively constant.

In the primitive atmospheres of the earth, oxygen probably was either scarce or absent, and most gases that were present were chemically reducing. The most probable constituents were methane, ammonia, water, and perhaps cyanogen (Pittendrigh, 1967). The two gases of principal importance to most present-day organisms, however, are oxygen and CO_2, and they have evolved in an atmosphere that probably has remained relatively constant over millions of years.

The atmospheric composition is not always the most important factor to organisms; rather, it is the partial pressure of a gas, especially oxygen and CO_2, which influences the existence of life, as illustrated by altitudinal studies on organisms. The partial pressure of a gas is the product of the total barometric pressure times the concentration of gas in dry air. For every 18,000 ft (5500 meters) the barometric pressure is reduced by half, but the composition of air is essentially unchanged. To illustrate this effect on animals, the following critical altitudes have been identified for most humans: above 10,000 ft, oxygen must be added to inspired air; at 35,000 ft, 100 percent oxygen must be inhaled; and at 50,000 ft, little

Table 2-2. Coefficients of Saturation of CO_2 in Water at Different Temperatures and Salinities (ml/liter in equilibrium with 760 mm Hg).

Temperature, °C	*Salinity, ‰*		
	0	28.91	36.11
0	1715	1489	1438
12	1118	980	947
24	782	695	677

SOURCE: After Sverdrup *et al.*, 1942.

oxygen reaches the alveoli even if 100 percent oxygen is breathed (Hitchcock, 1964).

The solubility characteristics of oxygen and carbon dioxide in water are different, for carbon dioxide is about 200 times more soluble than oxygen. Both temperature and salinity have pronounced effects on the solubility of these two gases; with either increased temperature or increased salinity, there is a decrease in solubility in water (Tables 2-2 and 2-3).

Table 2-3. Solubility* of O_2 from a Wet Atmosphere at a Pressure of 760 mm Hg, in Milligrams Per Liter, at Temperatures of 0° to 40°C, and at Different Salinities.

Temperature, °C	*a*	*b*
0	14.16	0.08405
1	13.77	0.08153
2	13.40	0.07908
3	13.05	0.07671
4	12.70	0.07440
5	12.37	0.07218
6	12.06	0.07002
7	11.76	0.06795
8	11.47	0.06595
9	11.19	0.06402
10	10.92	0.06217
11	10.67	0.06039
12	10.43	0.05869
13	10.20	0.05706
14	9.98	0.05551
15	9.76	0.05404
16	9.56	0.05263
17	9.37	0.05130

Temperature, °C	*a*	*b*
18	9.18	0.05005
19	9.01	0.04887
20	8.84	0.04777
21	8.68	0.04674
22	8.53	0.04579
23	8.38	0.04491
24	8.25	0.04410
25	8.11	0.04338
26	7.99	0.04272
27	7.86	0.04214
28	7.75	0.04164
29	7.64	0.04121
30	7.53	0.04085
31	7.42	0.04057
32	7.32	0.04036
33	7.22	0.04023
34	7.13	0.04018
35	7.04	0.04020
36	6.94	0.04029
37	6.86	0.04046
38	6.76	0.04070
39	6.68	0.04102
40	6.59	0.04141

* Solubility determined by use of the following formula: $C's = a - bS$, where $C's$ is solubility (parts per million) of O_2 in water of salinity S (parts per thousand), with values for a and b as tabulated. Example: If temperature of water sample is 30°C and its salinity is 15 ‰, then $C's = 7.53 - (0.04082 \times 15.00) = 6.918$ ppm. (After Truesdale, Downing, and Lowden, 1955.)

If gases are to be functionally important to an organism, there must be a mechanism whereby gases can enter and leave the body. Apparently the passage of oxygen across a membrane is accomplished by diffusion. The rate of diffusion can be calculated by Fick's law, which is based on concentration difference, amount of surface thickness of the membrane, and time, and is expressed as the diffusion coefficient (Davson, 1959). For oxygen, the diffusion coefficient is expressed as the number of milliliters per minute passing through a 1-cm² area with a 1-cm thickness when the pressure difference equals 1 atm (760 mm Hg pressure). This value is not the same for different substances. The rate of diffusion in animal tissue is 300,000 to several million times slower than in air. Some representative values are: muscle, 0.000014; connective tissue, 0.000011; water, 0.000034; air, 11.0; chitin, 0.000013. Not only is the diffusion rate different in various tissues, but the behavior of each gas also differs. Carbon dioxide, for example, diffuses through water and animal tissues 25 times faster than oxygen. Since diffusion is a very slow process, convection is also highly important to oxygen transport within an organism.

The importance of gases to the physiologic-ecologic interrelationships of animals will be discussed in detail in Chapter 4.

Light and Radiation

The radiant energy from the sun is the basic requirement for the existence of life on the earth. This source of energy is of fundamental importance to the production of food by plants, and as mentioned previously, the heat budget of the world is dependent on solar radiation. Although we generally think only in terms of visible light, the sun emits other radiations of different wavelengths (Fig. 2-3). Biologists have been primarily interested in only three regions near the center of the electromagnetic spectrum: (1) the infrared, (2) the visible light, and (3) ultraviolet regions. The infrared wavelengths, which are the longest of these three, are not visible to the human eye; they contribute to the warmth of the earth at high altitudes in the terrestrial atmosphere. Visible light, which is only a small fraction of the radiation spectrum, supplies most of the warmth the earth

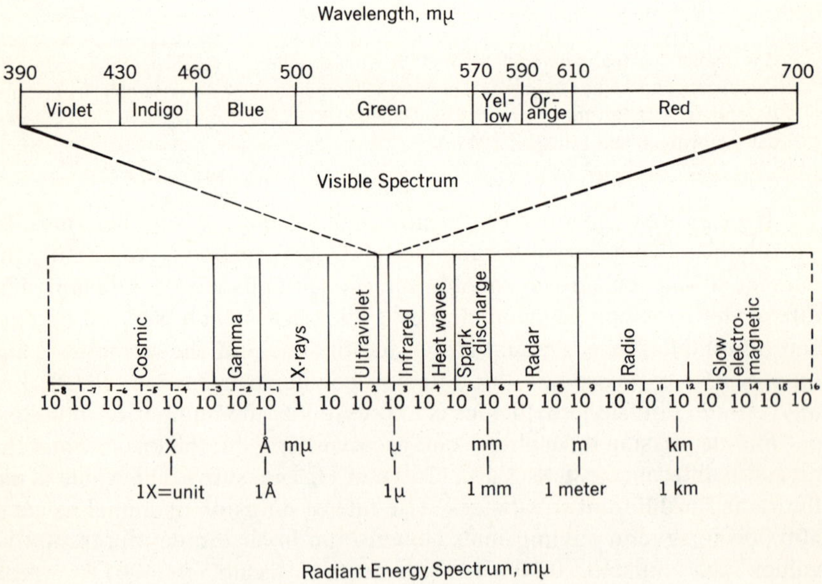

Figure 2-3. The spectrum of radiant energy.

receives from the sun and also provides the main source of energy. The wavelength of ultraviolet light from the sun is shorter than that of visible light, and it produces the upper levels of the earth known as the ionosphere. The ionosphere, which starts at an altitude of about 100 km and extends almost indefinitely into space, has a great effect on radio waves around the earth.

Two developments in recent years, nuclear technology and the invasion of outer space, have intensified interest in other forms of radiation as an environmental factor.

Pressure

Organisms living at sea level experience a pressure of about 15 psi, which is defined as 1 atm (760 mm of Hg). Pressure decreases as altitudes increase, and conversely, increases with depth so that with every 10 meters of submergence, the hydrostatic pressure increases by 1 atm. The average depth of the oceans has been estimated to be 3700 meters; at this depth the pressure is 370 atm (28,200 mm Hg); in contrast, at an altitude of 500 meters, where the highest permanent human settlement occurs, the pressure is only 0.54 atm (413 mm Hg). In addition to changes in barometric pressure with altitude or depth, the pressure at one location may change on a temporal basis; for example, at sea level a maximum of 1.06 atm (805 mm Hg) has been recorded. Pressure influences solubility, ionic dissociation, and surface tension, and water is slightly compressible with increased pressures. Since incompressible bodies are influenced very little by increased pressure, they will sink in the sea, and apparently this factor is important in food settling to the deep-water benthic fauna. Although gas-filled spaces and cavities are mechanically deformed by pressure changes, pressure acting alone does not significantly alter the viscosity of sea water. Increased pressure, however, can influence the velocity of certain chemical reactions, but not all; reaction rates that are slow at 1 atm in the absence of a catalyst may be greatly increased at higher pressures (Fawcett and Gibson, 1934).

Geomagnetism and Gravity

The earth may be described as a magnetized sphere in which the magnetic axis is inclined by about 11 deg relative to the geographical axis, although the angle between the geographical axis and magnetic axis of the earth changes slowly but constantly. At the present time the magnetic needle, while not pointing exactly to the north, deviates more or less from this position at different places on the earth (Elsasser, 1967). Considerable evidence from geological studies of rocks, however, indicates that the magnetic field may be and has been oriented toward the South Pole as well. Rocks may be formed either by cooling of fluid volcanic magma or by deposition of sediments in water, and in either process the direction of the earth's magnetic field is locked into the rock at the time of its formation. From such data it is known that the earth's magnetic field has been reversed nine times in the past 3.6 million years (Cox *et al.,* 1967). If the earth's magnetic field disappears or if it becomes greatly altered during a reversal of polarity, the earth loses some of its magnetic shielding against cosmic rays, and the consequent increases in radiation dosages could conceivably result in increased mutation rates. Using paleontological data as evidence, Uffen (1963) suggested that during these periods of reversal, the rates of evolution were exceptionally high.

In contrast to the variability of the earth's magnetic field, the direction of gravitational force toward the center of the earth is invariable. The pull of gravity varies, however, with the distance from the earth's center. Since gravitational pull is so constant and prevalent, experimentation is difficult, and therefore the influence of gravity on biological material is relatively unknown.

CHARACTERISTICS OF MAJOR DIVISIONS OF THE BIOSPHERE

The major divisions of the biosphere will be discussed here in terms of their distinctive environmental characteristics. The evolutionary strategies used by animals to meet the unique stresses encountered in these various habitats will be the subject of subsequent chapters.

The Ocean Environment

The earth is unique among planets of our solar system in that about 70 percent of its surface is covered by water. Furthermore, the quantity of sea water is so great that if the earth's surface were flattened, it would be covered with water to a depth of 12,000 ft. Although mapping of the sea floor is considerably unrefined in comparison to that of the "dry land," available data show irregular topography marked by mountain ridges, canyons, plains, and trenches. The mapping that has been done indicates that areas of great relief on the sea floor are as striking as those on land: the highest elevation on land, Mount Everest, is 29,028 ft high, and one of the deepest abysses in the ocean, the Mariana Trench in the Philippine Ocean, is 35,800 ft deep. In addition to being a unique and distinct ecologic subdivision, the sea exerts a strategic influence on the ecology of the total biosphere, since the ultimate source of water essential to all forms of life is the sea (see hydrological cycle, Fig. 2-1). Although relatively little is known of the actual cause-and-effect relationships, the physical properties of sea water allow it to moderate the climate, and fluctuation in oceanic currents can alter the weather of adjacent land masses. Thus climate is profoundly influenced by the sea.

The marine environment is not homogeneous but varies geologically, physically, chemically, and biologically. An indication of the fitness of the marine environment for biotic activity is that the greatest diversity of major animal types exists in the ocean. Furthermore, all life apparently originated in the sea, and the amount of living material (biomass) in the ocean far exceeds that on land. Unlike terrestrial or fresh-water systems, which may be isolated from one another, all the great oceans of the world are interconnected and the water flows continuously in a state of dynamic interchange. Theoretically, a small sample of sea water off the coast of South Carolina could, in time, reach any other part of the marine environment.

Sea water consists of a solution of inorganic salts, some organic substances, and dissolved atmospheric gases. The amount of salt in a sample of sea water is expressed as salinity and is written as parts per thousand or ‰ (grams per kilogram). Salinity of the open ocean waters is approximately 35 ‰, meaning 35 grams of dissolved solid material are contained in 1 kg of water. Salinity varies in different marine habitats, and the greatest fluctuations occur in the inshore waters, particularly in estuaries. Some intertidal ponds associated with salt flats in the tropics may have a salinity of 155 ‰ whereas the salinity of estuarine waters after extremely heavy rains may drop to 0.5 ‰, the level of fresh water. To help alleviate the confusion existing in scientific literature over the classification of marine waters according to salinity, the scheme in Table 2-4, known as

Table 2-4. Classification of Marine Waters (The Venice System).

Zone	*Salinity* (‰)
Hyperhaline	$> \pm 40$
Euhaline	$\pm 40 - \pm 30$
Mixohaline	$(\pm 40)\ \pm 30 - \pm\ 0.5$
Mixoeuhaline	$> \pm 30$ but $<$ adjacent euhaline sea
(Mixo-)polyhaline	$\pm 30 - \pm 18$
(Mixo-)mesohaline	$\pm 18 - \pm\ 5$
(Mixo-)oligohaline	$\pm\ 5 - \pm\ 0.5$
Limnetic (fresh water)	$< \pm\ 0.5$

the Venice system, was adopted by the 1958 Symposium on the Classification of Brackish Waters. Although the total amount of salt in sea water changes with dilution, the ratio of the major constituents to each other remains unchanged. Salinity influences other physical factors that have ecological significance. The saturation value of oxygen is lower in sea water than in fresh water (Table 2-3); the maximum density of sea water (about 36 ‰) is at −3.8°C but at 4°C for fresh water; and, the freezing point of sea water is lower than that of fresh water.

In the ocean, as in bodies of fresh water, more dense water will sink to the deeper regions, and water of similar densities will then tend to spread horizontally. Because of the differences in water densities, vertical gradients become more pronounced than lateral ones. The vertical mixing is essential if the nutrients required for primary productivity at the surface of the ocean are to be replaced.

Temperature fluctuations in the ocean are less extreme than on land. The temperature at the surface of the ocean is rarely higher than 27°C, and bottom-water temperatures range from 4 to −1.5°C. In contrast, surface land temperatures range from 71.5°C in Arizona to subzero levels in polar regions. Also, the rate of temperature change is slower in ocean waters than on land: at the surface of the sea a diurnal change of 0.3°C may be found, whereas on land at a depth of 0.4 cm, an extreme daily range of 56.6°C has been recorded. The ocean, which serves as an immense reservoir of heat, receives its input of heat energy mainly from the sun, and most of the radiation falls upon the surface of the ocean. In turn, water vapor is given off to the atmosphere, and the ocean contributes to the reservoir of heat in the atmosphere.

Based on various criteria, a classification of marine environments presented by Hedgpeth (1957) will furnish the basis for our discussion of the physiological ecology of marine animals (Fig. 2-4).

Intertidal Zone

Beginning from shore and progressing seaward, a number of distinctive habitat types may be observed. One of these types is the intertidal zone, which encompasses the area between the high-tide mark and the low-tide mark. This habitat has received much attention because it is easy to observe and reach, and does not require elaborate oceanographic equipment. In contrast to the more stable, unchanging, deep-sea habitat, however, it is an extremely complex environment to analyze because it is alternately exposed to aquatic and aerial factors that fluctuate on a daily and seasonal basis. Although tidal amplitude depends on such factors as the phase of the moon, the amount of tidal change varies greatly

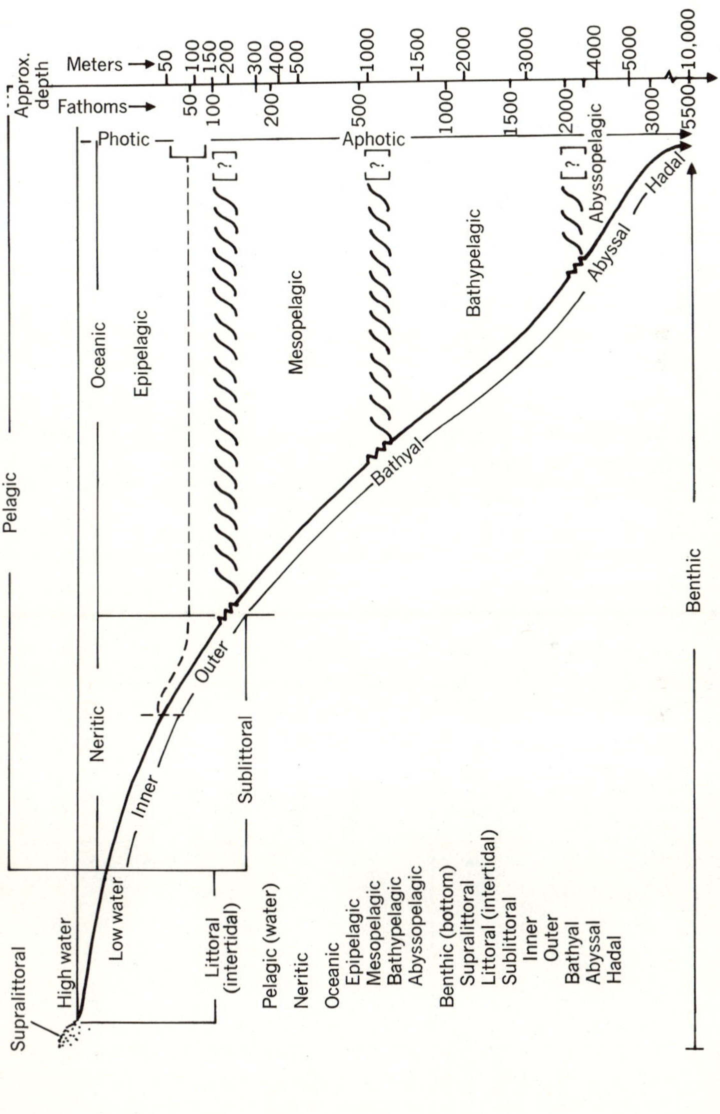

Figure 2-4. Classification of marine environments. (From Hedgpeth, 1957.)

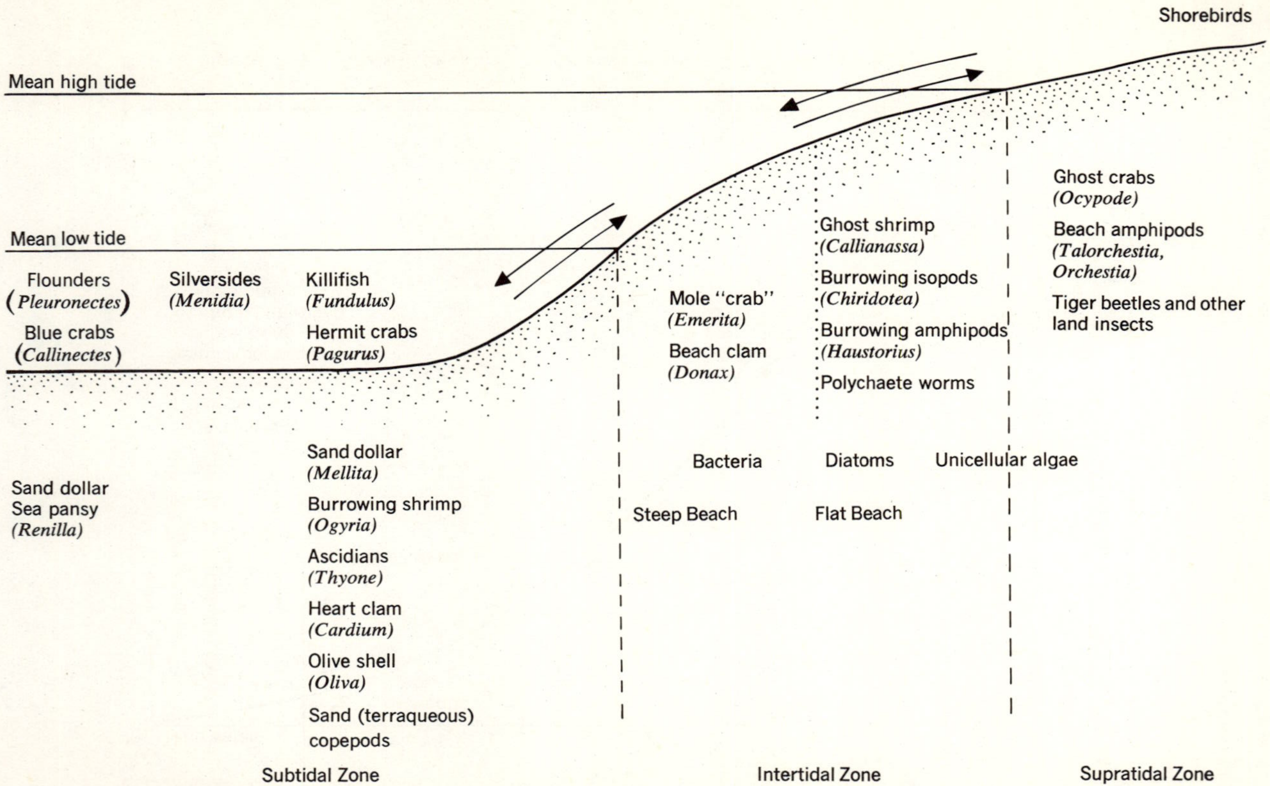

Figure 2-5. Tidal zones of a sandy beach. (From Odum, 1959, based on data from Pearse *et al.*, 1942.)

throughout the world. For example, in the Bay of Fundy the change may be as great as 15.4 meters, whereas in the Caribbean the change may be a matter of centimeters.

The substratum, which is very important to animals for attachment and shelter, forms the basis for the classification of intertidal-zone habitats. Thus we speak of sand beaches, mud flats, or rocky shores. Within each of these intertidal-zone habitats there is conspicuous zonation of organisms from the high-tide to the low-tide mark. Some animals (Fig. 2-5), such as the ghost crab, will be found primarily above or near the high-tide mark, while others, such as the box crab, are found only near or below the low-tide mark (Pearse *et al.,* 1942). These differences in zonation reflect the physiological and competitive capacity of organisms to respond to both abiotic and biotic factors in the environment.

One of the best-known intertidal zones is the sand beach. These beaches have fairly uniform topographical gradations, with few sharp irregularities such as rock jetties and with relatively little slope. Two general types have been described: (1) the open beach, which is unprotected from unrelenting wind and wave action; and (2) sand beaches in harbors and estuaries, which generally are protected from the pounding action of the surf. Unprotected beaches, particularly, are in a constant state of flux, and they may vary in extent from tide to tide, season to season, and from storm to storm. The biotic inhabitants of sand beaches must be able to adapt to a variable and harsh environment, for they are subject to desiccation, insolation, and wide temperature ranges.

The size of the sand grains present, which has a marked influence on the physical characteristics of the beach, may vary not only from beach to beach, but also at different regions of one beach. They therefore play an important role in the distribution of animals. Although the temperature regime of a beach depends on its geographical location, recent data on temperature variation on a beach in Tomales Bay, California, will give some idea of the daily, seasonal, and annual temperature changes at different depths. Animals living in the upper 1-cm stratum are exposed to a daily temperature range as great as three times that experienced by subtidal individuals of the same species. In contrast, animals at a depth of 10 cm or more are subjected to temperatures similar to that of the subtidal environment. The annual range and seasonal temperature changes at a 10-cm depth are similar to those in the open water of Tomales Bay (Johnson, 1965). This information is graphically represented in Fig. 2-6. Similar measurements of salinity of this same sandy beach (Fig. 2-7) were made (Johnson, 1967). During the summer months there is little variation in salinity associated with the slope and elevation of the beach, except in a few cases where evaporation produces change. In the winter, salinity

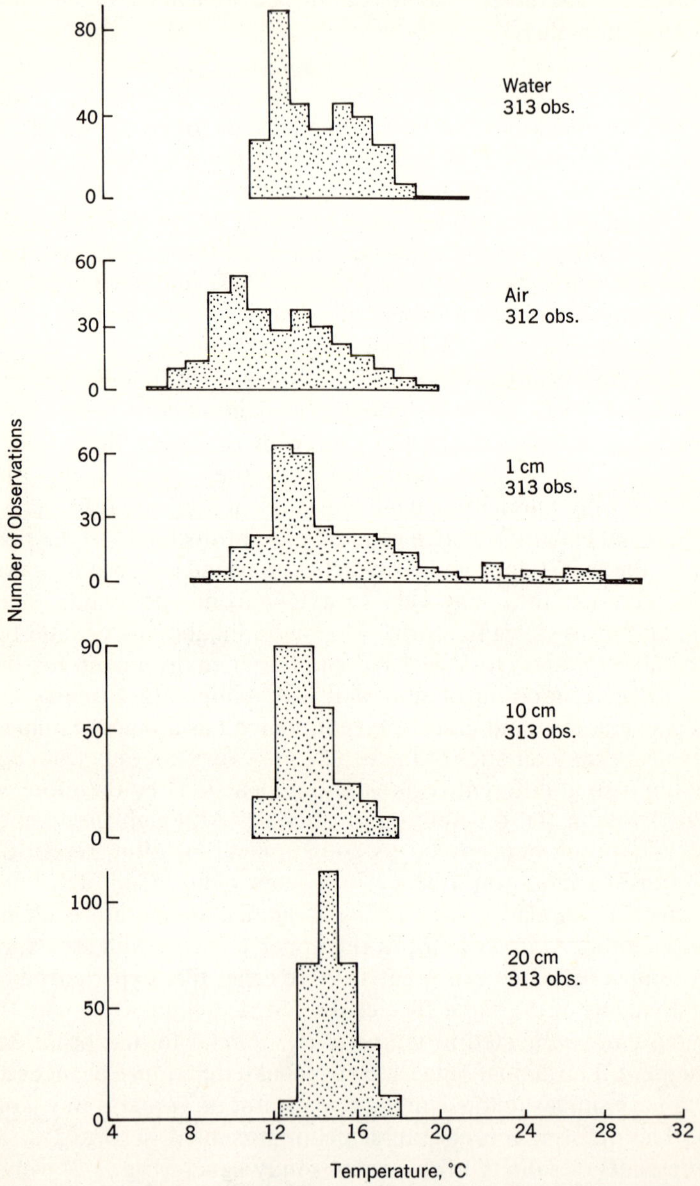

Figure 2-6. Temperature frequency distributions for air, water, and sediment at different depths. At a depth of 10 cm, temperatures are similar to those in open water. (From Johnson, 1965.)

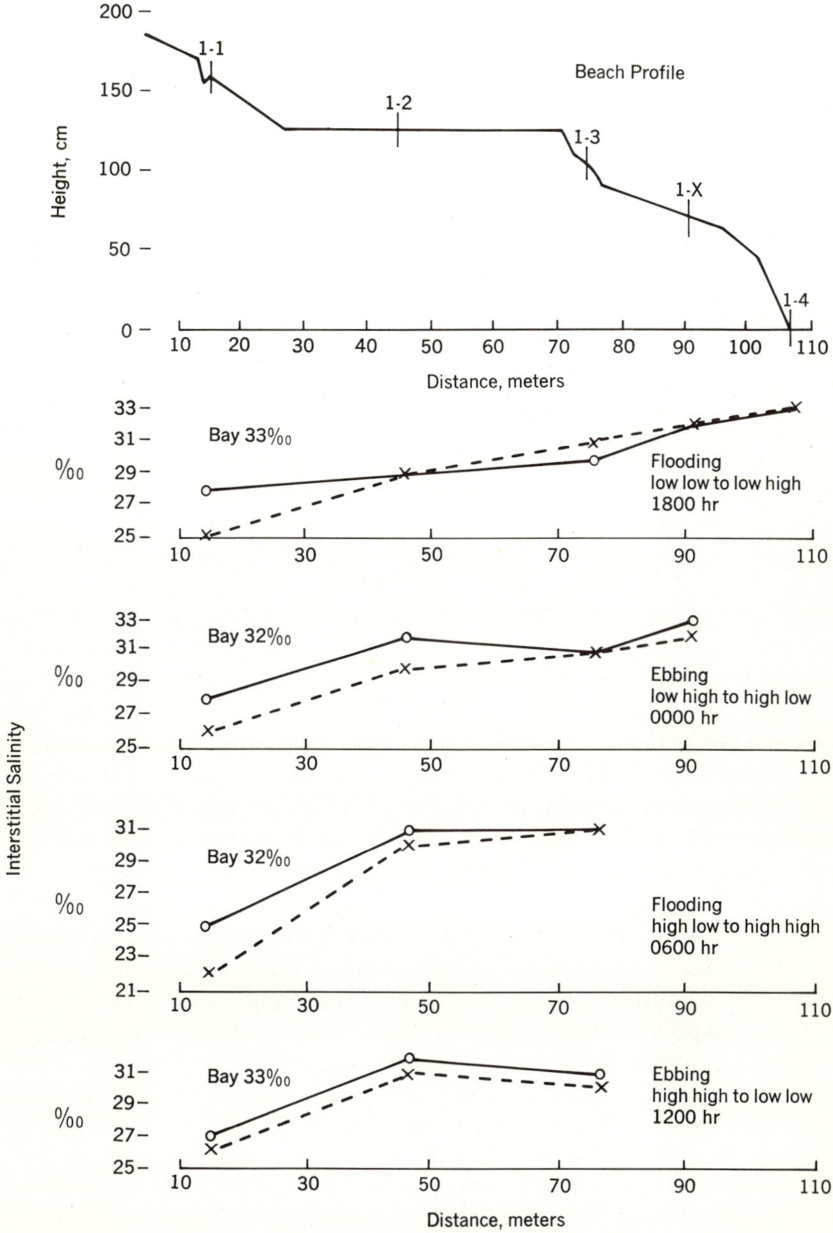

Figure 2-7. Beach profile and variation in interstitial salinity at 10 and 20 cm. (From Johnson, 1967.)

differences in the water are due to the increased flow of fresh water. High on the beach, the salinity at the 10- and 20-cm depths remains low throughout the tidal cycle. The salinity at the shallower depth of 10 cm remains higher than that at 20 cm. Below an elevation of 1 meter, however, the interstitial salinity varies less than in adjacent open waters of the bay. As a result, the environment of the infauna (animals that live in the sandy or muddy surface layers of the sea bottom) of the low intertidal zone is more stable than that of the epifauna (animals that live at the surface of aquatic bottoms and upon rocks, stones, pilings, and shells) in shallow water. The variations at the high perimeter of this intertidal zone must present a significant obstacle in preventing marine organisms from spreading into higher parts of the beach (Johnson, 1967).

Data from studies of the variation in oxygen content in sand beaches indicate that, in general, the vertical gradient in regions not agitated by waves has a steady decrease of oxygen availability from the surface downward, with some oxygen occurring at depths of 10 cm. Horizontally, differences in oxygen content gradate from land to sea, with the greatest concentration near the water's edge; however, during periods of calm, low values were also noted here. On protected beaches, a very irregular pattern of oxygen availability is found. Seasonal differences have been reported by Jansson (1967), who noted that the availability of oxygen is strongly governed by physical and chemical properties of the sand, including such features as porosity, permeability, and percentage air volume. Other factors that affect the oxygen concentration are the flow of interstitial water in the sand and the wave action.

In the intertidal mud flats, another dominant intertidal-zone area, gradations from pure sand to pure mud result in transitional intertidal zones that have varying degrees of sand composition. Typically, mud cannot exist where there is significant wave action, and this type of habitat generally occurs in protected waters such as in estuaries. Since mud is of fine texture and has a high colloidal content, there is little interstitial water, but there is likely to be a high organic content, indicating a rich food supply for the organisms in this area. Owing to this high food content and poor circulation, however, anaerobic conditions are not uncommon on the mud flats. Usually, a black layer of mud a short distance below the surface indicates low oxygen and high hydrogen sulfide content. Since the substratum is soft, most organisms living here are burrowing animals, and a few attached organisms may be present, except in those areas of a mud flat where shell or other hard materials are mixed in. Those organisms that require a hard surface on which to crawl are absent. Figure 2.8 shows that because of the poor circulation of interstitial water, physical features of the mud change little during tidal cycling and during exposure to air (Sanders *et al.*, 1965).

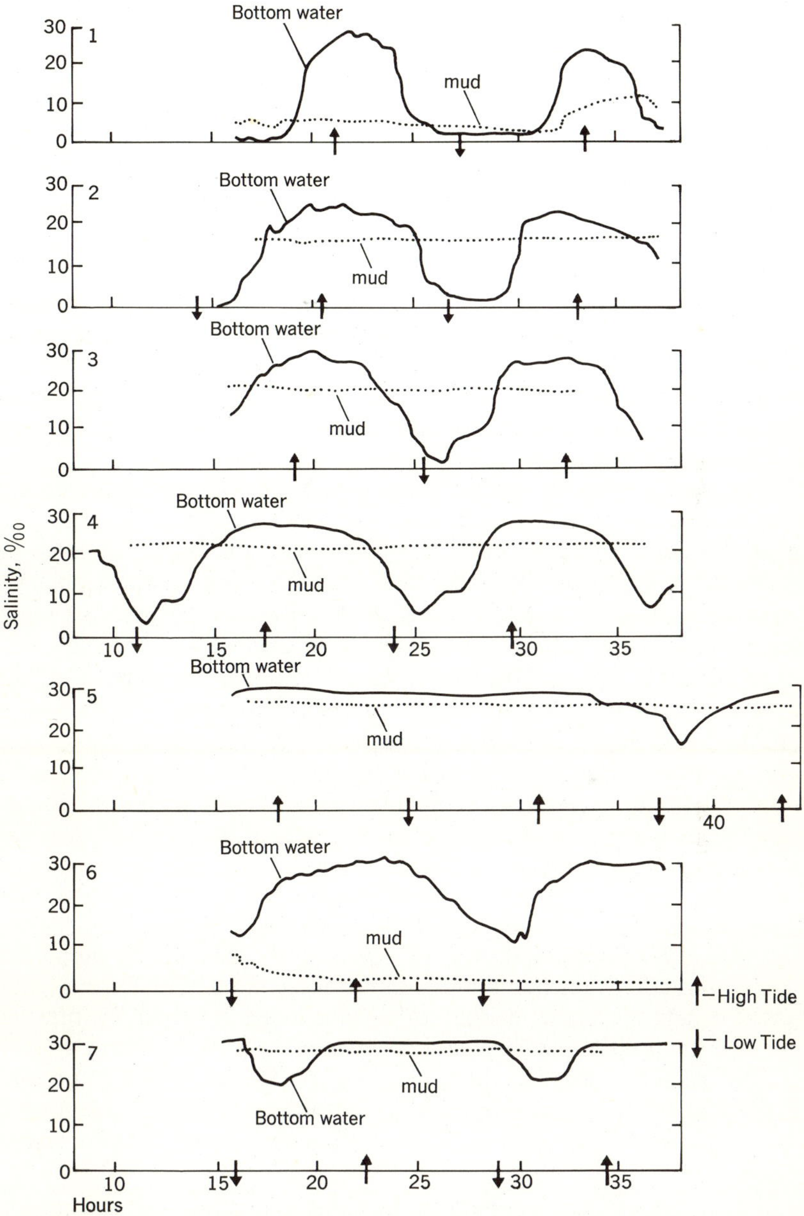

Figure 2-8. Sediment and bottom water salinities in an intertidal mud flat. (From Sanders *et al.*, 1965.) Each of the seven graphs represents a different sampling area.

Rocky shores also characterize the intertidal-zone habitat. Rocky intertidal habitats—where the substrate is a hard, solid surface—vary greatly in that some are uniformly flat from sea to land, whereas others are characterized by boulders varying in size and having irregularly shaped surfaces. In this latter type there are many microenvironments accompanied by a rich variety of fauna and flora. Owing to marked differences in various environmental factors, organisms present on rocky shores exhibit a high degree of zonation. Although considerable information on zonation is available throughout the world, a uniform classification has not been universally accepted by biologists. Two types of habitats are associated with the rocky intertidal zone: (1) tide pools, and (2) the splash zone, a region above the high-tide mark which is splashed with sea water as a result of considerable wave action. Heavy wave action generally occurs only in those rocky intertidal zones that are not fully protected from the direct impact of the open ocean.

Organisms living on rocky shores must contend with problems of both an aquatic sea environment when the tide is "in," and an aerial environment when the tide is "out." Several factors are apt to influence the ecology of organisms in the rocky intertidal zone, among them a marked temperature gradient which may be observed on the surfaces of rocks. This gradient is correlated with the length of exposure of the rocky surface to the direct action of the sun. Typically, those areas nearest the high-tide mark will be exposed for a longer period of time and will show more extreme temperature variation throughout the year than those areas closer to the low-tide mark. Animals inhabiting the underside of rocks, however, will not necessarily be exposed to marked temperature fluctuations; they are not exposed to the direct effect of the sun and therefore will not experience the same degree of temperature change that occurs on the surface. Salinity will not vary excessively in a rocky intertidal zone, although salinity in tide pools associated with this zone may be very high as a result of evaporation, or may be dramatically reduced during periods of heavy rainfall. Since fluctuations in water level may be very rapid and of great magnitude in this habitat, organisms living in the rocky intertidal zone face the problem of desiccation. Generally, the desiccatory effect is less on organisms closer to the low-tide mark than on those at the high-tide mark, but some animals live at the high-tide mark on the undersurface of rocks where there is some residual water and where the relative humidity is high.

The degree of light will vary in the same magnitude and direction as does temperature; that is, the organisms from the high-tide mark may be exposed to direct sunlight for a longer period of time, and therefore will be exposed to strong effects of ultraviolet rays. The direct effect of the sun can result in an increased heat load on the animal or can increase its body temperature and cause desiccation. To avoid these harmful effects, the sessile animals tend to have thicker shells or body coverings; the non-sessile animals commonly remain underneath or on the side of rocky surfaces during the day, and are active on the top surface only at night. Tidal currents also influence the distribution of animals on rocky shores, and the magnitude of the tidal current determines the degree of settling of animals on rocky surfaces.

In all these factors, time is exceedingly important because the length of time an animal can withstand a stressful factor greatly influences its vertical distribution. This relationship between exposure and submergence is graphically represented in Fig. 2-9.

Pilings of piers and other such structures offer another intertidal-zone habitat. Many of the same stresses that act upon the rocky shore habitat are also operative upon organisms found on pilings. The type of piling determines the types of animals that will be associated with it. For example, wood pilings commonly attract many wood-burrowing organisms, whereas concrete pilings attract only those animals that can attach to the surface or the few very specialized forms that can burrow into concrete.

Irrespective of the type of intertidal zone, debris accumulates above the high-tide mark when the tide recedes. Although the size, shape, and composition of this debris varies, a relatively complex community of organisms lives here in what is appropriately called the drift-line habitat. The type of intertidal zone determines the drift-line habitats. For example, the microclimate of the sand beach drift line does not approach the extremes of the exposed sand surface, but it is drier and hotter than the drift line present in *Spartina* marshes (Barnes and Barnes, 1954).

Estuaries

Wherever water from a fresh-water system pours into the sea, an unstable environment is created. This region of dynamic encounter is referred to as an *estuary*. Recently, Pritchard (1967) defined an estuary as "a semi-enclosed coastal body of water which has a free connection with the open sea and within which sea water is measurably diluted with fresh water derived from land drainage." Other definitions have been proposed to describe all different types of estuarine areas throughout the world, but one feature is common to all: estuaries are restricted to the mouths of rivers in an area of oceanic tidal fluctuation. The upper limit of an estuary is difficult to pinpoint, since in some estuaries (such as the Amazon), tidal influences are observed several hundred miles upstream and well beyond the presence of saline water. In some estuaries there is relatively little

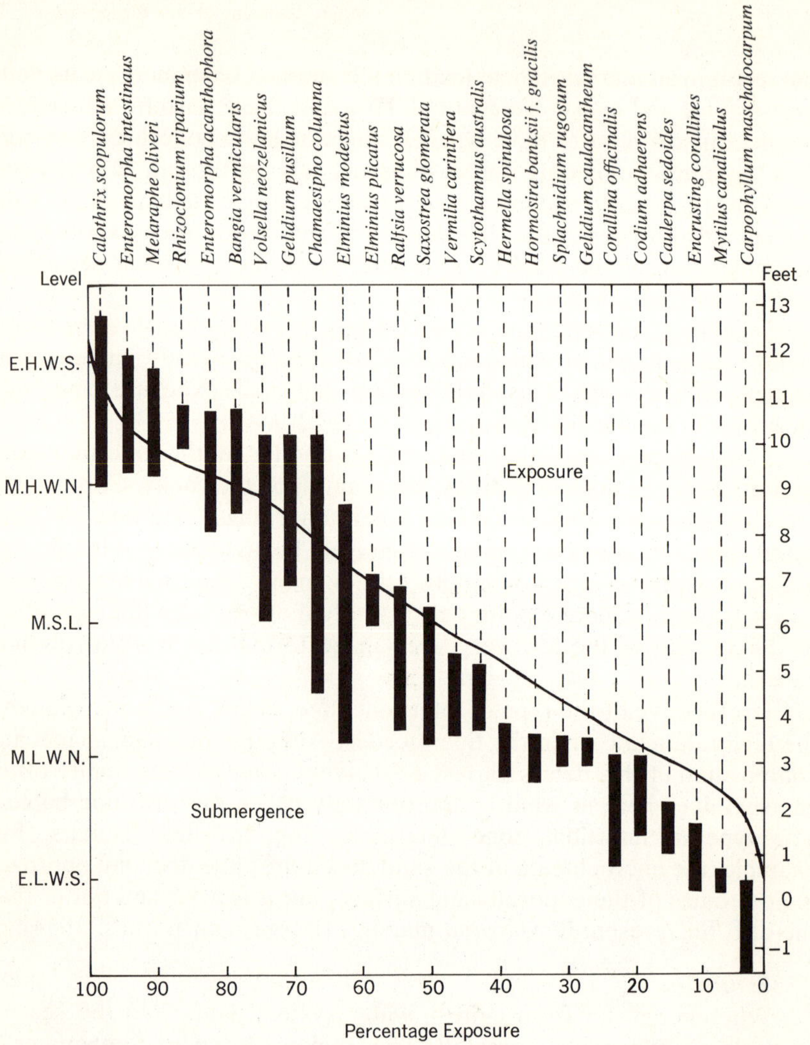

Figure 2-9. Vertical distribution of various organisms at Aukland, New Zealand. E.H.W.S.—extreme high water of spring tides; M.H.W.N.—maximum high water of neap tides; M.S.L.—mean sea level; M.L.W.N.—minimum low water of neap tide; E.L.W.S.—extreme low water of spring tide. (From Dellow, 1950.)

salinity change, depending on fresh-water inflow. Using relative salinities as a basis for distinction, two general types of estuaries have been described: (1) a normal, or "positive," estuary in which the salinity is reduced upstream as a result of adequate inflow of fresh water, and (2) a hypersaline, or "negative," estuary in which the salinity is higher than that of the ocean because of inadequate runoff of fresh water. Although the estuary is chiefly characterized by a marked change in salinity, the complex interaction of many fluctuating environmental factors must be considered when studying the physiological ecology of estuarine animals.

Tidal amplitude varies temporally and geographically, and the development of intertidal areas bordering the estuarine shore depends upon the degree of tidal flux. Vast amounts of silt borne by fresh-water runoff and deposited in the estuary form extensive tidal flats. The water bordering these tidal flats tends to be shallow, and temperature fluctuations may be extremely pronounced during one tidal cycle or may be seasonally controlled by temperature changes in the open sea. The circulation within an estuary is also a factor in determining the amount of environmental fluctuation. Since many physical factors influence circulation patterns, no one pattern is typical. The principal types have been characterized by Bowden (1967) and are listed in Table 2-5 and are shown in Fig. 2-10.

The generally high concentration of nutrients in estuaries may account for their productive contribution to the economy of the sea. Not only do resident populations of commercially important animals live here, but also many neritic species (species that live in the shallow-water zone

Table 2-5. Types of Estuarine Circulation.

Type	*Physical Processes*	*Forces*
1. Salt wedge	River flow dominant	Pressure gradients, field accelerations, Coriolis effect, interfacial friction
2. Two-layer flow with entrainment, including fjords	River flow modified by tidal currents	Pressure gradients, field accelerations, Coriolis effect, entrainment
3. Two-layer flow with vertical mixing	River flow and tidal mixing	Pressure gradients, field accelerations, Coriolis effect, turbulent shear stresses
4. Vertically homogeneous (a) with lateral variation (b) laterally homogeneous	Tidal currents predominating	Pressure gradients, field accelerations, turbulent shear stresses, Coriolis effect in (a)
5. Exceptional cases: intensive mixing in restricted sections, tributary estuaries, sounds, straits, etc.		

SOURCE: Bowden, 1967.

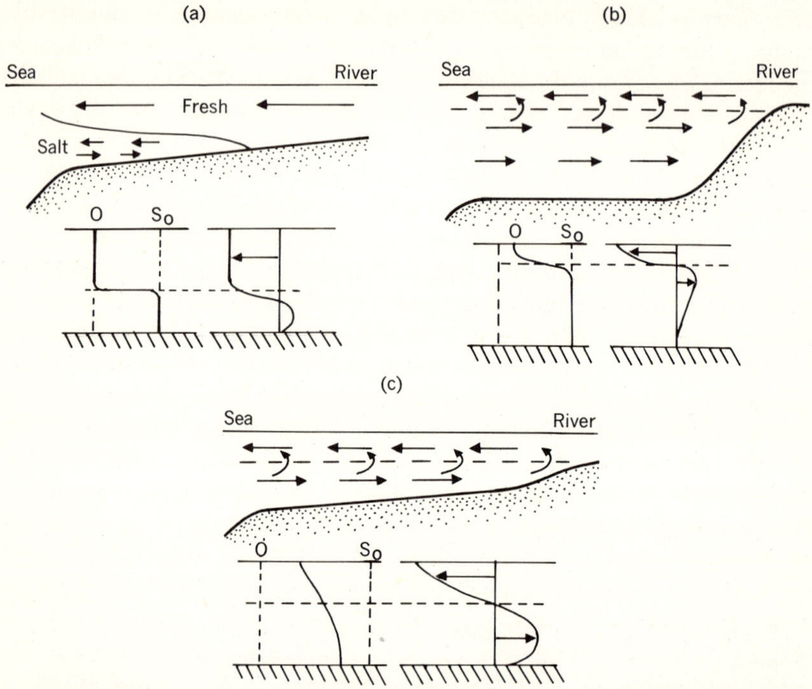

Figure 2-10. Examples of types of estuarine circulation: (*a*) salt wedge estuary; (*b*) two-layer flow with entrainment; (*c*) partially mixed estuary with entrainment and mixing. In each of the drawings the upper part represents a section along the estuary, and the lower part represents typical salinity and velocity profiles. (After Bowden, 1967. Copyright 1967 by the American Association for the Advancement of Science.)

on the continental shelf), which use the estuary as a nursery ground before migrating to the open sea.

Estuaries have extreme biological importance in themselves and also have numerous distinctive types of habitats along their borders. Because these habitats represent a transition between the aquatic and terrestrial zones, a dynamic interaction exists between them. For example, these regions have a characteristic level of biological production of organic material in excess of what is consumed. Much of this excess organic material is transplanted to the estuary, where it is utilized by resident animals. In addition to a resident fauna and flora, aquatic and terrestrial animals may venture into these border regions alternately, depending on tidal fluctuation.

Because of the complex nature of climatic conditions influencing this border region, different habitat types have evolved in various parts of the world. For example, mangrove swamps predominate in the tropics, whereas *Spartina* marshes are a dominant vegetational type in the temperate zone.

Tidal Marshes. Tidal marshes border every estuary in temperate and tropical regions except in areas subjected to surf or where rocky shores are present. Tidal marshes develop when plants that are resistant to the harsh environmental conditions of an intertidal zone invade higher elevations of bare mud and sand flats. Under favorable conditions of tide and wave action, these plants trap sediments, and with the subsequent accumulation of sediments and increased land elevation, environmental changes take place which are favorable to the invasion of different species. The resultant zonation pattern is graphically represented in Fig. 2-11. Distinctive species of animals that are adapted to the unique environmental characteristic of each zone are associated with each of these different succesional stages. The physical conditions encountered by animals living in this habitat type are generally extremely variable: salinities in excess of 80‰ may be found in marsh areas temporarily cut off from the estuary during a period of high temperature, whereas salinities less than 1‰ may be present after excessive rains. High hydrogen sulfide levels and anaerobic conditions occur in regions where a large amount of decaying organic matter exists.

Oyster and Mussel Aggregation. One of the most conspicuous communities bordering estuaries is formed by the aggregation of oysters or

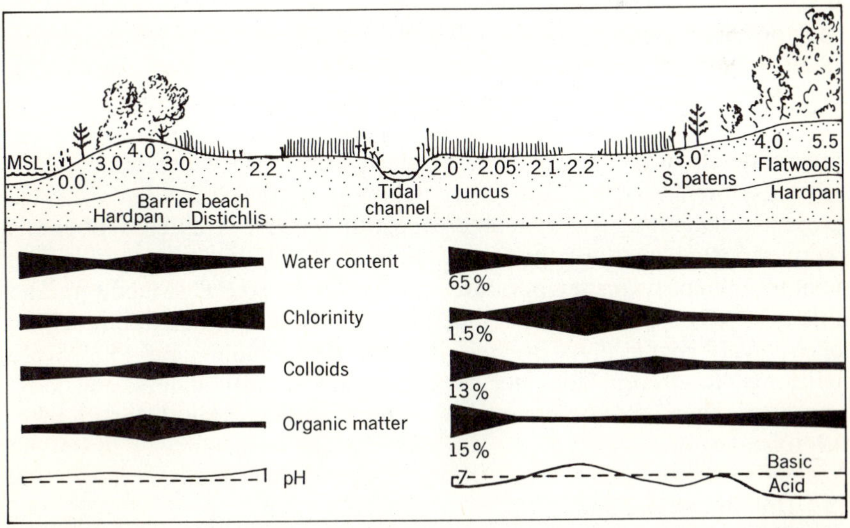

Figure 2-11. Generalized transect of a salt marsh. (From Kurz and Wagner, 1957.)

mussels. These molluscan beds provide a hard substrate for attachment of and protection for plants and animals. One indication of the biotic diversity of this habitat type can be found in the study of Wells (1961), who recorded 303 species of animals associated with various oyster beds located in Newport River, North Carolina. The number of species associated with these aggregations decreases at the lower end of the salinity gradient. These oyster beds vary in size from multilayered aggregations, forming a reef, to small clusters of animals scattered in the intertidal zone. Mussels also aggregate to form beds that vary in size and complexity.

Both mussel beds and oyster reefs influence circulation patterns and sedimentation characteristics in estuaries. These structures may be parallel to or transversely positioned across the path of the current. In either case, since the oysters and mussels feed by filtering material from the medium, they are dependent on water movements. Animals associated with this habitat may experience marked fluctuations in environmental factors, depending in part on the size of the molluscan aggregation and its location in the estuary. Parts of large reefs permanently submerged are more environmentally stable than those exposed to air at low tides.

Coral Reefs

Coral reefs are widespread throughout the tropics and abound with an extremely diverse fauna and flora (Fig. 2-12). Coral reefs, which originate from biotic activity rather than geologic processes, are made up primarily of corals and calcareous algae. By the very nature of their architecture they provide numerous microhabitats for other reef organisms. A rich resident infauna of mobile and sedentary species are present, and many active free-swimming species are periodically attracted to the reef.

Environmental conditions necessary for reef corals to live and grow are relatively well known. In general wherever the water temperature remains above 23–25°C and rarely below 16°C, reef-building corals will grow if sufficient light is available. Recently however tropical reef corals were reported on the continental shelf off North Carolina where the winter ambient water temperatures are as low as 10.6°C (Macintyre and Pilkey, 1969). In addition to temperature stability, a sufficient circulation of water is required to prevent thermal stratification and to provide nutrients. Reef coral does not survive in low-salinity waters, and most species exposed to 50 percent sea water will die within 24 hr (Yonge, 1940).

The Pelagic Zone

The pelagic zone encompasses the entire water mass covering the bottom, or benthic zone. It is arbitrarily divided into two major sub-

Figure 2-12. Soft coral reef. (Russ Kinne/Photo Researchers.)

divisions: the neritic zone, or coastal waters; and the oceanic, or open-ocean, waters. The division line between these two zones is not well defined, but generally a depth of 200 meters is considered to be the boundary (Sverdrup *et al.,* 1942). Typically, a greater variation in physical factors occurs in coastal water than in oceanic waters, and it is particularly pronounced in those features most influenced by runoff from land and fresh-water systems, such as increased sediments, reduced salinity, and fluctuation in nutrients. The chemical composition of waters near shore varies more than that of oceanic water. Another differentiating feature between oceanic and coastal waters is the influence of a relatively shallow bottom in coastal areas, which allows greater opportunity for mixing and recycling of nutrients when bottom sediments are disturbed by wave action. Furthermore, the interaction between a rich benthic fauna and surface-dwelling species occurs more readily because of spatial proximity to oceanic waters.

Since the oceanic mass is not entirely separated from the coastal water mass, the edge of one frequently will invade the other. This intrusive water may retain its physical and biotic characteristics for some time. For example, the edge of the Gulf Stream frequently invades coastal water in the dynamic region of Cape Hatteras, North Carolina, where it may alter the thermal characteristics of the region, depending on the season of the year.

Both the neritic and the oceanic regions have a surface photic zone and a deeper, dark (aphotic) zone. At different latitudes the transition point between them seasonally varies with such factors as the amount of particulate matter present and whether this matter is inanimate. Since the quantity of light decreases with depth, there are qualitative changes in the penetration ability of different wavelengths. Generally, the blue-green wavelengths reach deep oceanic water, whereas the yellow-green wavelengths penetrate inshore waters.

The pelagic regions are dominated by plankton. Without the photosynthetic activity of phytoplankton, life in the sea would cease to exist except for restricted inshore populations of animals feeding on food originating from the land or a fresh-water system. Plankton is not evenly distributed in pelagic waters, and its presence varies temporally and spatially with fluctuations in environmental factors. Knowledge of certain of the physiological capabilities of plankton under different environmental conditions and certain aspects of population dynamics makes it possible to construct formulas to predict the density and seasonal occurrence of plankton (Riley *et al.,* 1949). The depth to which phytoplankton can survive for any period of time is determined in part by the point at which sufficient light is available for the photosynthetic rate to be equal to respiration. This depth, called the *compensation point,* varies with many conditions, including temperature and turbidity.

Zooplankton and some macrofauna show marked diurnal vertical migrations. These tend to be more pronounced in oceanic water than in the shallower coastal waters. Although migration is achieved by complex interaction of various factors, light is one of the most important requirements for maintaining plankton at a certain level. By means of ultrasonic depth indicators, oceanographers have detected a distinctive layer (or layers), at various depths in the ocean, which moves vertically with changes in light. This is called the *deep scattering layer* (DSL). Little is known about the species composition of the DSL or the ecological interrelationships of these organisms.

A striking feature of oceanic waters is the occurrence of an *oxygen minimum layer* at an intermediate depth between higher oxygen concentrations above or below this layer. Despite an oxygen content of only 0.1 ml per liter, rich collections of organisms have been taken from the oxygen minimum layer. Comprehensive studies on the ecological dynamics of this layer are lacking, however. The DSL biota at times overlaps that of the residual oxygen minimum layer.

The Abyssal Zone

It is difficult to define the limits of the abyssal zone, owing to lack of agreement on the depth at which it begins. Hedgpeth (1957) states that it begins between 2000 and 3000 meters and extends downward to about 6000 meters, but he adds that a better definition would stipulate the abyssal zone to be the region in which temperature is never above 4°C. A universally accepted definition based on temperature and/or depth is difficult to find because these factors may not be uniform throughout the abyssal zone, as demonstrated by the following data: below 2000 meters the temperature is 14°C in the Mediterranean; in the Caribbean, it varies from 6°C to 4°C; and in the Antarctic and Arctic it is less than −1.2°C. Nonetheless, it is recognized that the abyssal benthic zone is the largest benthic habitat type in terms of geographical area, occupying approximately 90 percent of the ocean floor, while oceanic waters cover about 70 percent of the earth's surface. This habitat is characterized by the relative uniformity in the intensity of physical factors: salinity is about 34‰ throughout the year in the Atlantic; light is absent, except for bioluminescence, below 1000 meters; temperature, except in a few cases as cited above, varies with latitude over the range from 3.6 to −0.6°C; and little or no seasonal variation occurs. Sediment type, however, varies from pteropod ooze, which is composed of shells of pelagic molluscs, to red clay.

Great hydrostatic pressure is one of the unique features of the abyssal benthic habitat. For each 10 meters of depth, the pressure increases by 1 atm. Although the environmental conditions are relatively constant, the abyssal zone appears to vary geographically with respect to sediment type, topography, and characteristics of water masses. Since light does not penetrate to this depth, there is no primary production of food by photosynthetic activity, and organisms living in this zone must depend on an input of food from the upper layers of the water.

Figure 2-13. Crater Lake, Oregon. (Dr. P. J. DeCoursey.)

Fresh-Water Habitats

Fresh waters are distributed over the surface of the land in a series of different systems, all of which ultimately join the ocean waters. These fresh-water systems are generally divided into two groups based on the activity or rate of movement of the water: (1) standing quiet (lentic) waters, which include lakes, ponds, swamps, and marshes; and (2) flowing (lotic) waters, which are the streams and rivers. Each of these presents a different set of environmental parameters.

Standing Water

A lake can be defined as any inland depression in the surface of the earth, ranging in depth from a few meters to about 1800 meters (Fig. 2-13). Smaller bodies of shallow, standing water with rooted plants growing over most of the bottom are commonly referred to as ponds (Fig. 2-14). Most lakes and ponds have outlet streams, and neither are static and permanent features of the landscape because many forces operate to level the shoreline, fill in the basin with sediments, and erode the outlets.

Although natural lakes may form in different ways, the lakes in any given area generally have originated as a result of one particular force. For example, most lakes in central North America are formed by the

Figure 2-14. Upland pond, Stonington, Connecticut. (Mary M. Thacher/Photo Researchers.)

action of continental glaciers, whereas those in Florida occupy basins dissolved from a soluble, porous limestone. Other lakes have been formed in craters of extinct volcanos or after landslides have blocked off a large area. Lakes can also be formed following a shift in the crust of the earth, where mountains are uplifted or rock strata are broken up or displaced. Man-made lakes have been formed by damming rivers or streams, and many small ponds have been constructed. Ponds can also form when abandoned strip mines and rock quarries fill with water or when beavers dam up streams.

The temperature of a lake varies with season and depth, and even in the tropics some thermal stratification is present in the deeper lakes. Most temperate-zone lakes exhibit an annual temperature cycle and pronounced thermal stratification (Fig. 2-15), and in these lakes the relationship between temperature and fresh water is important. The maximum density of water is reached at 4°C; if water temperature is either warm or cool, the density falls. Sometime in the spring, the lake is isothermal at maximum density (4°C); surface winds mix the lake and chemical conditions throughout the lake become relatively uniform. At this time, the entire lake is in motion, and this condition is termed the

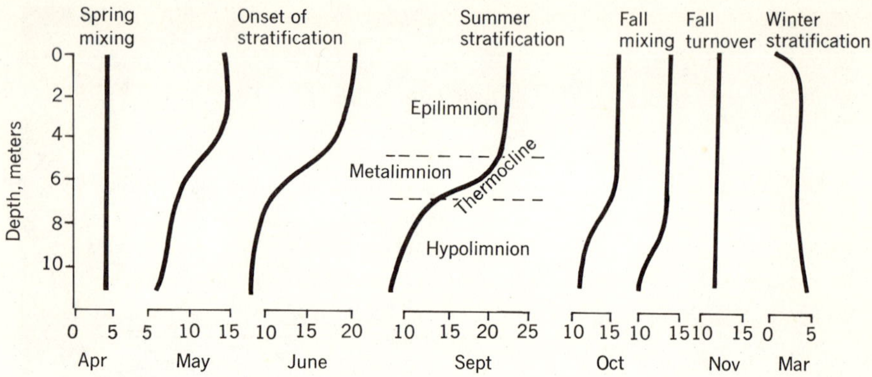

Figure 2-15. Annual temperature cycle and thermal stratification of lakes. (From English, 1962.)

spring turnover. As the atmospheric temperatures increase, the surface temperatures of the lake also increase, but the depth to which the temperature increases throughout the lake depends upon exposure to wind, sheltering topography, and duration of the warming period.

With the arrival of a calm, warm period, when the wind action no longer thoroughly mixes the lake, a temperature profile with three distinct layers is observed: a uniformly mixed upper layer of higher temperature, termed the *epilimnion;* a region of abrupt temperature change, termed the *metalimnion;* and a lower, uncirculating part of the lake with even lower temperatures, called the *hypolimnion.* The region of abrupt temperature change within the metalimnion is called the *thermocline.* This temperature change at the thermocline (the drop can be 6 degrees per meter) and the corresponding density gradient produce a very stable condition, referred to as *summer stagnation.* The fact that the epilimnion and hypolimnion exist as two nearly separate bodies of water, with little mixing and interchanging substances, often accounts for population fluctuations.

With autumn cooling, heat is lost faster than it is absorbed, the epilimnion temperature decreases, and differences in density between the strata become less. As the wind generates currents, which cause increased circulation of the cooler waters, the thermocline sinks farther until it reaches the bottom of the lake. Subsequently, the lake once again mixes entirely, and this state is called the *fall turnover.* Here again the nutrients from the lower levels are replenished, and often there is a reoxygenation of the lower layers from the upper waters.

When surface temperatures decrease with the onset of winter, this colder, denser water sinks to the bottom, and within a short time the entire lake is again at maximum density. With freezing surface temperatures, an "inverse" stratification may be observed in which 0°C occurs at the ice cover and 4°C lies just below the surface. If the entire lake becomes covered with ice and no external forces mix the lake, it is said to be in a state of *winter stagnation.*

Gradations also occur in the amount of oxygen and light in different parts of a lake. The depth of light penetration depends upon the turbidity of the water. Generally, the stratification of oxygen is more or less parallel to that of temperature during the summer, with the greatest amount near the surface. During the spring and fall turnovers, the deeper regions of the lake are reoxygenated. If the lake does not freeze over during the winter, the oxygen content will remain relatively constant, but if ice forms, the amount of oxygen beneath it may be seriously depleted.

The biotic zones in a lake can be divided into several layers (Fig. 2-16). At the upper layer is the lake surface, with the interface between the air and the water. The organisms associated with this zone, referred to as the *neuston,* consist mainly of snails, flat worms, and some insect larvae. The second zone is that region of open water that extends vertically from the surface to the bottom, and it is referred to as the *limnetic zone.* Since the top is a lighted zone, it supports a plant-based animal community. Animals associated with the upper limnetic zone tend to be free-swimming animals such as fish, and are referred to as *nekton.* In the lower part of the limnetic zone only bacteria and plankton are present. The *littoral zone* extends from the shoreline where rooted plants predominate to the area where rooted plants are no longer present. The lowest level is called a *profundal zone.* Here light is absent, oxygen is low, and the animals must depend on the limnetic zone for food. Mainly bacteria and fungi occur in this area, along with some chironomid larvae and annelids.

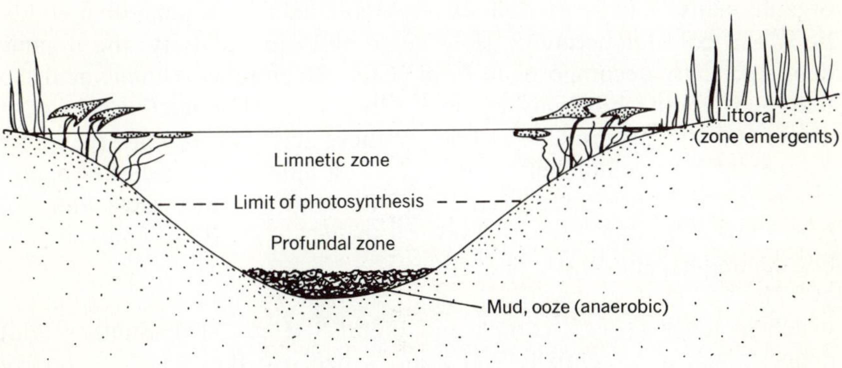

Figure 2-16. Biotic zones in a lake.

In contrast to the marked stratifications of lakes, environmental conditions in ponds are almost entirely uniform. Ponds have a relatively large littoral area and small or absent limnetic and profundal regions. Since ponds may be temporary areas, they offer only a transient existence for the animals living there. Some of these temporary ponds will recur regularly and some quite irregularly, depending on the precipitation in the area. Both types, however, support a unique form of life, for in order to live in a temporary pond, an organism must be able to reproduce very rapidly when favorable conditions are present, and it must have some means of surviving in the absence of water. Numerous invertebrates have evolved mechanisms that permit survival for many months, or even years, without any water being present. The eggs of rotifers, for example, can survive many months in an absolutely dry atmosphere, and as soon as water becomes available again, the eggs hatch and the cycle is repeated. Other animals, such as amphibians and adult insects, are able to live in this environment because they can move in and out of the pond at will.

Although a lake is transitory, the various successional stages in its life are definable and can be studied as an ecosystem. In its early stages, when a lake is deep and clear and has relatively low productivity, it is called *oligotrophic*. Gradually the bottom begins to fill up, added nutrients appear, productivity in the euphotic zone increases, and the shore areas are invaded by hydrophytes; now the lake is said to be *eutrophic*. Although this aging of lakes is a naturally occurring process, man has done much to hasten it through pollution (see Chapter 12).

Bogs, Swamps, and Marshes. Bogs, swamps, and marshes develop during the late fill-in stages in lake succession—with the exception of some swamps that develop on the flood plains of southern North America. Bogs are present in humid areas, particularly in the cold, northern, forested regions of North America, Europe, and Asia (Fig. 2-17). They generally develop where the drainage is blocked and little organic matter can be carried away. A combination of anaerobic conditions and cold temperatures slow down biological activity; the organic debris partially decomposes to form peats, which release humic acids so that the water in a bog is always decidedly acidic. Although few species of plants and animals are able to tolerate these acidic conditions, they may be present in large numbers. Bogs support a number of species of plankton, particularly phytoplankton, along with some aquatic insects, amphibians, a few fish, some birds, and several species of small mammals such as bog lemmings, muskrats, and shrews.

The marshes are wetlands characterized by grasslike vegetation, which includes sedges, cattail, and bullrush (Fig. 2-18). Smith (1966) defines them as essentially wet prairies because they are low, poorly

Figure 2-17. Sphagnum Bog, Hope Lake, Wisconsin. (Dr. P. J. DeCoursey.)

drained lands that have standing water for several months of the year. Marshes support a much richer fauna than do bogs, particularly, a larger number of invertebrates and fish and many birds.

Swamps are wetlands that support large trees and shrubs (Fig. 2-19). They offer a wide variety of environmental conditions, both aquatic and semiaquatic, which in turn support a rich variety of animals such as many species of aquatic insects, reptiles, and birds. More terrestrial animals such as deer, bear, and squirrel also are commonly associated with swamps.

Flowing Waters

A stream has been defined by Reid (1961) as "a mass of water with its load moving in a more or less definite pattern and following the course of least resistance toward a lower elevation." Streams may begin as outlets of ponds or lakes, or they may arise from seepage areas or springs. The

Figure 2-18. Marsh, Mines County, South Dakota. (USDA Photo.)

Figure 2-19. Florida swamp.

outstanding characteristic of a stream is that the water is continuously moving, and as it flows from the source in the direction of least resistance, it cuts a channel by eroding small furrows that grow into gullies. The greater the volume of water and the steeper the slope, the faster the land is eroded away. As the water moves along, carrying a load of erosional debris, the channel is cut deeper and wider. Thus the character of a stream changes as it flows from the source to the mouth. Up in the hills at the source, the headwater streams are small and shallow, the current is swift, and water temperatures are cold. As the water moves downstream, the current slows and the water spreads wider and is deeper. By the time the water reaches the lowlands, the current is very slow, and the bottom is covered with silt and mud (Fig. 2-20).

Streams offer a diverse range of habitats that support a rich and varied fauna; species near the source of a stream differ from those farther downstream, as illustrated graphically in Fig. 2-21.

Figure 2-20. Yellowstone River, Yellowstone National Park, Wyoming. (Dr. P. J. DeCoursey.)

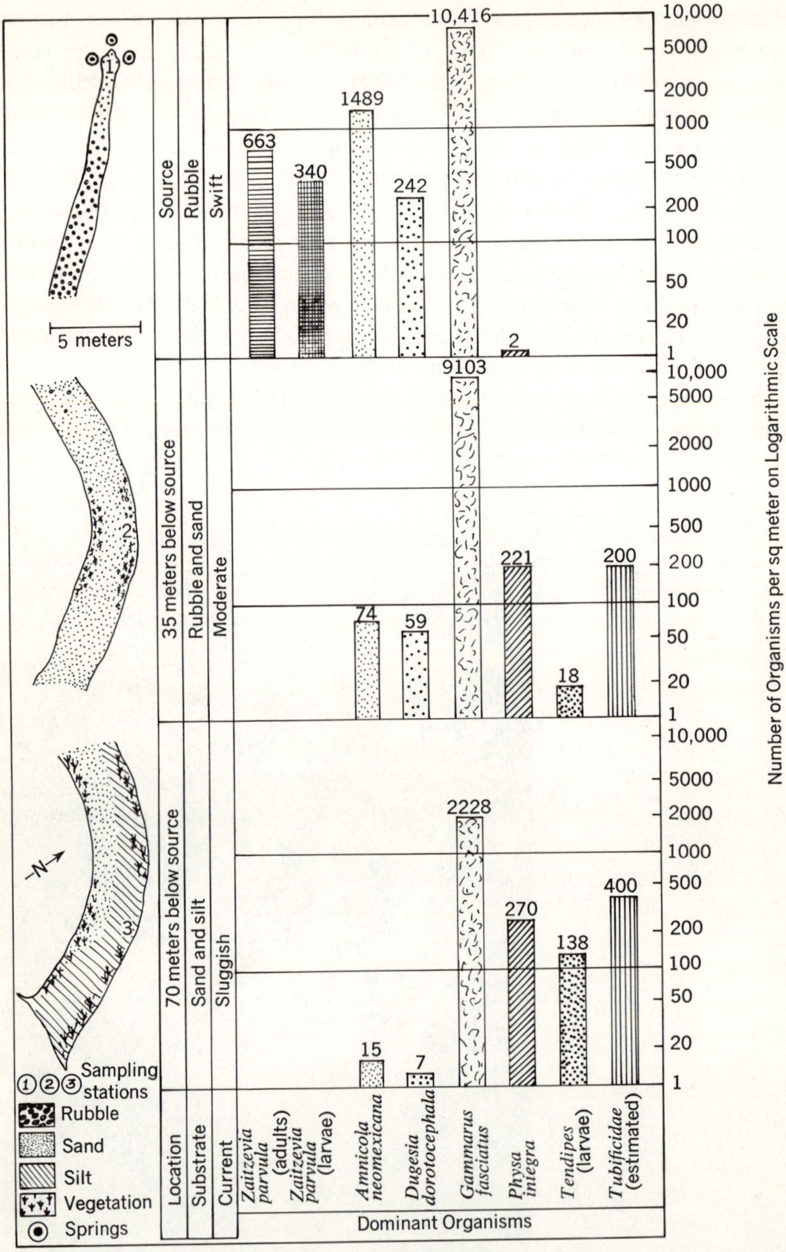

Figure 2-21. Habitat characteristics and populations of animals in Lauder Springbrook, New Mexico. (From Reid, 1961.)

The current of streams produces enough turbulence to prevent the stratification that is characteristic of lakes. Furthermore, the oxygen content in unpolluted streams is generally higher than that in lakes. Turbidity, however, may be greater, especially during certain seasons.

One of the most important environmental features that an organism encounters in streams is the continuous flow of water, and animals must be able to maintain position if they are not to be caught up in the current and carried to less suitable environments. Stream animals have evolved many morphological adaptations that make this possible; streamlined bodies lessen resistance to water current, and flattened bodies enable the animals to cling to undersurfaces of stones. Some organisms are able to attach themselves temporarily to the substrate, and others protect themselves by constructing cases which are cemented to the sides and bottoms of stones.

Terrestrial Habitats

Tundras

Tundras, which are nearly level and treeless regions covered with low vegetation, offer an extremely harsh environment for both plants and animals. Above the 65-degree latitude, tundras can be found at sea level; below this latitude they are confined to mountain sides and are generally described as either arctic or alpine, depending on their location. In the northern hemisphere, a large part of the area north of 57°N Lat. is arctic tundra, starting as a ring around the Arctic Ocean and extending southward to Labrador, Iceland, and southern Greenland, and across Canada as far south as northern Manitoba. Arctic tundras also surround the Bering Sea from Alaska to Siberia and from the Bering Straits to the Aleutians. In the southern hemisphere, there are only widely isolated tundralike areas on some subantarctic islands; there is no counterpart of the arctic tundra in the Antarctic because this region is almost entirely covered by ice. In the Arctic the climate is controlled by the annual cycle of solar radiation so that in areas located north of the Arctic Circle there are long periods during the winter when the sun does not rise above the horizon. The climate is dry and very cold, and air temperatures dip down to −70°F or below. Snowfall is limited and the soil water is frozen; wind velocity can be very great, and during the summer the sun does not set. Near sea level, solar radiation melts some ice and snow, but much of the soil is underlain by permafrost, and air temperatures seldom rise above 60°F.

Although alpine tundras are similar in many respects to the arctic tundras, alpine areas have more snow, rain, and wind, and the photoperiods are different. In alpine areas there is some sunlight in the winter and some darkness in the summer. Since alpine tundras are found on the tops of mountains, the oxygen pressure is low. Neither the flora nor fauna of these areas is very extensive. The flora consists mainly of low and compact herbaceous, perennial plants. Winds are very strong on the tundras, and the lowness and compactness of the plants offer a definite advantage. Because the growing season is short, annual plants would be at a distinct disadvantage because they depend on heat to ripen seeds, whereas perennial plants can store food during one summer, which permits rapid growth during the first part of the following summer. Since the plants are herbaceous, the buds are under the soil surface or near the surface under the snow, and thus are protected from adverse winds and weather. The number of species adapted to live under such conditions, however, is limited to a few hundred, in contrast to the more than 100,000 species found in the tropics (Billings, 1964).

In the arctic tundra, the permafrost may extend to more than 1000 ft in depth, and only the upper 6 to 8 in. thaw during the summer. This limits the habitats available to those species that dig burrows or dens. Animals inhabiting the arctic tundra tend to be small in size; Morrison (1964) reported that 70 percent of the 33 species of mammals found on the Siberian tundra weigh less than 1 kg and 60 percent are under 100 grams. Since the tundra vegetation has a low aboveground productivity, the herbivores must depend on the total production of large areas for food. Thus, many of these herbivores, such as lemmings, caribou, and reindeer, move from place to place in order to obtain food. Metabolically, tundra animals must also be able to acclimate quickly to very low temperatures. Because of the nature of the light and dark cycle, arctic animals generally do not depend on photoperiod to initiate seasonal physiological changes.

During the short summer season, bird life is abundant and varied, but during the winter the birds migrate to more tolerable climes. Numerous insects, many of which overwinter as larval stages, also are present. There are, however, no amphibians or reptiles, probably because of the dearth of habitats where these animals can overwinter (Hock, 1964).

Forests

Generally, a forest is characterized by the location and type of tree that predominates; on this basis, forests can be arbitrarily divided into three different types: the boreal coniferous forest, the temperate deciduous forest, and the tropical rain forest.

The boreal coniferous forests are also known as the north woods; since this type of forest is so common in Siberia, the Russian word "taiga" is sometimes used. The trees are needle-leaved evergreens; the dominant

Figure 2-22. Eldorado National Forest, California. (USDA Photo.)

species are mainly spruces, but there are also fir, pine, poplar, birch, and aspen. Many lakes are also associated with this type of forest. Since the climate is severe, the number of species that can thrive under these conditions is limited; winters are long, cold, and snowy, and the summers are brief. In North America, the taiga extends in a belt from Alaska southeastward across Canada to Newfoundland. The northern boundary is distinct and forms the arctic timberline with the tundra. The southern boundary is less distinct, but in North America it extends down into northern Minnesota, northern Michigan, and northern New England. In Eurasia, the taiga extends from Norway across Sweden, Finland, Russia, and Siberia.

The boreal forest is relatively simple in structure; compared to other forests, there are few species of trees (Fig. 2-22). The forest is stratified into four layers: an evergreen tree stratum, which is 40–70 ft high; a short shrub stratum, 2–4 ft high; a layer of low herbaceous plants; and finally a very low moss and lichen layer (Billings, 1964). Much of the ground has a thick layer of needles and dead twigs. The animal species that live in the forest tend to be shy and retiring, and they must be able to tolerate the cold snow. Many of the animals present are closely linked to the vegetation: the moose feed on willows and water plants, beavers feed on the aspen, rodents feed on the conifer seed. In one group of birds indigenous to these forests, the crossbills (*Loxia*), the beak is structurally specialized for picking seeds from the cones. During the winter the invertebrate populations lie dormant, but during the summer months there is much activity by a large invertebrate population, particularly the spruce budworm and the tent caterpillar. Many migratory birds and mammals are present only during the more favorable summer months.

In North America, many areas of the boreal forest have been lumbered or destroyed by a number of disastrous fires that have occurred within the past century. As a result, many of the large mammal populations have been depleted (Shelford, 1963).

The temperate deciduous forests of North America extend from the center of the Great Lakes region south to the Gulf of Mexico, west to the Ozark Mountains, and over the northern two-thirds of the Florida peninsula. In 1600, this entire region was forested, except for a few areas in the northeastern states and in Virginia, where there were large Indian populations. As the country was settled by Europeans, however, the forest was gradually destroyed by agriculture or modified by logging; at the present time, the areas in which no trees have been removed is less than 0.1 percent of the original forest (Shelford, 1963). Deciduous forests also occur in eastern Asia and in the temperate regions of Europe.

In all these areas, the temperate deciduous forests are dominated by broad-leaved deciduous trees, which shed their foliage in fall, are bare over winter, and grow new foliage in the spring (Fig. 2-23). There are many kinds of dominant trees here, typified by elms, maples, beeches, oak, and hickory. The climate is very seasonal, with intermittent warm moist weather and a considerable amount of precipitation throughout the year. Two factors limit the distribution of deciduous forests to other areas: low temperatures and lack of precipitation. These forests have many layers. The upper canopy, which consists of deciduous trees that can attain a height of 75–175 ft, shades nearly 90 percent of the ground. Many invertebrates live on the bark, leaves, and upper trunk of these trees; in the lower part of the canopy, birds and small mammals live in trees cavities produced by decay when limbs break off. Below the upper canopy is a second story that is 20–35 ft high, and below this is a shrub layer of deciduous trees and evergreens that are from 1–8 ft high. Spiders, snails, centipedes, birds, and insects live in these shrub layers. On the ground, a herbaceous layer is present along with moss and lichen on fallen logs and rocks.

Figure 2-23. Deciduous forest. (Dr. E. R. Degginger.)

Before large areas of the forest in North America were cleared, the wolf, mountain lion, and wapiti were common in these forests, but now very few are left (Shelford, 1963). In areas that have become reforested, there are some large mammals such as the Virginia deer and the black bear. Small mammals are well represented, along with birds, amphibians, and reptiles.

Tropical rain forests occur in three major areas on either side of the equator: Central Africa, southeastern Asia, and South America. These areas are characterized by a relatively high, uniform temperature throughout the year and by abundant rain. The trees in these forests are mainly broad-leaved evergreens, but there are so many species that no one species is predominant. In a single square mile in a Brazilian forest, for example, as many as 300 species were noted (Billings, 1964). The fauna of rain forests is also rich in species. One area that has been studied extensively, and where the fauna is well known, is Barro Colorado in the Isthmus of Panama. This region, which is about 6 sq miles in size, is almost entirely covered with a rain forest. Among the animals living here are 58 species of mammals, 306 species of birds, and 66 species of reptiles. The invertebrate fauna is equally rich, and it has been estimated that probably over 20,000 species of insects are present (Aubert de la Rüe *et al.*, 1957).

A tropical rain forest is stratified into many layers. The dominant trees generally are 80–125 ft high, but some may extend as high as 200 ft. Below the dominant trees there may be as many as five other stories, all of which must tolerate deep shade (Fig. 2-24). Each level shelters plants and animals that may be entirely different from those living above or below them. The most abundant and varied fauna occurs in the crown of the

Figure 2-24. Tropical rain forest, Rhodesia. (Carl Frank/Photo Researchers.)

trees, where birds are the dominant vertebrates, most of which are fruit and vegetable eaters rather than predators. Among the numerous mammals present are the monkeys and apes, which feed almost entirely on leaves and fruit and descend only rarely to the ground. Arboreal squirrels, rats, some marsupials and sloths, and many species of reptiles, frogs, and salamanders are also present. These animals have evolved a number of morphological adaptations to arboreal living, such as claws, adhesive pads, sucker organs, opposable digits, and prehensile tails. The incredibly rich insect fauna tends to be brightly colored.

Epiphytes, which are plants without roots in the soil, provide numerous habitats for a wide variety of animals. Some epiphytes occur near the ground, but others grow in the higher branches of trees where leaf mold accumulating between the roots of these epiphytes forms a "suspended" soil that permits the development of a typical terrestrial microfauna high in the trees. In some plants, water derived from rain or condensation accumulates in the axils of the leaves and provides a suitable habitat situation for animals such as amphibians, which must have water to complete their life cycle (Aubert de la Rüe *et al.,* 1955).

Other habitats near and on the ground around the tree trunks support many species of birds, mammals, lizards, snakes, and amphibians. In contrast to the brightly colored insects of the tree tops, invertebrates living on the tree trunks are noted for their dull color and protective resemblance to their habitat background.

The areas where the dense primary rain forest is still intact are continually decreasing, and the destruction of the forest can almost always be attributed to the work of man. Once the land has been cleared, the soil deteriorates rapidly without its cover because it is then exposed to the effects of high temperature. Consequently, the proportion of humus diminishes, and in some areas the ground hardens into a laterite and becomes completely useless. Moreover, tropical rain, which can be torrential, causes massive erosion of the soil.

Grassland

Grasslands extend over vast areas, commonly in the interiors of continents, and tend to look very similar, whether they are present in the tropics or in the temperate-zone areas (Fig. 2-25). Many different terms have been applied to them: in Eurasia, the grasslands are called steppes; in Venezuela, the llanos; in Brazil, the campos; in Argentina and Uruguay, the pampas; and in North America, the prairies and the plains. In these areas the vegetation is dominated by grasses, legumes, and composites. Since many other families are present, however, the flora is diverse.

Figure 2-25. Midgrass prairie of central Wisconsin, showing a prairie chicken performing the courtship dance. (Dr. P. J. DeCoursey.)

Climate is the main factor controlling the invasion of the grasslands by trees, since there is not enough rainfall to support trees, but there is enough to keep deserts from forming. On the American prairies, fires also seem to serve as a limiting factor, for if a prairie is protected from fire, trees then tend to invade the more moist areas.

It has been estimated that probably 45,000,000 bison and a similar number of pronghorns lived in the grassland of North America in 1600 (Shelford, 1963). Today, most of the temperate-zone grassland areas are under cultivation or are used for controlled grazing so that the populations of bison and pronghorn have been reduced to a small fraction of earlier numbers. The fauna is dominated by large grazing herbivores, which occur in vast numbers along with the attendant carnivores, but there are also many burrowing animals, particularly among the mammals and insects.

Deserts

Deserts occupy approximately one-fifth of the earth's surface, and all have one common characteristic: lack of water (Fig. 2-26). Most deserts do receive some rain every year, but in some desert areas the rain falls only once in several years.

Deserts may be classified as being either warm or cold. Warm deserts are tropical or subtropical, and winter is either absent or very mild. In these deserts the vegetation is tropical. There is a great diurnal variation in temperature, and although temperatures are much lower at night than

Figure 2-26. White Sands National Monument, Alamagordro, New Mexico. (Dr. P. J. DeCoursey.)

during the day, frost occurs only rarely. Cold deserts, on the other hand, are subject to great seasonal extremes in temperature, ranging from −40 to 110°F, and some snow may occur. The vegetation is derived from mountain flora and from the saline marshes and sand dunes present along seacoasts in the colder areas.

Animal life is interesting and quite diversified, although the number of species and their population sizes are small. Many of the adjustments made to desert life are physiological or behavioral rather than morphological. Many animals are nocturnal, and others estivate during the most severe periods of drought. Large mammals are scarce in the desert areas, but smaller mammals, particularly rodents, are plentiful. Birds, reptiles, amphibians, and some invertebrates have become adapted to desert life, but productivity generally is low and the weight of animals per unit area is low.

SUMMARY

All biological systems must interact within the abiotic complex, which includes such factors as temperature, water, gases, light and radiation, pressure, geomagnetism, and gravity. Many factors in this abiotic complex where life has evolved have remained unchanged for millions of years, although modern man has shown capabilities for modifying them.

Of all the areas of the earth, the ocean, where life is thought to have originated, offers one of the most stable environments. All the oceans are interconnected and support a varied and abundant fauna, but the oceans are not homogeneous and vary geologically, physically, chemically, and biologically. Inshore areas of the sea, such as the intertidal zone and estuaries, are more complex and their environmental factors fluctuate more than in the open ocean.

Fresh waters, which ultimately join the ocean, are not so stable an environment as the ocean, nor do they support so rich a fauna. Standing waters, or the lakes and ponds, are transitory in time. Lakes, especially those in temperate-zone regions, are thermally stratified and can be divided into several biotic zones, but environmental conditions in ponds are almost entirely uniform. In flowing waters there is no stratification, but the animals living there must adapt so that they can maintain position in the continuously moving water.

Terrestrial habitats are infinitely varied, and environmental conditions range from the extremely harsh environments of the tundra or the desert to the uniformly favorable environment of the tropical rain forest. Terrestrial environments, especially forests and grasslands, have undergone constant change, and man has played the major role in their modification.

We know that animal life abounds throughout all these diverse environments of our biosphere. Now our goal is to gain an understanding of how an organism meets the challenge of its particular environment.

References

Aubert de la Rüe, E., F. Bourlière, and J.-P. Harroy. 1957. *The Tropics*. Knopf, New York.

Barnes, B. M., and R. D. Barnes. 1954. The ecology of the spiders of maritime drift lines. *Ecology*, 35:25–35.

Billings, W. D. 1964. *Plants and the Ecosystem*. Wadsworth, Belmont, California.

Bowden, K. F. 1967. Circulation and diffusion. In G. H. Lauff (ed.), *Estuaries*, pp. 15–36. A.A.A.S. Publ. No. 83, Washington, D.C.

Cox, A., G. B. Dalrymple, and R. R. Doell. 1967. Reversals of the earth's magnetic field. *Sci. Am.*, 216:44–54.

Davson, H. 1959. *Textbook of General Physiology*, 2d ed. Little, Brown, Boston, Mass.

Dellow, U. 1950. Intertidal ecology at Narrow Neck Reef, New Zealand (studies in intertidal zonation 3). *Pacific Science,* 4:355–374.

Elsasser, W. M. 1967. The earth's magnetism. *In* H. Odishaw (ed.), *The Earth in Space,* pp. 77–87. Basic Books, New York.

English, S. M. 1962. A limnological study of Sandy Lake, Pennsylvania, with special reference to algal stratification and productivity. Unpublished Ph.D. Thesis, Univ. of Pittsburgh.

Fawcett, E. W., and R. O. Gibson. 1934. The effect of pressure on a number of organic reactions in the liquid phase. *J. Chem. Soc.,* 1934:386–395.

Glueckauf, E. 1951. The composition of atmospheric air. *In* T. F. Malone (ed.), *Compendium of Meterology,* pp. 3–9. Am. Meteorol. Soc., Boston, Mass.

Hedgpeth, J. W. 1957. Classification of marine environments. *In* J. W. Hedgpeth (ed.), *Treatise on Marine Ecology and Paleoecology,* Vol. I, pp. 17–28. Geol. Soc. Am. Mem. 67, Washington, D.C.

Henderson, L. J. 1913. *The Fitness of the Environment.* Macmillan, New York.

Hitchcock, F. A. 1964. Animals in high altitudes: the aerosphere. In *Handbook of Physiology, Section 4, Adaptation to the Environment,* pp. 835–839. Am. Physiol. Soc., Washington, D.C.

Hock, R. J. 1964. Terrestrial animals in cold: reptiles. In *Handbook of Physiology, Section 4, Adaptation to the Environment,* pp. 357–359. Am. Physiol. Soc., Washington, D.C.

Jansson, B. 1967. The availability of oxygen for the interstitial fauna of sandy beaches. *J. Exp. Mar. Biol. Ecol.,* 1:123–143.

Johnson, R. G. 1965. Temperature variation in the infaunal environment of a sand flat. *Limnol. Oceanog.,* 10:114–120.

Johnson, R. G. 1967. Salinity of interstitial water in a sandy beach. *Limnol. Oceanog.,* 12:1–7.

Kurz, H., and K. Wagner. 1957. Tidal marshes of the Gulf and Atlantic coasts of Northern Florida and Charleston, South Carolina. *Florida State Univ. Studies,* No. 24.

Macintyre, I. G. and O. H. Pilkey. 1969. Tropical reef corals: tolerance of low temperatures on the North Carolina continental shelf. *Science,* 166:374–375.

Morrison, P. R. 1964. Adaptation of small mammals to the Arctic. *Federation Proc.,* 23:1202–1206.

Odum, E. P. 1959. *Fundamentals of Ecology,* 2d ed. Saunders, Philadelphia.

Pearse, A. S., H. J. Humm, and G. W. Wharton. 1942. Ecology of sand beaches at Beaufort, N. C. *Ecol. Monographs,* 12:136–190.

Pittendrigh, C. S. 1967. Life on other planets. *In* H. Odishaw (ed.), *The Earth in Space,* pp. 297–307. Basic Books, New York.

Pritchard, D. W. 1967. What is an estuary: physical viewpoint. *In* G. H. Lauff (ed.), *Estuaries,* pp. 3–5. A.A.A.S. Publ. No. 83, Washington, D.C.

Regnault, M. V. 1852. Recherches sur la composition de l'air atmosphérique. *Ann. Chim. Phys.,* 36:385–405.

Reid, G. K. 1961. *Ecology of Inland Waters and Estuaries.* Reinhold New York.

Riley, G. A., H. Strommel, and D. F. Bumpus. 1949. Quantitative ecology of the plankton of the western Atlantic. *Bull. Bingham Oceanog. Coll.,* 12:1–169.

Sanders, H. L., P. C. Mangelsdorf, Jr., and G. R. Hampson. 1965. Salinity and faunal distribution in the Pocasset River, Massachusetts. *Limnol. Oceanog.,* 10:(Suppl., Nov.) R216–R229.

Schwarzschild, M. 1967. The sun as a star. *In* H. Odishaw (ed.), *The Earth in Space,* pp. 14–24. Basic Books, New York.

Shelford, V. E. 1963. *The Ecology of North America.* University of Illinois Press, Urbana, Ill.

Smith, R. L. 1966. *Ecology and Field Biology.* Harper & Row, New York.

Sverdrup, H. U., M. W. Johnson, and R. H. Fleming. 1942. *The Oceans, Their Chemistry, and General Biology.* Prentice-Hall, New York.

Truesdale, G. A., A. L. Downing, and G. F. Lowden. 1955. The solubility of oxygen in pure water and sea-water. *J. Appl. Chem., London,* 5:53–63.

Uffen, R. J. 1963. Influence of the earth's core on the origin and evolution of life. *Nature,* 198:143–144.

Wells, H. W. 1961. The fauna of oyster beds, with special reference to the salinity factor. *Ecol. Monographs,* 31:239–266.

Yonge, C. M. 1940. The biology of reef-building corals. *Sci. Repts. Great Barrier Reef Exped.,* 1:353–391.

3

The Zone of Resistance

Environmental factors are not constant. They fluctuate temporally and geographically, and the stresses on animals resulting from these environmental alterations affect the complex biotic structure of our biosphere. Learning to predict the consequences of these stresses may be important to the continuing survival of many forms of life, including that of man himself. Beforehand, however, it is essential to understand the nature of environmental gradients and the adaptive methods used by organisms to tolerate a changing environment.

The gradients established in an environment often vary widely; they may be steep or sharply discontinuous, as in the thermocline, or they may be gradual and of relatively small magnitude, as the annual temperature change in tropical coastal waters. Organisms persisting in a specific geographical area must be able to survive environmental extremes found there. However, if an environmental factor reaches a given intensity for a sufficient period of time, it may limit species distribution. For example, populations of the cold-water mussel, *Mytilus edulis,* often become established during the winter in the area south of Cape Hatteras, but if they are subjected to prolonged summer water temperatures in excess of 27°C, the mussel populations cannot survive (Wells and Gray, 1960).

Biologists have recognized for years that environmental factors can be limiting. Over a hundred years ago, Liebig (1840) noted that plants need a minimal amount of specific constituents to survive. This generalization is known as the Law of the Minimum. Later, Shelford (1913) observed that animals are limited at both maximal and minimal points in an environmental gradient, and he proposed the more general concept of tolerance limits (Law of Toleration). Between the upper and lower tolerance limits, the adaptations of organisms are termed *capacity* adaptations; the adaptations at the tolerance limits are called *resistance* adaptations. Organisms are not exposed to one environmental factor at a time, and the interaction of two or more factors nearing tolerance limits generally produces a more drastic effect on organisms than would any single factor.

What are the toleration limits of organisms living, for example, in an estuary that is being polluted or altered by dredging operations? What happens to organisms when man attempts to modify the weather? To answer questions like these, it is essential to study resistance adaptation of organisms to various environmental stresses; it is necessary to know not only toleration limits of organisms from different habitats, but also the basic mechanisms of adaptation. In addition to considering the traditional climatic factors such as temperature, salinity, and oxygen, the present-day student must investigate man-inspired disruptive factors, such as radioactive and chemical pollutants.

TEMPERATURE

Temperature is one of the most obvious changeable environmental factors. It varies not only between climatic regions, but also in the temporal changes of all habitats. Since environmental biologists frequently study the obvious, and since temperature can be controlled and measured with ease and precision, it is not surprising that the data available on the influence of temperature on organisms are much greater than for any other environmental factor. As a result, much is known of thermal effects on organisms at all levels of biological organization, from submolecular to ecosystem, including the variables that govern organismal responses to temperature. Despite the wealth of valuable data, there are still serious gaps in our understanding of thermal biology. In this chapter, only one aspect of this broad topic will be emphasized, namely, the thermal resistance adaptations of organisms. We shall use this information to answer such questions as: How does temperature limit organisms? What organismic functions are involved? What strategies have evolved to allow organisms to occupy thermally distinct habitats?

To determine how temperature limits an organism, various experimental procedures have been used to describe its thermal lethal limits. In general, these techniques involve subjecting an organism to a given controlled temperature for a period of time and then judging the animal to be either dead or alive. This, the quantitative expression of a lethal factor, usually involves both time and a specific point along a thermal gradient. For example, an individual belonging to species Z dies after 20 min exposure to 44°C. One must subsequently judge if this response is typical of other individuals of species Z or if other specimens can survive in this temperature for shorter or longer times. To circumvent this problem,

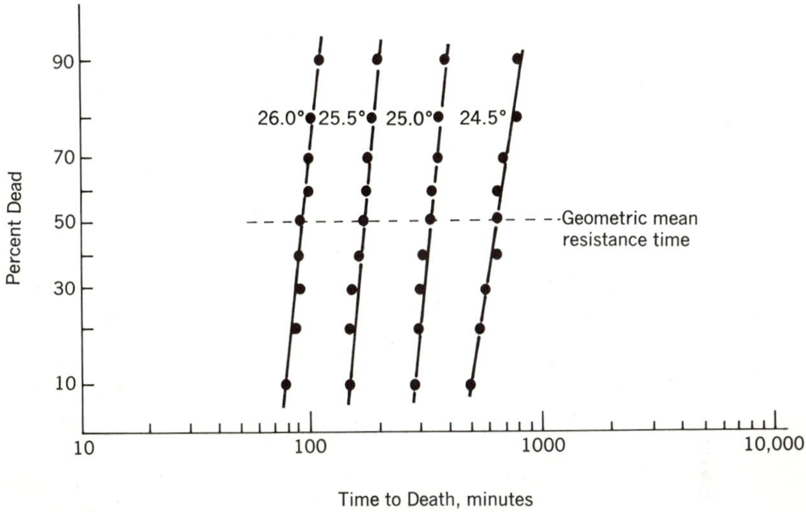

Figure 3-1. Survival time at various upper lethal temperatures of young spring salmon, *Oucorhynchus ishawytscha,* acclimated to 10° in fresh water. (From Brett, 1952.) The LD_{50} at 26° is 90 minutes but at 24.5° it is almost 650 minutes.

many research workers quantitatively express their results as the time that it takes at a certain temperature to produce a stated percentile mortality in a population sample. Although various percentile values have been used, the value of 50 percent is generally accepted. This value is termed the LD_{50} (lethal dosage), or the median tolerance temperature. An example of LD_{50} curves is given in Fig. 3-1. If the LD_{50} values are plotted against a series of temperatures, the resulting curve graphically represents time/temperature relationships. These curves are of particular value when responses of one species are compared under different experimental conditions or when a number of different species are compared (Fig. 3-2).

Many experimental variables must be considered in determining thermal lethal limits of organisms—for example, whether or not organisms should be subjected immediately to a new experimental thermal level, or whether experimental temperature should be increased or decreased at a gradual and predetermined rate until the desired temperature is reached. The criterion of death, or other indication of the lethal limit, is often different for each species and with each experimenter. Moreover, if one is more interested in the survival of a population of animals than of individuals, this same criterion may not be valid. First, although an individual may survive a given thermal exposure, the exposure may make it reproductively sterile because of permanent damage to

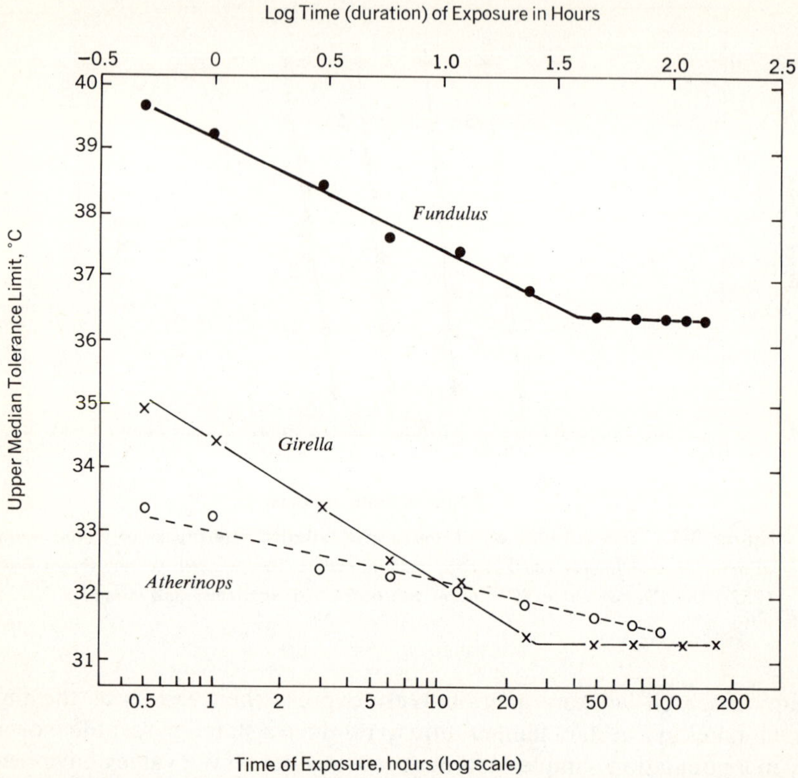

Figure 3-2. Thermal tolerance of different species of marine fish. (From Doudoroff, 1945.)

gonadal tissue. Such an individual will not contribute to the perpetuation of the species, and it may therefore be considered genetically dead. Second, many species require a certain range of temperature for reproduction, and this does not kill the adult or injure its gonadal tissue. But if the environmental temperature is above or below this required range, the animal will not reproduce so that there will be no addition of young animals to the population. If this trend persists, the population will disappear. Obviously, absolute values obtained by means of these different methods may not be comparable, but there are common indicators that help in understanding the ecology of animals.

In poikilotherms, the body temperature varies with surrounding environmental temperature (Fig. 3-3). The temperature range that a cold-blooded animal tolerates can be correlated closely with the environmental temperature normally encountered in nature: tropical animals cannot

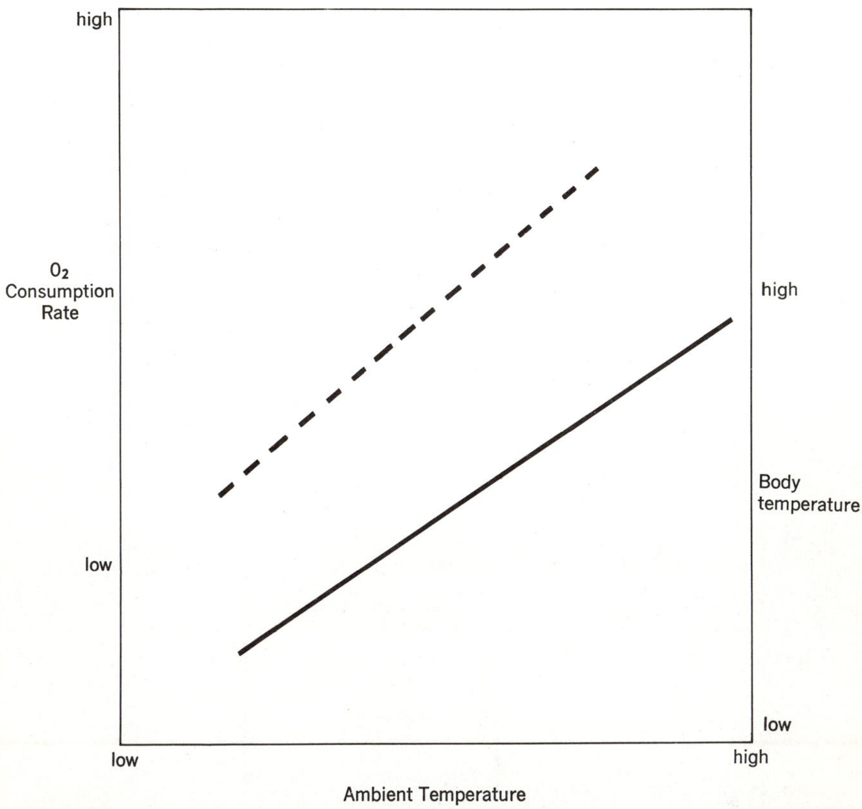

Figure 3-3. Generalized response of body temperature and oxygen consumption rate of a poikilotherm to a changing ambient temperature.

withstand low temperature; polar animals cannot withstand high temperature; temperate-zone animals survive a wide temperature range (Fig. 3-4). Many animals, both vertebrate and invertebrate, show a rather precise discrimination of temperature. If the temperature decreases or increases appreciably, these animals may tolerate and adjust to change, may avoid the change by seeking a less exposed environment, or may become inactive. Some animals respond by becoming dormant during periods of extreme temperatures. Such dormancy is termed *estivation* at high temperatures and *hibernation* at low temperatures.

Many poikilotherms exercise a degree of thermoregulation by either physiological or behavioral mechanisms. For example, when temperatures drop, lizards will bask in the sun to achieve the desired body temperature; once this thermal level is attained, they will divide their time

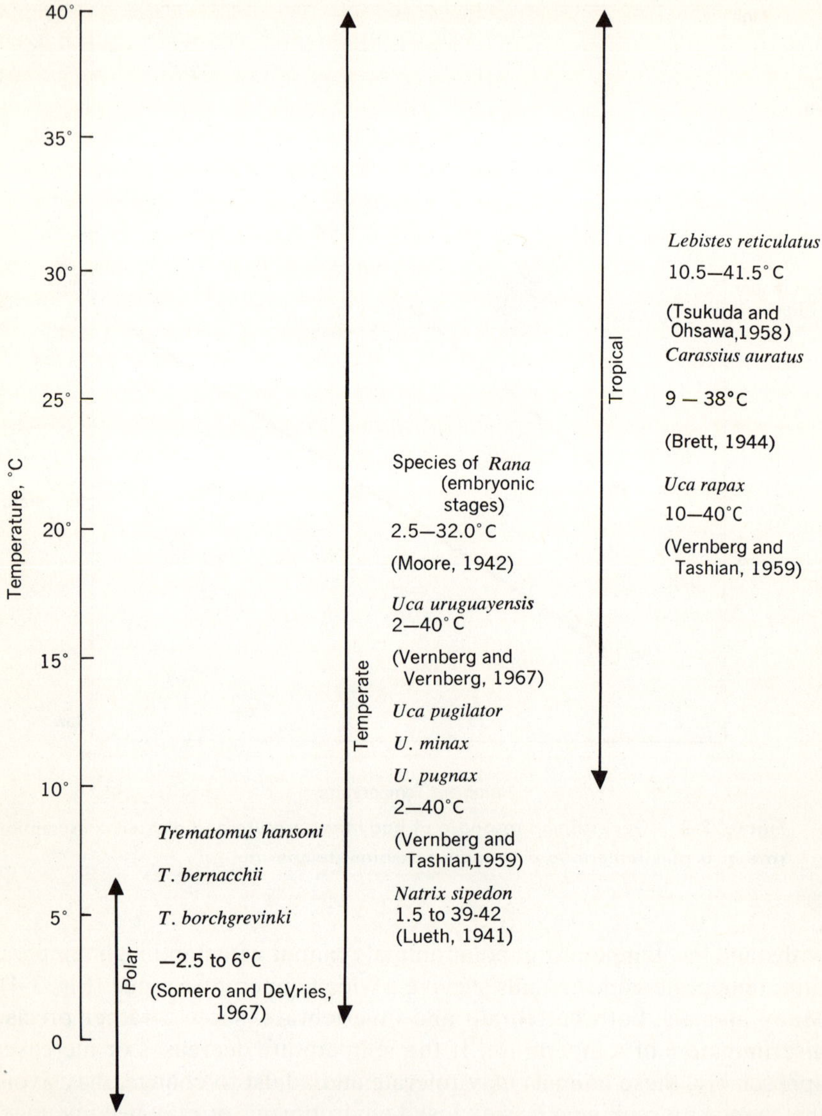

Figure 3-4. Comparative upper and lower thermal limits of tropical, temperate, and polar zone animals.

between sun and shade to maintain it. Poikilotherms such as desert dwellers that live in environments where the temperature is apt to be very high are often nocturnal and thus avoid the highest temperatures of the day. Other groups of animals, such as social insects, will aggregate to raise the thermal level to a point compatible with survival. If a temperature drop is not too stressful, many poikilotherms are able to achieve an initial transient increase in metabolic rate resulting from increased nervous activity, followed by a decrease in metabolic rate, with a subsequent leveling off of oxygen consumption. If the temperature drop is too great, however, chill coma results, followed by death. Seasonal metabolic changes do occur, and a temperature that would be lethal in midsummer could be well within the tolerance range of an animal in mid-January. Some poikilotherms, both vertebrates and invertebrates, lower their body temperatures slightly by evaporative cooling. In frogs and reptiles, evaporative cooling can occur through the skin or via the respiratory tract by panting (Schmidt-Nielsen and Bentley, 1966; Warburg, 1967). Among the invertebrates, evaporative cooling has been reported in tropical intertidal-zone animals (Lewis, 1963).

In contrast to poikilotherms, the body temperature of homeotherms changes little with the environmental temperature (Fig. 3-5). In general, the deep-body temperature of homeotherms like arctic birds and mammals is the same as that of homeotherms living in the tropics or in the desert. These organisms are able to maintain such constancy of body temperature by a combination of several factors: (1) a thermoregulating

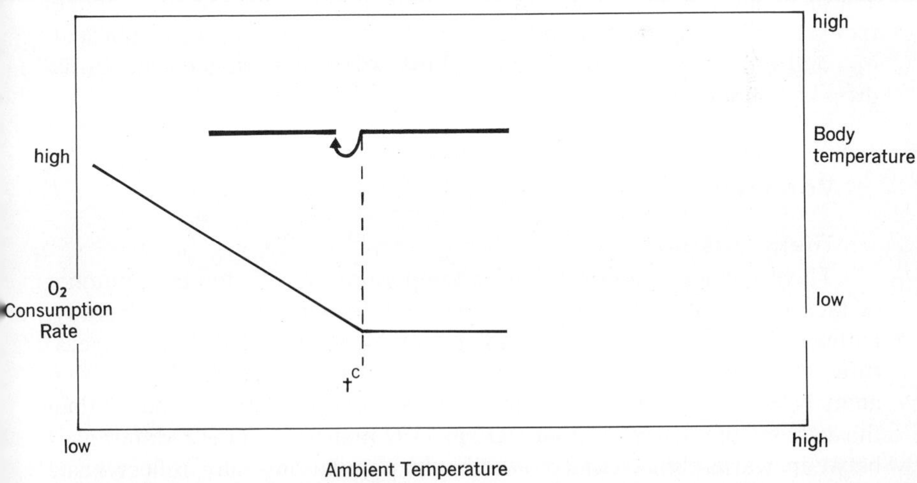

Figure 3-5. Generalized response of body temperature and oxygen consumption rate of a homeotherm to a changing ambient temperature. At a critical temperature (t^c) the body temperature drops and the oxygen consumption rate increases. As a result, the body temperature returns to the previous level.

center in the brain; (2) insulation; (3) a peripheral vascular response to ambient temperature; and (4) metabolic compensation.

Among these factors, insulation is highly important, particularly in enabling large arctic animals to withstand very low temperature. If the temperature is lowered, the oxygen consumption rate of tropical mammals increases, whereas that of arctic mammals remains basal. This response pattern is due to differences in insulation; some arctic animals are fur-insulated, whereas others are insulated with a layer of blubber. Many small animals, however, are not well insulated and must seek heat retention by other means, often by huddling together. Some animals compensate for a decrease in temperature with an increased metabolic capacity, although this capacity is subject to seasonal changes.

Homeothermic animals living in very hot environments cannot tolerate greatly elevated body temperature any better than those living in cold environments, and they must utilize methods that will facilitate heat transfer to the environment, including an increased peripheral blood flow and surface cooling by sweating or panting. Birds may accomplish this by rearranging their plumage so that more skin is exposed. There is, however, no reduction in metabolic rate. Animals subjected to high temperature may exhibit diurnal patterns of behavior—that is, they may reduce locomotor activity during the heat of the day or may move into the shade to avoid direct sunlight.

Various species shift between homeothermy and poikilothermy. These include animals such as the pigmy mouse and the little pocket mouse, which respond to temperature extremes by estivating or hibernating, and others such as the hummingbird, which experience a nocturnal drop in temperature.

Variables

Thermal History

The response of an organism to temperature is affected by a number of factors, one of the most important being thermal history. Generally, animals found in warmer environments can withstand higher temperatures than animals from colder situations. It is also generally true that animals from colder environments tolerate lower temperatures better than those from the warmer climes (Figs. 3-6 and 3-7). These differences between warm-climate and cold-climate populations may reflect basic genetic differences or may be phenotypic expressions resulting from different thermal histories. Thus, within the genotypic framework of a population, the lethal limits may be labile and can be shifted by previous thermal conditioning. For example, the thermal limits of fiddler crabs

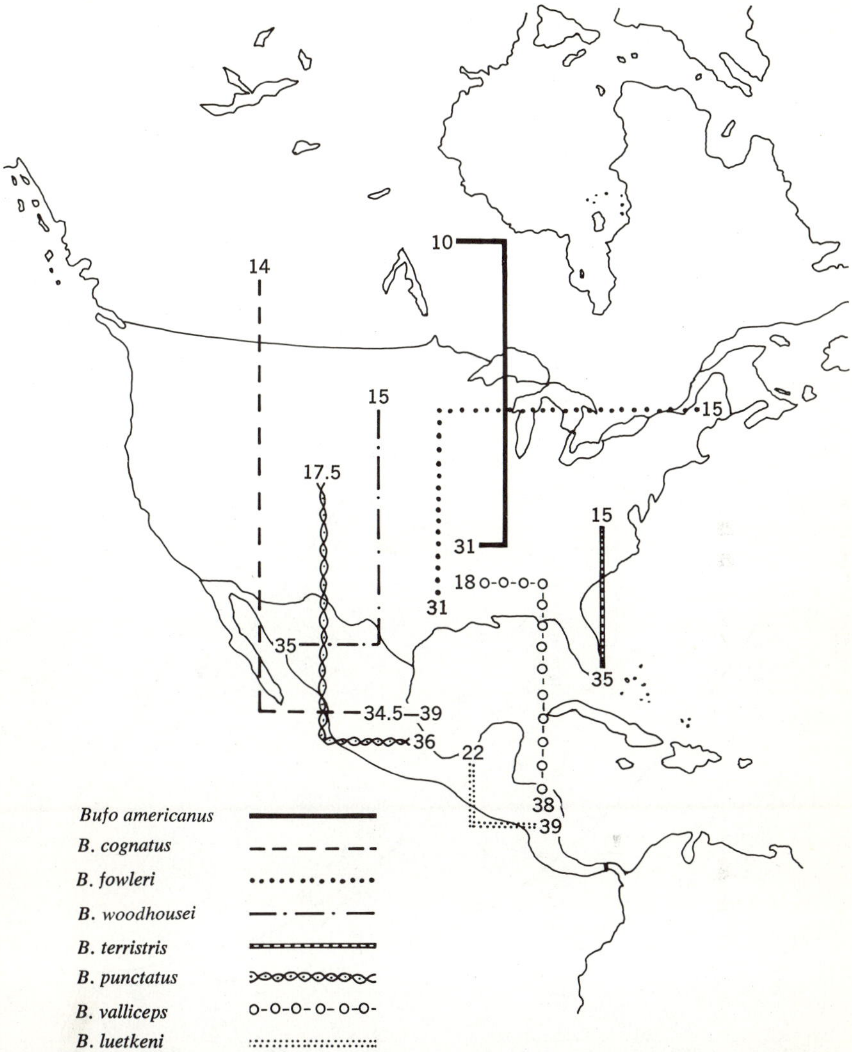

Figure 3-6. Thermal lethal limits of various species of toads from different geographic regions. The numbers indicate upper and lower thermal lethal temperature (°C) for each species. (Based on data of Ballinger and McKinney, 1966.)

(*Uca*) from the temperate zone are a function of the maintenance temperature existing previous to experimentation. Cold-acclimated animals can withstand low temperature longer than warm-acclimated animals, while the reverse response is observed at high temperatures.

There are, however, interspecific differences, and crabs from the tropics cannot survive low temperatures. They show limited genetic

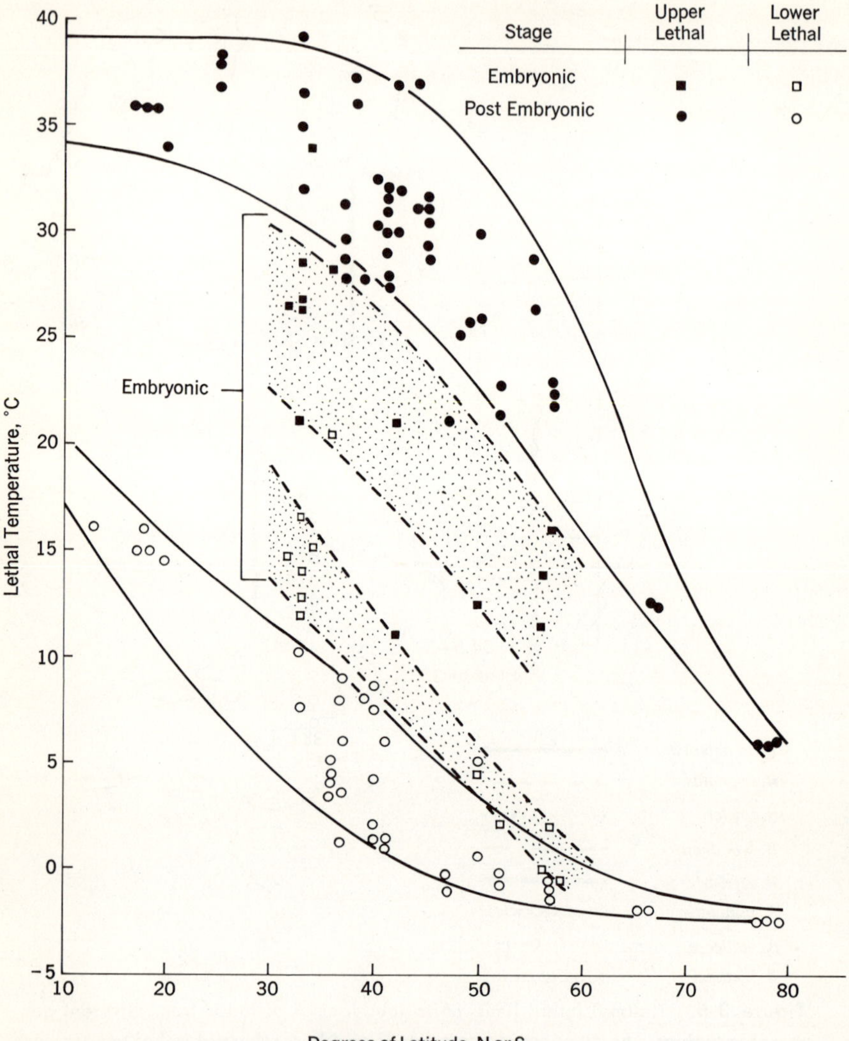

Figure 3-7. The correlation of upper and lower lethal thermal temperature limits of embryonic and post embryonic fishes with latitude. (Brett, 1970.)

lability; that is, their response is changed only slightly by thermal acclimation. These interspecific differences may be correlated with the seasonal thermal regime each species encounters. Tropical species live in a relatively constant, warm environment, whereas temperate-zone species are confronted with marked daily and seasonal temperature fluctuations. Shifting of lethal limits may be studied by acclimation work in the

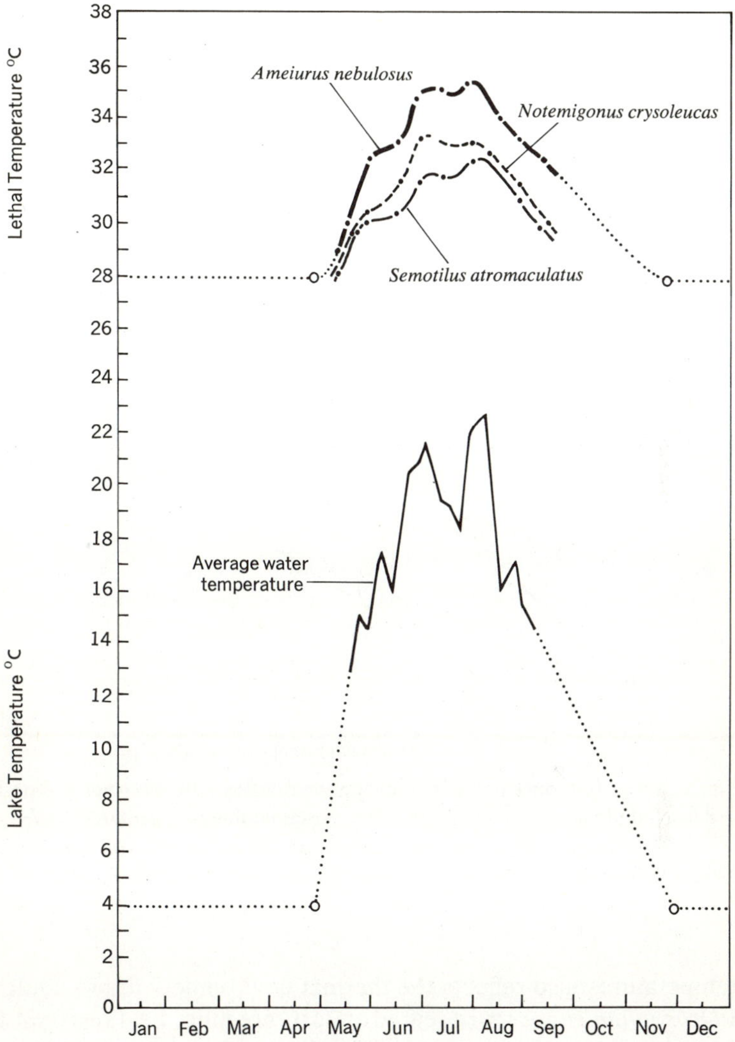

Figure 3-8. Changes in the lethal temperature of certain fresh water fish in response to seasonal changes in ambient temperature. (From Brett, 1944.)

laboratory, as well as by using animals that are acclimated in the field. For example, seasonal shifts in the upper lethal temperature of fresh-water fish is directly correlated with monthly changes in their habitat temperature (Fig. 3-8). Generally, temperate-zone animals during the summer, or animals acclimated to summer temperatures, can withstand higher thermal stress than winter animals. Survival time of adult organisms at

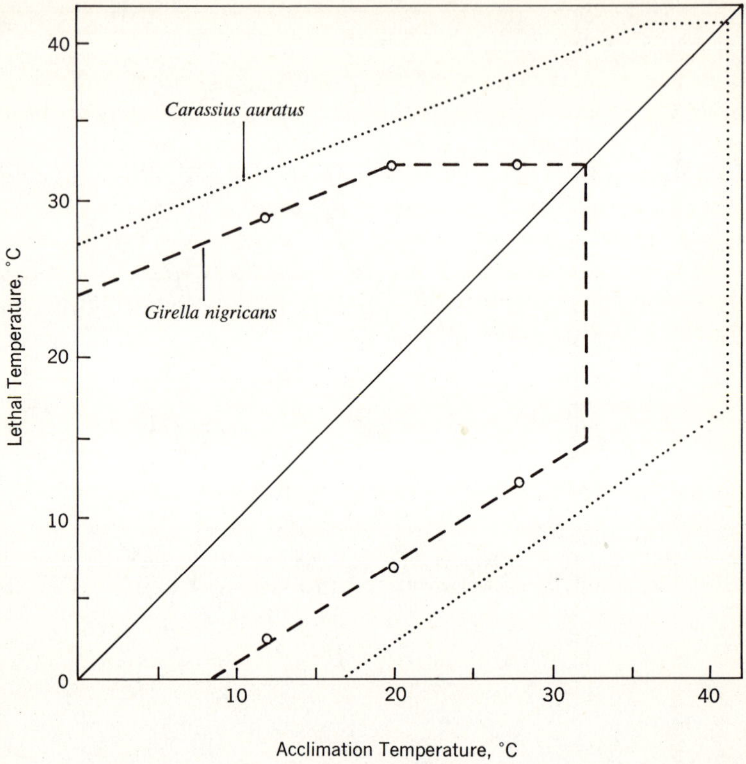

Figure 3-9. Tolerance polygons showing the relation between median high lethal and low lethal temperatures as a function of acclimation temperatures in two species of fish. The limits of the tolerance polygon of *Carassius* are wider than those of *Girella* indicating that *Carassius* is more eurythermal. (After Brett, 1944.)

high temperatures also reflects the thermal environment of pre-adult life. Adult *Drosophila subobscura* reared at 25°C are more heat resistant than flies reared at 15°C (Smith, 1957). Excellent methods for graphing the influence of acclimation on lethal limits have been developed by Fry and his co-workers. One example is the tolerance polygon shown in Fig. 3-9.

Genetic Differences

The thermal resistance of a species may be closely identified with the genetic composition of the parent stock, as illustrated by hybrid development in two anuran species, *Bufo valliceps* and *B. luetkeni* (Ballinger and McKinney, 1966). The lower lethal temperature for development of *B. valliceps* is 18°C; for *B. luetkeni,* 22°C. In hybrid crosses between these two species, the lower lethal limit is between that of the two parents at 19.5°C.

The lethal development temperature of hybrid organisms reflects not only the genetic composition of their parents, but also the thermal limits of the adult organisms. Hybrids resulting from a cross between a warm-water species of speckled trout, *Salvelinus fontinalis,* and a cold-water species of lake trout, *S. namaycush,* were acclimated to 10°C and then placed under high thermal stress; the response was intermediate between that of the two parents. When the hybrid fish were acclimated at 20°C, however, they assumed the characteristic response of the more heat-tolerant parent species (Fig. 3-10).

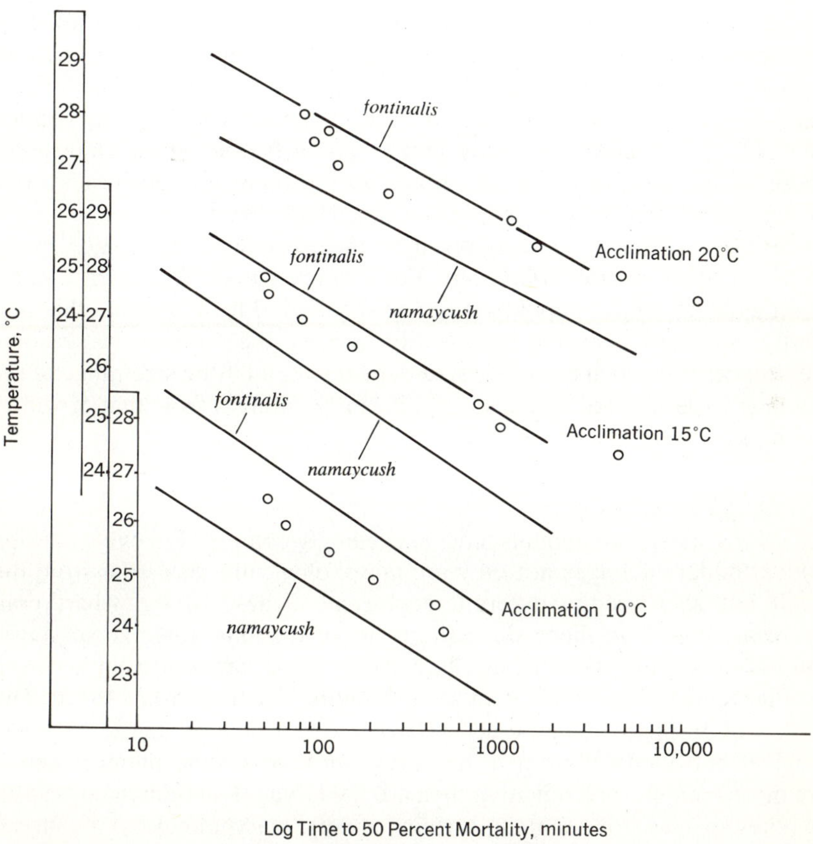

Figure 3-10. Resistance of hybrid fish to thermal stress in comparison with that of the parents, *Salvelinus fortinalis* and *S. namaycush,* when acclimated to different temperatures. The o is the lethal point of the hybrid. (From Fry and Gibson, 1953.)

Diet

No consistent relationship between the total intake of food and resistance to thermal stress has been demonstrated. Some species are more sensitive to elevated temperatures when they are starved for even short periods of time, while the thermal response of others is not influenced by starvation. On the other hand, goldfish showed increased resistance to high temperature when placed on a high fat diet (Hoar and Cottle, 1952).

Size

Size is one of the most complex factors to be considered when thermal lethal limits of animals are being compared. In some species, the smaller animals are more resistant to higher temperatures than the larger ones (Todd and Dehnel, 1960); in some species the reverse is noted; and in other species size is not an apparent variable (McLeese, 1956). Furthermore, the response of large- and small-sized animals at high- and low-temperature extremes apparently is not the same. Some smaller animals die faster at low temperatures than do larger animals of the same species, but at high temperatures body size is not a factor. These differences indicate that the cause(s) of death are not the same at opposite ends of a thermal gradient (Barron and Tahmisian, 1948; Brett, 1952).

Body size must be considered in intraspecific comparisons under different experimental conditions. For example, the LD_{50} at 42°C for a summer and a winter population of one species is 3 hr. However, this similarity in values does not necessarily indicate the absence of seasonal differences if thermal limits are size-dependent and if the size range of the summer experimental animals is significantly different from that of winter animals.

Stage in Life Cycle

Many species of animals have exceedingly complex life cycles during which the larval stages not only are morphologically dissimilar from the adult but also occupy different ecological niches. In the wharf crab *Sesarma cinereum,* thermal requirements of the planktonic zoeal stages are different from those of adults because these larvae are limited to a smaller temperature range than are the adults (Costlow *et al.,* 1960). The larvae of tropical fiddler crabs (*Uca*) and the pelagic larvae of some tropical bottom-dwelling fish, however, show a reverse pattern: adults are more temperature-sensitive than are the larvae (Kuthalingham, 1959). In other species (for example, certain parasitic platyhelminths) the larvae reflect the thermal limits of the adult stage despite their morphological dissimilarity and different ecological situations (Vernberg, 1961; Davies and Walkey, 1966).

In fish reared in the laboratory under constant thermal conditions, the responses to temperature during the life cycle were different (Morris, 1962). Larvae could not compensate for either high or low temperatures, but as the fish attained a larger size, the animal could metabolically compensate for low temperatures. Metabolic acclimation to high temperatures occurred as the animal became even larger, but was accompanied by loss of ability to compensate for low temperatures.

Other Factors

In addition to the factors discussed above, several others have reportedly affected the thermal lethal limits of organisms. The sex of the animal may be important, for according to a few reported studies, the female tends to be more tolerant than males (Tsukuda and Ohsawa,

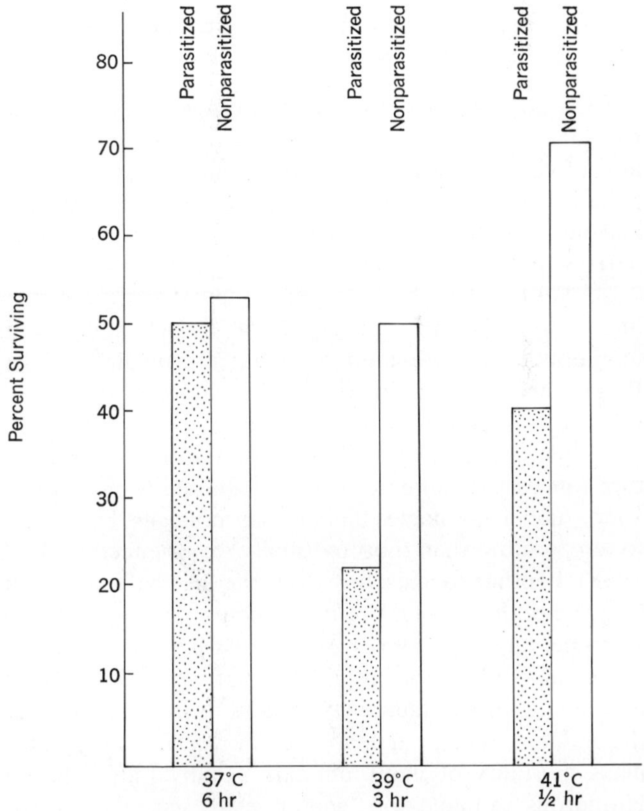

Figure 3-11. Survival time of parasitized and nonparasitized *Nassarius obsoleta* (Say) at different temperatures. (Based on data from Vernberg and Vernberg, 1963.)

1958). Among invertebrates, molting adds yet another variable. In some crabs molting adversely affected heat resistance, but had no effect on other species (Spoor, 1955; Todd and Dehnel, 1960).

Parasitism is also a factor that should be considered. Mud-flat snails, *Nassarius obsoleta,* which are heavily infected with trematode larvae, cannot withstand high temperatures as well as nonparasitized snails (Fig. 3-11).

Hormones have been shown to influence cold and warm resistance as well as acclimation to temperature, especially in mammals and a few other invertebrates. These interactions are complex, and the student is referred to endocrinology textbooks, journals, and review articles for further details (Fisher, 1958; Gorbman and Bern, 1962; Rao, 1967; Rose, 1967; Hart, 1969).

Temperature as a Limiting Factor in the Distribution of Animals

In both the northern and southern hemispheres, many species extend from the warm, temperate areas far into the colder regions. Species that are active over a wide temperature range are said to be *eurythermal.* As a general rule, terrestrial animals are more apt to be eurythermal than marine or aquatic species. In contrast, other species are limited to a very narrow range of environmental temperature, and these are termed *stenothermal.* For the most part, tropical forms are stenothermal animals limited to high temperatures, although some stenothermal animals in the temperate zone are limited to an intermediate range of temperature. Poikilotherms inhabiting arctic and antarctic areas tend to be stenothermal forms limited to low temperatures.

Most organisms are able to survive the "normal" seasonal temperature ranges encountered in their habitats. Some species in the intertidal zone, such as barnacles and certain molluscs, have wider tolerance limits to temperature than those normally experienced in that area (Southward, 1958). In other species, however, the thermal lethal limits are very similar to those of the "normal" environmental temperature range, and when environmental temperatures deviate sharply, mass mortality occurs. An excellent example of the lethal effects of unusually high temperature may be found in reports of faunal investigations in the Peru coastal current. Here, the sudden irregular appearance of a warm countercurrent resulted in the mass mortality of many animals usually found there (Gunther, 1936). Numerous examples of "winter kills," especially among Florida fishes, have been reported following unusually severe winter temperatures.

Any difference between the laboratory-determined lethal limits and the temperatures normally encountered in nature reflects a margin of safety for survival. At the distributional limits of a species, this margin of safety may be less than that of a centrally located population. Evidence to support this generalization comes from studies on fiddler crabs, *Uca rapax*. During the severe winter of 1957, the populations of this fiddler crab at its northernmost habitat in Florida were killed, and their northern survival areas retreated southward by approximately 100 miles.

In marine environments there are animals living at nearly all environmental temperatures. A good example are fish that live in the Antarctic under a nearly permanent ice cover in water continuously at freezing temperature (Wohlschlag, 1964). As noted earlier in this chapter, the animals from colder geographical areas are generally less heat tolerant and more cold tolerant than those animals from warmer regions; for example, specimens of *Aurelia*, a jelly fish, from Nova Scotia die at a water temperature of 29–30°C, while *Aurelia* from Florida can tolerate temperatures up to 38.5°C. Similarly, the horseshoe crab *Limulus* at Woods Hole dies at 38.5°C, whereas specimens from Florida survive up to 46.2°C (Mayer, 1914). Whether these differences reflect different genetic constitutions of the two populations or simply physiological adaptation is not known.

Not only lethal effect of temperature in the geographical distribution of animals but also temperature effects on breeding, feeding, and other biological activities must be considered. Generally, the distribution of shallow-water marine species can be assigned to four types of zonation. In the first type, northward distribution is dependent on thermal lethal limits during the winter months, and southern distribution is dependent on summer temperature limits. In a second type, the thermal limits required for repopulation determine the north to south distribution. In the third type of zonation, the thermal requirements for repopulation determine the poleward habitat in summer, and the maximum temperature determines the equatorward survival area. Finally, the minimum temperature for survival determines the poleward limit in winter and temperatures limiting repopulation determine the southward range (Fig. 3-12). Terrestrial invertebrates, particularly arthropods, generally are distributed in all thermal environments where life is found. Many arthropods that have invaded the colder areas have one stage in their life cycle which is very resistant to cold, enabling them to overwinter until warmer weather returns (Salt, 1964). Birds and mammals also are adapted to live in nearly all thermal environments. The distribution of amphibians and reptiles, however, is limited to the relatively warmer thermal climates. Hock (1964) has listed three factors that limit the invasion of reptiles into cold environ-

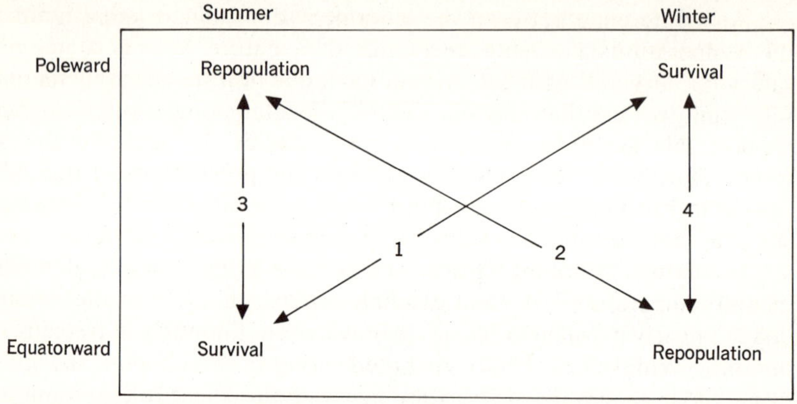

Figure 3-12. Diagrammatic representation of the four basic types of limitation to extension of geographic range of organisms. Arrows indicate the ways in which pairs may combine. See text for explanation. (After Hutchins, 1947.)

ments: the daily ambient temperature must be high enough to allow activity; the daily ambient temperature must be high enough and long enough to allow breeding and to allow adults and young to acquire food for "overwintering"; and there must be adequate sites for hibernation.

With this background, the question may be asked why a temperature that is lethal to one species is compatible with life in another species.

Cause of Death

Since animals from the polar regions usually die from heat death at temperatures below the thermal lethal levels of most tropical species, any interpretation of these diverse temperature relationships becomes very involved. It is known that temperature continuously influences each level of biological organization from the submolecular to the organismic, although not necessarily in the same manner. If all temperature effects are integrated and the integrity of the organism is maintained, the organism survives. The thermal lethal limits of the whole animal are the expression of the irreversible failure of some process or a number of processes to operate properly. Little is known of the actual causes of death at either high or low temperatures, however.

When animals are exposed to extremes in temperature, death may occur within minutes or it may be delayed for hours or days, which suggests that death may be due to various causes, each operating at a different combination of temperature and time. It is reasonable to expect that the resistance capacity of the most sensitive tissue necessary for life determines the lethal point of the entire organism. There is good evidence to show that the nervous system in poikilotherms is a primary site of thermal injury at both high and low temperatures (Brett, 1956; Konishi and Hickman, 1964). In frogs, the whole organism is more thermally sensitive than is the neuromotor junction, followed by muscle, heart and nerves, respectively, whereas the nervous tissue of skates is inactivated before somatic and smooth muscles. Other evidence suggests a correlation between the lethal temperature of the whole organism and failure of the central nervous system. At high temperatures, respiratory movements in fish cease entirely with the approach of death, suggesting inactivation of the respiratory center, and brain tissue metabolism is markedly depressed at approximately the same thermal level that is lethal to the intact organisms. There is also evidence to indicate that brain tissue adapts metabolically in animals living in low-temperature habitats. At a temperature of 10°C, the respiration rates of brain tissue from a warm-water species of fish were reduced, but brain tissue from a cold-water species was able to maintain a consistent rate over the thermal range of 0–20°C (Peiss and Field, 1950).

Studies on invertebrates have demonstrated a similar interrelationship of tissue and whole-animal lethal limits. The LD_{50} values of isolated pieces of gill tissue from several species of lamellibranchs are similar to those of the intact organism. Furthermore, the thermal limits of both tissues and intact animals can be correlated with habitat preference; for example, tissues from intertidal-zone animals survive at higher temperatures than do tissues from species living in cold, deep water.

Various theories have been advanced to explain the lethal effect of temperature. One suggests that key enzyme systems are inactivated by heat; then, after a certain point, the inactivation is irreversible. Thermal inactivation of an enzyme system has been demonstrated in bivalved molluscs. It has been suggested that the thermostabile aspartic glutamic transaminase found in some of these molluscs enables them to live in high-temperature environments (Read, 1963). This, however, may not be a valid correlation, since the aspartic glutamic transaminase could be of importance only in facultative anaerobic bivalves rather than being an important enzyme in temperature tolerance per se.

Work on antarctic fishes indicates that enzyme inactivation probably cannot account for thermal deaths in all animals. Brain tissue of three species of cold-water fishes were inactivated at temperatures as low as 5°C, which is indeed a very low temperature for enzyme inactivation. Furthermore, gill tissue from the same fish was not similarly affected. Instead, it has been suggested that thermal death is in some way related to lipid metabolism. At such low temperatures, death of the antarctic fishes may be related to biophysical changes in the cellular lipids in the brain, a suggestion that is consistent with the theory advanced several years ago by Heilbrunn (1952) and Somero and DeVries (1967).

Still another theory stresses the importance of the viscosity of protoplasm in limiting the metabolic rate. Essentially, this theory is as follows: Protein molecules are held in the intramicellar spaces by capillary action. Bound water is held by these protein molecules and to a greater or lesser degree it is impermeable to solutes. This relationship of the nonsolute-bearing space to the solute-bearing space governs the ease with which the metabolites can be mobilized, thus influencing the metabolic rate. The absolute amount of bound water within a cell is taken as the factor that governs the stability of a protein molecule, and this stability determines molecular resistance to temperature extremes. As a result, heat denaturation occurs at elevated temperatures, owing to the breakdown of the micellar structure and not to the coagulation of proteins. Withdrawal of vital water from the cell causes death at low temperature.

Comprehensive references on the causes of thermal death may be found in the works of Bělehrádrek (1935), Johnson *et al.* (1954), Precht and Hensel (1955), and Hoar (1966).

SALINITY

There are two general types of responses to salinity: animals that can adjust to wide ranges of salinity are referred to as *euryhaline,* and those restricted to very narrow limits are termed *stenohaline.* These are arbitrary terms, and many animals, especially marine forms, are intergrades between these two extremes.

The amount of salt in the water has a pronounced effect on the salt concentration within the body of an animal. The ability to control the total amount of salt is termed *osmoregulation.* Organisms that are able to maintain a constant internal osmotic pressure independent of the external medium are said to be regulators and are *homoiosmotic.* Those animals in which the osmotic pressure varies with the external medium are termed adjustors and are *poikilosmotic.* An euryhalinic organism may be either homoiosmotic or poikilosmotic.

Whole-Animal Response

As a general rule, animals living in fresh or open ocean waters are much less tolerant of salinity fluctuations than those living in estuarine situations. Some animals can live in water ranging from fresh to concentrated sea water, such as the brine shrimp *Artemia salina*. Even the eggs of this shrimp are able to develop in a wide range of salinity. Other marine species, such as oysters and the blue crab, can withstand dilute sea water. In contrast, corals found in tropical waters cannot withstand salinity reductions of 20 percent, while many fresh-water species cannot withstand salinity increases over 5–8 percent. At the other extreme, a number of animals can tolerate hypersaline conditions. Two species of crab, *Pachygrapsus crassipes* and *Hemigrapsus nudus,* living in a hypersaline pond cut off from the ocean waters, were active until the salinity reached about 66 ‰ (Gross, 1961). *Uca rapax* from the salt flats of Puerto Rico can withstand even higher salinities than *P. crassipes,* tolerating salinity up to 90 ‰.

Although the salinity of different oceans varies to a certain extent, the open-ocean waters tend to be fairly constant in salinity, approximately 35 ‰. Near the shore, however, large river systems open into bays, sounds, and estuaries where the salinity is much less constant than in open-ocean waters. Unusually heavy rains may reduce the salinity almost to the level of fresh water and thereby cause mass mortality of many of the sedentary animals. Still, not all marine animals are killed when salinity is suddenly reduced. Some are able to survive for prolonged periods at low salinity, while others migrate considerable distances to avoid them. The number of species of marine animals, however, declines in direct proportion to decreasing salinity, with the greatest number of species found near shore in undiluted sea water. Salinities ranging from 5–8 ‰ represent a faunistic barrier between marine and fresh-water species (Khlebovich, 1969). For example, along the Texas coast, of a total of 112 species of fish, 109 species were found in salinities above 30 ‰, 73 species were found in salinities of 20 ‰ and less, and only 39 species in salinities below 5 ‰ (Gunter, 1945). Similar results have been reported for marine invertebrates. In an area where the average salinity was 31.8 ‰, 220 species were tallied, but where the mean salinity was 19.0 ‰, only 56 species were recovered (Wells, 1961). Specific examples of the effects of decreasing salinity on distribution are given in Table 3-1.

Table 3-1. Numbers of invertebrate species collected in relation to salinity from the Beaufort, North Carolina area. (From Wells, 1961).

	Shark Shoal	*Sluice-way*	*Gallant Point*	*White Rock*	*Cross Rock*
Mean salinity ‰	31.8	30.9	28.2	24.0	19.0
Total number of species	220	203	105	92	56

Although open-ocean water is quite stable with respect to salinity, zooplankton from the open ocean have been found to withstand greater salinity changes than those normally encountered, so that salinity does not appear to limit their distribution into estuaries (Hopper, 1960). But salinity has been suggested as the principal physical factor in limiting the distribution of the planktonic zoeal stages of *Sesarma cinereum,* an intertidal estuarine brachyuran crab. The larvae of this crab can develop to the crab stage in salinities of 20.1–26.5 ‰, but at salinities as low as 12.5 or as high as 31.1 ‰, the zoea will not go beyond the megalopa stage (Costlow *et al.,* 1960). Although the larvae of the intertidal crab can withstand lower salinities than the oceanic forms, the oceanic species are better able to tolerate high salinities.

On a broad zoogeographical basis, salinity interacts with temperature to influence distribution. Tropical animals seem to survive low salinity best in combination with high temperature, while northern species tend to tolerate low salinity better at low temperatures. In some marine species, populations are limited to brackish water in the northern range of distribution; in low latitudes they are found in fresh water (Vernberg, 1962).

Tissues

Intact organisms have been used in most of the studies on salinity lethal limits. A few observations, however, have been reported on isolated tissues. Gill tissue from intertidal species generally can withstand a wider range of salinity than can tissue from sublittoral species; among the sublittoral species, tissues from deep-water forms were more stenohaline than were tissues from shallow water species (Fig. 3-13). These responses correlate well with the salinity regimes of these species and appear to be adaptive, since the intertidal-zone animals experience the greatest fluctuations in salinity.

Acclimation

Salinity tolerances of marine animals can be a function of the salinity of their environment. Populations of the polychaete worm *Nereis diversicolor* from France were found to have a more rapid weight regulation in a salinity of 7 ‰ than populations of the same species of worms from Great Britain. On this basis, worms from these two areas were regarded as "physiological races." But in a series of laboratory experiments, Smith (1955) was able to show that there were no genotypic differences in salinity tolerance between populations if the animals were adapted slowly to a series of salinities ranging from sea water to fresh water. Thus he demonstrated the phenotypic lability of the adults of *N. diversicolor.*

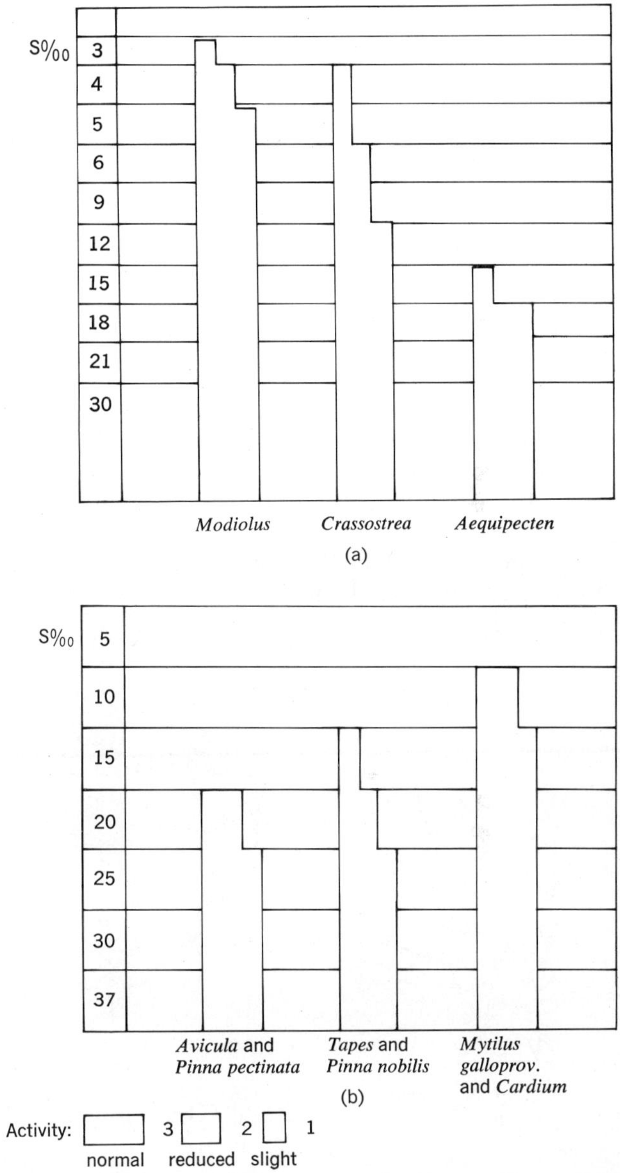

Figure 3-13. Comparison of the cellular osmotic resistance is isolated gill pieces of marine bivalves occupying different habitats. (*a*) Estuarine species: *Modiolus* and *Crossostrea* may be found in low saline waters and *Aequipecten* is restricted to salinities generally above 15 ‰. (After Vernberg *et al.*, 1963.) (*b*) Oceanic species: *Mytilus* and *Cardium* are intertidal; *Tapes* is intertidal and *Pinna nobilis* is found at depths of 10 meters; *Avicula* and *Pinna pectinata* occur at depths of 100 meters. (After Schlieper *et al.*, 1960.)

There is some evidence that the salinity tolerance of the progeny of marine invertebrates is dependent upon the salinity to which the parent stock is adapted. The offspring of adult amphipods, *Gammarus duebeni,* reared at a salinity of 2 ‰, tolerate a transfer to fresh water better than do offspring from a different stock grown at either 30 ‰ or 40 ‰. Thus there appears to be a persistent adaptation to the salinity in which the parent stock live and the embryos develop.

Salinity stress can also modify structural and physiological characteristics of an animal, as in the brackish water hydroid *Cordylophora caspia.* Animals of the same colony—which would mean, of course, that they would all have the same genotype—can survive a wide range of salinity if these changes in salinity are made gradually. A few of the structural modifications resulting when hydroids are grown in salinities of 30 ‰, 15 ‰, and fresh water are represented in Fig. 3-14. Many of these changes have physiological significance. For example, the surface/volume ratio of the hydroid is lowest in fresh water and increases with increasing salinity. Reduction of free-surface area is an adaptation to hypotomic stress, and if the animals are transferred abruptly from fresh water to 30 ‰, and thus placed under osmotic stress, the growth rate is inhibited.

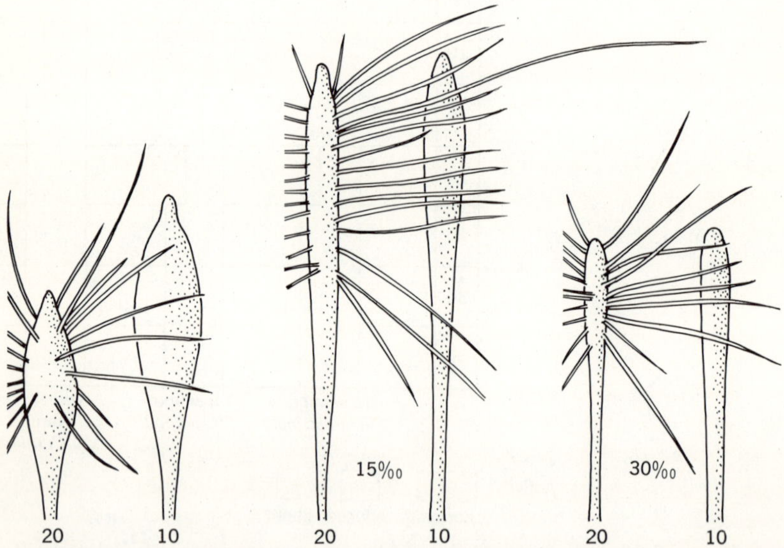

Figure 3-14. *Cordylophora caspia.* Structural alterations in hydranths adapted to fresh water, 15 % and 30 ‰ salinity both at 20° and at 10°C. Shapes drawn from average values of 25 hydranths. Tentacles omitted from 10°C shapes and shown for 20°C on the right side in full length (except 15 ‰ where only one tentacle appears in full length). (From Kinne, 1958.)

Ions

The relative abundance of certain ions is equally as important as the total osmotic pressure of the external medium, since either a deficiency or an excessive amount of an ion can be lethal to organisms. The mechanisms for ionic and osmoregulation are discussed in Chapter 5, but a few examples of the lethal effect of different ions are cited here.

Although the concentration of potassium in sea water is quite low, it has long been recognized that potassium is necessary for normal heart activity of many invertebrates. Sodium, in contrast to potassium, is the cation found in greatest concentration in sea water. Excessive amounts, however, are toxic to many organisms. There has been very little experimentation on solution without sodium because it has proved difficult to replace this cation with osmotically active substances that will not affect the protoplasm. Then, too, other cations exert an antagonistic effect, further confusing the exact role of sodium in living systems (as in sea urchin eggs, where both magnesium and calcium tend to prevent the toxic effects of sodium and potassium). Although the magnesium ion is necessary for the activation of certain enzymes (for example, isocitric acid dehydrogenase), in excess it has an anesthetic effect on many marine organisms.

Calcium is essential for all animal life. It is relatively abundant in sea water, and is found in lesser amounts in brackish water. Apparently it is this reduction in calcium that is responsible for the shorter survival time of some marine animals in low salinities. For example, the thermal-resistance time of gill tissue from *Mytilus edulis* is twice as long from animals maintained in sea water with a salinity of 30‰ at 15°C as in animals from brackish water. Yet, when the calcium content is increased to equal that of sea water, the thermal-resistance time is equal to that of tissue in sea water (Schlieper and Rudolf, 1960).

GASES

Lack of oxygen is seldom a critical factor in restricting the distribution of terrestrial animals. The supply of oxygen available to terrestrial animals is adequate and constant. Only in the alpine regions is there a significantly reduced oxygen tension, and even here the distribution of invertebrates apparently depends on low temperatures and lack of food rather than on decreased amounts of oxygen. Insects generally are very resistant to reduced oxygen pressures and to rapid changes in pressure, and there are several orders of insects that can survive in a vacuum tube down to 1/10,000 mm Hg (Lutz, 1929). Vertebrates are less resistant to reduced pressure; for example, frogs cannot survive when the total pressure of gases is less than 100 mm Hg. Man is even less resistant, dying at about 350 mm Hg.

Animals living in marine situations are more apt to be influenced by a reduced oxygen supply than are terrestrial organisms. Fresh water has an extremely variable supply of oxygen. In mountain streams the amount is high. In most deep-water lakes, in shallow lakes with muddy bottoms, or in lakes with a thermocline or chemocline, however, the oxygen content is very low or nonexistent. Polluted waters are also quite low in oxygen content. The availability of dissolved oxygen in marine environments ranges from the extreme anaerobic conditions, such as in the deeper parts of the Black Sea, to saturated or supersaturated situations, as reported in some marshes. Although no animals have been reported living in the completely anaerobic areas of the Black Sea, there is abundant marine life in other areas, even where the oxygen content is permanently low. Bottom muds frequently are anaerobic except for small amounts of oxygen at their surface, but animals there are able to survive prolonged periods of anaerobiosis without injury. There is also a relatively abundant fauna at various depths in sandy beaches where oxygen concentrations are quite low (Gordon, 1960). There are occasions, however, when mass mortality of oceanic species occurs, apparently caused by oxygen deficiency and/or by poisonous gases, particularly H_2S. Brongersma-Sanders (1957) reported eight incidences in various oceans, and Tulkki (1965) attributed the disappearance of the benthic fauna from the southern portion of the Baltic Sea to an oxygen deficiency.

In other marine environments the oxygen content normally may fluctuate greatly over a relatively short period of time (Fig. 3-15). Intertidal areas generally have widely varying concentrations of oxygen. Eelgrass beds have been reported to vary from completely anaerobic conditions to a supersaturated situation within a 12-hr period. In rock pools there is a similar situation: Active photosynthesis occurs during the day, accompanied by a large output of oxygen and followed by a marked reduction in oxygen at night due to the respiratory activity of animal organisms. Adaptation to fluctuation in oxygen content is not a characteristic restricted only to intertidal animals. Migrating animals that exhibit great vertical migration, such as copepods and euphausiids, must be able to cope successfully with a wide oxygen gradient (see Vernberg, 1970, for recent review of tolerance limits of marine animals).

Only a very few animals such as some intestinal protozoans are obligate anaerobes; that is, they cannot survive in the presence of oxygen. Otherwise, animals may be divided roughly into three groups with respect

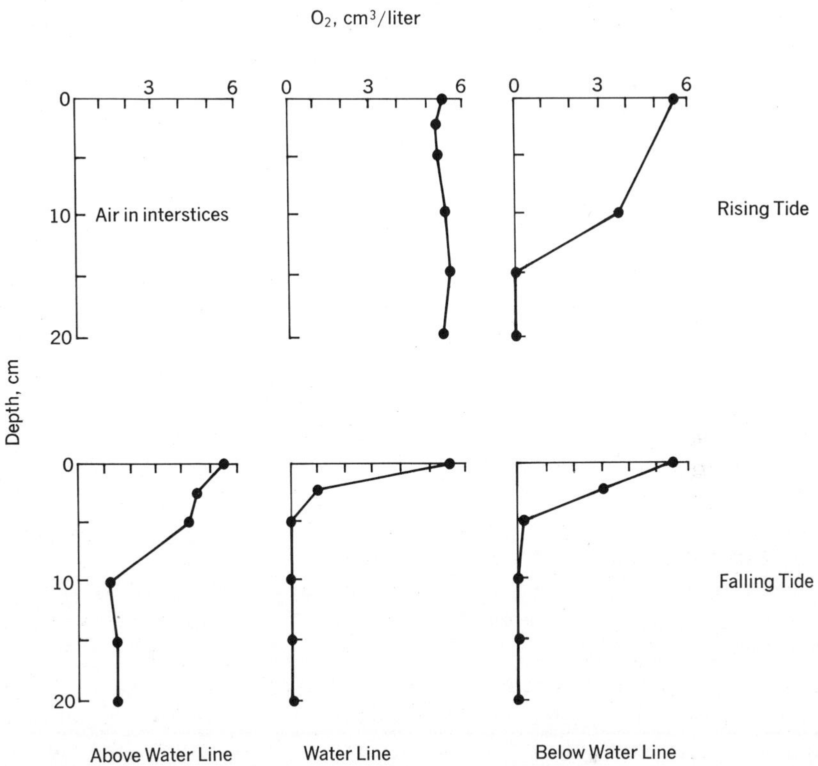

Figure 3-15. Oxygen concentrations at various depths in three areas of a sand beach. Above water line: highest point reached by wash of breaking waves; water line: point covered by wash about 50 percent of time; below water line: lowest point exposed to air by receding wash. (After Gordon, 1960.)

to their demands for oxygen. Many of the animals living in environments permanently low in oxygen content can survive anaerobically for several weeks, although normally these animals are aerobic. Another group of animals comprises those that live in a habitat with fluctuating amounts of oxygen present, and these organisms can survive relatively short periods of time in anaerobic conditions, but not nearly so long as the first group. Finally, some animals require a constant, relatively high oxygen content in the environment. Active swimming animals generally require more oxygen in water than do sluggish or sessile species. Thus the fast swimming teleost, the scup, dies of suffocation even when oxygen tension is relatively high, while the sluggish, bottom-dwelling toadfish survives well and can remove almost all oxygen from the water. Young salmon and brown trout are able to detect and avoid low oxygen concentrations that are well above their lower incipient limiting level (Bishai, 1962).

The need for molecular oxygen is not necessarily the same throughout the life cycle of an animal. In trematodes the free-living cercarial stage dies quickly in the absence of oxygen, but the adult stages can withstand prolonged periods of anaerobiosis.

Some mud-dwelling animals are actually sensitive to atmospheric oxygen. *Chironomus plumosis, Tubifex,* and the ostracod *Heterocypris* survive longer in 4 percent oxygen than in 21 percent oxygen (Fox and Taylor, 1955). Although few animals are truly anaerobic, some animals are killed or injured by high concentrations of oxygen. Intestinal flagellates of termites, for example, are killed when exposed to pure oxygen. Cultures of parasitic protozoans, *Trichomonas,* survive better anaerobically than in the presence of oxygen, and pure oxygen kills the nematode parasite *Ascaris* within 30–60 min. Not all parasitic helminths, however, are sensitive to 100 percent oxygen. The trematode *Gynaecotyla adunca* survived in the presence of pure oxygen for 24 hr without apparent injury (Hunter and Vernberg, 1955).

Different life processes of an organism may be influenced by various oxygen tensions. Polychaetes (*Capitella capitata*) suspended in plastic cages in field test sites of different oxygen content survived for several days, but did not feed in regions of little or no dissolved oxygen. With a median-dissolved oxygen content of 2.9 ppm, the worms did feed, and at a median oxygen content of 3.5 ppm the experimental population sample was able to reproduce and to complete its life cycle (Reish and Barnard, 1960).

In many habitats the disappearance of oxygen often is accompanied by an increase in the concentration of hydrogen sulfide (H_2S). Waters that generally have low oxygen and high H_2S concentrations usually have few permanent residents, and the sudden introduction of oxygen-deficient, H_2S-rich solutions into a body of water will cause mass mortality among animals living there (Brongersma-Sanders, 1957). Active and mobile species are more sensitive to H_2S than are less mobile ones, and animals living in soft substrates are more H_2S-resistant than animals living in sandy or hard substrates (Jacubowa and Malm, 1931; Theede *et al.*, 1969).

DESICCATION

Water economy is one of the major problems confronting terrestrial and semiterrestrial animals. The amount of water necessary to maintain life varies from species to species. It also varies with the stage in the life cycle of a single species. Many mechanisms have evolved which enable animals to maintain the critical level of water balance, thereby permitting some species to live in deserts, some in water, and others in habitats intermediate in moisture content. Water is used in so many metabolic processes that it is impossible to define which mechanism or mechanisms are disrupted during periods of desiccation and which of these ultimately lead to the death of the organism.

To determine the lethal limits of water loss, animals are exposed for varying lengths of time to dry air or to air of different humidities expressed as saturation deficiency. Usually, this is done by placing the animals in a glass container, such as a desiccator, along with a desiccant or various chemicals used in controlling moisture levels (see Buxton and Mellanby, 1934). Modifications of these techniques have been described by Edney (1957). Some workers base the length of survival on the number of hours necessary to kill all experimental animals, while others express the results in terms of 50 percent survival level (LD_{50}). Along with length of survival time, rate of water loss by the experimental animal often is determined and is expressed as percent of initial body weight. Water loss is determined by wiping off the excess water on the outside of the animal, weighing the animal at the beginning of the experiment, and then reweighing the animal at regular intervals of time; thus an indication of the rate of water loss will be obtained as well as the total amount of water lost at the time of death. There is evidence that body size, sex, and temperature are important factors to be considered in the design of an experiment.

The habitat preference of the animal can often be correlated with tolerance to aridity. Adaptations of animals to meet the problems of dryness are behavioral, anatomical, and physiological, and since these adaptations are often closely allied, it is difficult to separate them distinctly.

Two of the main effects of desiccation are loss of body water and the drying out of the respiratory epithelium. The body wall of some animals, such as land planarians, earthworms, and amphibians, is an especially sensitive epithelium, which also functions as a respiratory membrane; if this membrane is subjected to drying, the cells are permanently injured. Therefore, to survive, these animals must always remain in moist, humid conditions. To help maintain a moist situation, many secrete a mucus coating, and in unusually dry situations they may resort to behavioral adaptations such as burrowing, coiling, or forming aggregates.

Arthropods have an external covering over the respiratory surfaces, and some of the predominantly terrestrial forms of marine crustacea have an actual reduction in the respiratory surface (Gray, 1957). Insects and myriopods have a particularly well-developed covering over the respiratory surface as well as a tracheal system. Vertebrates have stratified skin with several cellular layers in addition to a well-protected respiratory surface. Many of the structural modifications that prevent water loss are devices for effectively closing off the animals from the external environment. This is illustrated particularly well by the operculum of gastropod molluscs. The efficiency of the operculum in preventing water loss depends upon the size of the shell mouth in relation to the size of the animal and how well the operculum fits the shell mouth. Similarly, the close-fitting plates of barnacles are especially effective in preventing water loss. The serpulid polychaetes have one or more of the gill filaments enlarged at one end, which are utilized to close the mouth of the tube when the animal withdraws into it during periods of stress. In insects a wax layer makes the integument relatively impermeable. The integument of terrestrial crustacea is less permeable to water than that of predominantly littoral crustacea, although a wax layer has not been demonstrated.

Some animals apparently have the ability to regulate the rate of loss of body water, an example is the amphibians, that are able to slow the rate of water loss when they are first exposed to dry air (Thorson, 1955, 1956). In general, those animals that can survive desiccation best can also tolerate wide ranges in salinity; in intertidal-zone animals the loss of water is accompanied by osmotic adjustments and regulation. Control of the transpiration rate also is an important factor in the regulation of the loss of body water, since a reduction in the transpiration rate tends to conserve water. In the crab *Carcinus maenas,* the transpiration rate can be reduced by a third through increased calcification of the exoskeleton. Some evidence also indicates that a decreased transpiration rate accompanies increased terrestrialism, as for example in *Ligia oceanica,* a littoral isopod that has a much greater transpiration rate than *Porcellio,* a more terrestrial isopod (Edney, 1957).

Some animals obtain needed water from food as partial replacement of water lost through excretion. Although excretion will be discussed in more detail in a later chapter, it might be appropriate here to note that the excretory process has evolved differently for the more terrestrial animals. Aquatic species tend to excrete more nonprotein nitrogen than do terrestrial species. With the approach of terrestrialism, there also appears to be a general suppression of nitrogen metabolism rather than conservation of water, and the species shifts from ammonotelic to uricotelic excretion so that little or no water is excreted with uric acid. In the more terrestrial species the fecal pellets tend to be drier.

The resistance of animals to desiccation can often be correlated with the habitat of the organism; for example, animals living in moist areas tend to be less resistant to dry air than animals living in drier or arid habitats.

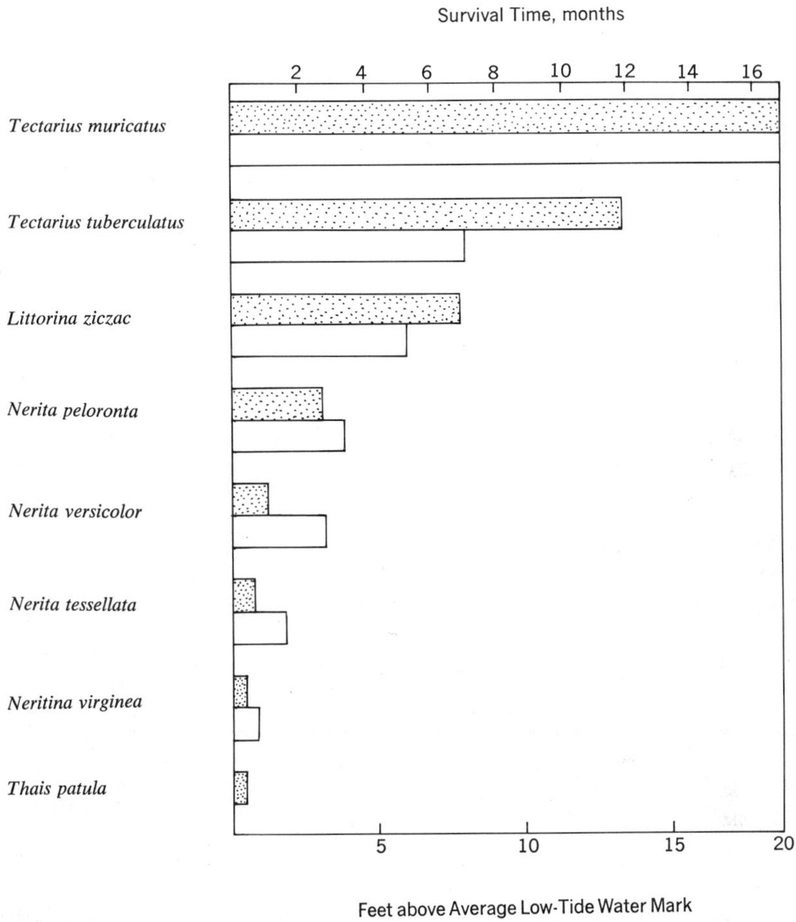

Figure 3-16. Correlation of survival time in exsiccation and intertidal distribution of marine snails. (Modified from Mattox, 1949.)

Among the intertidal-zone animals, those from the lower regions of the intertidal zone cannot survive desiccation as well as animals found near the high-tide mark (Fig. 3-16). Some of the less mobile animals living near the high-tide mark can survive without environmental water for incredibly long periods. Barnacles living in this area are covered with water only 5 percent of their lifetime, and a barnacle can live completely out of water for at least 44 days. In contrast, some of the more mobile intertidal-zone animals are not so resistant to desiccation as the less mobile ones such as some isopods, which survive only 2.5 hr at a relative humidity of 75 percent (Edney, 1957).

INTERACTION OF MULTIPLE ENVIRONMENTAL FACTORS

In nature an organism is subject to many environmental factors, each of which may exert an influence on the organism and some of which are more important than others. In laboratory studies, only one factor is usually considered at a time; it may be altered while others are kept constant. Even in field studies where data are collected on several physical factors, one factor usually is singled out for particular attention. Thus, in view of the complex nature of environment, it is difficult to know which variables are significant.

Physical factors may interact to kill an organism at intensity levels which, if taken separately, would not be lethal. Thus the limits of tolerance for a given factor are set in part by the extent to which the total environment is favorable. This can be illustrated by numerous examples, but the following three will serve to show the importance of this concept:

1. In the American lobster *Homarus americanus,* the thermal lethal limit is 32°C when the salinity is 30 ppt and the oxygen content is 6.5 mg oxygen per liter. If, however, the oxygen content is decreased to 2.9 mg oxygen per liter, the thermal lethal limit is lowered to 29°C. A similar shift

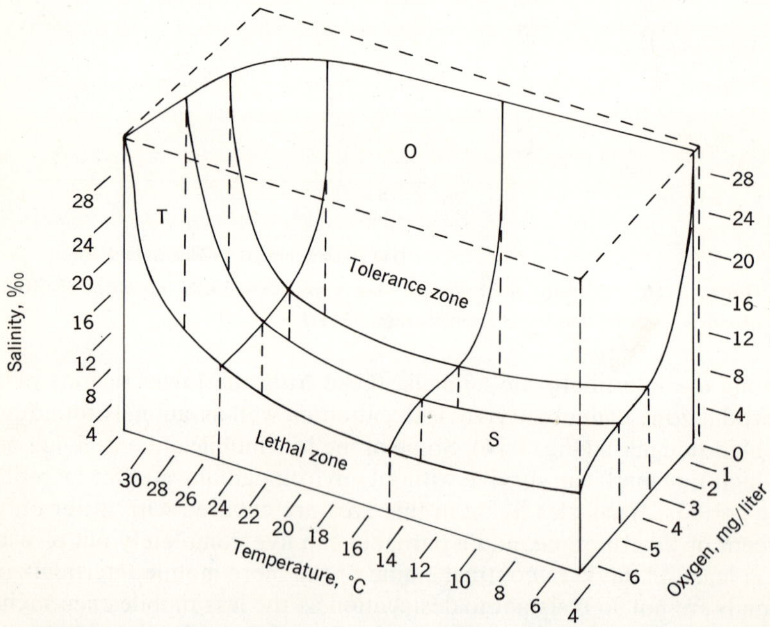

Figure 3-17. Three dimensional representation of the boundary of lethal conditions for lobsters for various combinations of temperature, salinity and oxygen. (From McLeese, 1956.)

in the temperature tolerance was observed when animals were maintained at a reduced salinity and a high oxygen content. In all, lobsters were subjected to 27 combinations of temperatures, salinity, and oxygen. From these results, McLeese (1956) was able to construct a three-dimensional diagram to represent the interaction of these factors on the lethal limits of a species (Fig. 3-17).

2. Shrimp (*Crangon vulgaris*) that have been cold-acclimated to 5°C have a greater tolerance to low salinity than shrimp that have been warm-acclimated at 15°C.

3. At the tissue level, the influence of temperature on the survival of isolated pieces of gill tissues of lamellibranchs in low salinity is correlated with the habitat of the species; for example, deep, cold-water forms can survive low salinities at low temperatures, but cannot survive correspondingly low salinities at higher temperatures. Conversely, many intertidal animals can tolerate low salinities better at higher temperatures than can cold-acclimated animals. See Table 3-2.

When some organisms are measured individually, their tolerance to a number of environmental factors is much greater than the animal will normally experience in its habitat. The functional significance of this phenomenon may be that if various combinations of these factors are studied

Table 3-2. Osmotic Resistance of Isolated Gill Tissue from Lamellibranchs Subjected to Various Salinities for 24 Hr. (Determined March 1961)

Salinity	*Modiolus*		*Crassostrea*		*Aequipecten*	
(ppt)	*W-A*[a]	*C-A*[b]	*W-A*	*C-A*	*W-A*	*C-A*
30	3	3	3	3	3	3
27	3	3	3	3	3	3
24	3	3	3	3	3	3
21	3	3	3	3	3	3
18	3	3	3	3	2·6	2·7
15	3	3	3	3	1·7	2·4
12	3	3	3	3	0	1·0
9	3	3	2·8	2·5		
6	3	3	2·1	1·6		
5	3	3	1·9	1·4		
4	2·6	2·4	1·2	0·4		
3	1·6	0·7	0	0		
2	0·2	0	0	0		

SOURCE: Vernberg *et al.*, 1963.
[a] W-A = warm-acclimated animals.
[b] C-A = cold-acclimated animals.

simultaneously, the resultant zone of tolerance, as represented by McLeese in Fig. 3-18, approximates more nearly the concomitant fluctuation in environmental parameters experienced by the organism in the field.

SUMMARY

Each species can tolerate the middle sector of a gradient for any one environmental factor; at either end of this gradient, however, a point is reached where the animal can no longer survive. An analysis of the lethal limits of an animal to a factor operating independently and in concert with other factors is imperative in assessing the role of environmental factors in its ecology and distribution. These data not only have intrinsic value to basic zoologic studies, but also have practical value in predicting the survival of animals in man-disturbed habitats.

The determination of lethal limits is not simple, since both intrinsic factors (such as body size and the sex of the animal) and extrinsic factors, including interaction of multiple factors, can profoundly influence the organism's response to the zone of resistance.

The lethal limits to temperature, salinity, gases, and desiccation are emphasized in this chapter because, of all the environmental factors, they have been most intensively studied. Different stages in the life history of one species occupying different habitats, as well as species from different habitats, show varying degrees of adaptation to these factors. With the development of more sophisticated laboratory equipment for controlling multiple factors, and with computer programs to analyze the data, it should be possible to have a more meaningful basis for predicting the biotic success of organisms under different combinations of environmental stress.

References

Ballinger, R. E., and C. O. McKinney. 1966. Developmental temperature tolerance of certain anuran species. *J. Exptl. Zool.*, 161:21–28.

Barron, E. S. G., and T. N. Tahmisian. 1948. The metabolism of cockroach muscle (*Periplaneta americana*). *J. Cell. Comp. Physiol.*, 32:57–76.

Belehradek, J. 1935. Temperature and living matter. *Protoplasma Monographien*, 8:1–229.

Bishai, H. M. 1962. Reactions of larval and young salmonids to water of low oxygen concentration. *Extrait du Journal du Conseil International pour L'Exploration de la Mer,* 27:167–180.

Brett, J. R. 1944. Some lethal temperature relations of Algonquin Park fishes. *Univ. Toronto Biol. Ser.,* 52:1–49.

Brett, J. R. 1952. Temperature tolerance in young Pacific salmon, genus *Oncorhynchus. J. Fish Res. Bd. Canada,* 9:265–323.

Brett, J. R. 1956. Some principles in the thermal requirements of fishes. *Quart. Rev. Biol.,* 31:75–87.

Brett, J. R. 1970. Temperature and fish. *In* O. Kinne (ed.), *Marine Biology,* Vol. 1, Interscience Publishers, London.

Brongersma-Sanders, M. 1957. Mass mortality in the sea. *In* J. W. Hedgpeth (ed.), *Treatise on Marine Ecology and Paleoecology,* Vol. I, pp. 941–1010. Geol. Soc. Am. Mem. 67, Washington, D.C.

Buxton, P. A., and K. Mellanby. 1934. The measurement and control of humidity. *Bull. Ent. Res.,* 25:171–175.

Costlow, J. D., Jr., C. G. Bookhout, and R. Monroe. 1960. The effect of salinity and temperature on larval development of *Sesarma cinereum* (Bosc) reared in the laboratory. *Biol. Bull.,* 118:183–202.

Davies, P. S., and M. Walkey. 1966. The effect of body size and temperature upon oxygen consumption of the cestode *Schistocephalus solidus* (Müller). *Comp. Biochem. Physiol.,* 18:415–425.

Doudoroff, P. 1945. The resistance and acclimatization of marine fishes to temperature changes. II. Experiments with *Fundulus* and *Atherinops. Biol. Bull.,* 88:194–206.

Edney, E. B. 1957. The water relations of terrestrial arthropods. *Monographs Exptl. Biol.,* Cambridge University Press, London and New York.

Fisher, K. C. 1958. An approach to the organ and cellular physiology of adaptation to temperature in fish and small mammals. *In* C. L. Prosser (ed.), *Physiological Adaptation,* pp. 3–49. Am. Physiol. Soc., Washington, D.C.

Fox, H. M., and A. E. R. Taylor. 1955. The tolerance of oxygen by aquatic invertebrates. *Proc. Roy. Soc. B.,* 143:214–225.

Fry, F. E. J., and M. B. Gibson, 1953. Lethal temperature experiments with speckled trout × lake trout hybrids. *J. Heredity,* 44:56–57.

Gorbman, A., and H. A. Bern. 1962. *A Textbook of Comparative Endocrinology.* Wiley, New York.

Gordon, M. S. 1960. Anaerobiosis in marine sandy beaches. *Science,* 132:616–617.

Gray, I. E. 1957. A comparative study of the gill area of crabs. *Biol. Bull.,* 112:34–42.

Gross, W. J. 1961. Osmotic tolerance and regulation in crabs from a hypersaline lagoon. *Biol. Bull.,* 121:290–301.

Gunter, G. 1945. Studies on marine fish of Texas. *Univ. Texas Inst. Mar. Sci., Publ.,* 1:1–199.

Gunther, E. R. 1936. A report on oceanographical investigations in the Peru coastal current. *Discovery Repts.,* 13:107–276.

Hart, J. S. (ed.). 1969. Proc. of international symposium on altitude and cold. *Federation Proceedings, Am. Soc. Exp. Biol.,* 28:919–1321.

Heilbrunn, L. V. 1952. An *Outline of General Physiology.* Saunders, Philadelphia.

Hoar, W. S. 1966. *General and Comparative Physiology.* Prentice-Hall, Englewood Cliffs, N.J.

Hoar, W. S., and M. K. Cottle. 1952. Dietary fat and temperature tolerance of goldfish. *Can. J. Zool.,* 30:41–48.

Hock, R. J. 1964. Terrestrial animals in cold: reptiles. In *Handbook of Physiology, Section 4, Adaptation to the Environment,* pp. 357–359. Am. Physiol. Soc., Washington, D.C.

Hopper, A. F. 1960. The resistance of marine zooplankton of the Caribbean and south Atlantic to changes in salinity. *Limnol. Oceanog.,* 5:43–47.

Hunter, W. S., and Vernberg, W. B. 1955. Studies on oxygen consumption in digenetic trematodes. II. Effects of two extremes in oxygen tension. *Exp. Parasitol.,* 4:427–434.

Hutchins, L. W. 1947. The bases for temperature zonation in geographical distribution. *Ecol. Monographs,* 17:325–335.

Jacubowa, L., and E. Malm. 1931. Die Beziehungen einiger Benthosformen des Schwarzen Meeres zum Medium. *Biol. Zbl.,* 51:105–116.

Johnson, F. H., H. Erying, and M. Polissar. 1954. *The Kinetic Basis of Molecular Biology.* Wiley, New York.

Khlebovich, V. V. 1969. Aspects of animal evolution related to critical salinity and internal state. *Mar. Biol.,* 2:338–345.

Kinne, O. 1958. Adaptation to salinity variations—some facts and problems. *In* C. L. Prosser (ed.), *Physiological Adaptation,* pp. 92–106. Am. Physiol. Soc., Washington, D.C.

Konishi, J., and C. P. Hickman, Jr. 1964. Temperature acclimation in the central nervous system of rainbow trout (*Salmo gairdnerii*). *Comp. Biochem. Physiol.,* 13:433–442.

Kuthalingham, M. D. K. 1959. Temperature tolerance of larvae of ten species of marine fishes. *Current Science,* 28:75–76.

Lewis, J. B. 1963. Environmental and tissue temperature of some tropical intertidal marine animals. *Biol. Bull.,* 124:277–284.

Liebig, J. 1840 (4th ed. 1847). *Chemistry in its Application to Agriculture and Physiology.* Taylor and Walton, London.

Lueth, F. X. 1941. Effects of temperature on snakes. *Copeia,* 3:125–132.

Lutz, F. E. 1929. Experiments with "wonder creatures." *Nat. Hist.,* 29:160–168.

Mattox, N. T. 1949. Effects of drying on certain marine snails from Puerto Rico. *Ecology,* 30:242–244.

Mayer, A. G. 1914. The effects of temperature upon tropical animals. *Pap. Tortugas Lab.,* 6:3–24. Carnegie Institution, Washington, D.C.

McLeese, D. W. 1956. Effects of temperature, salinity and oxygen on the survival of the American Lobster. *J. Fish. Res. Bd. Canada,* 13:247–272.

Moore, J. A. 1942. Embryonic temperature tolerance and rate of development in *Rana catesbeiana. Biol. Bull.,* 83:375–388.

Morris, R. W. 1962. Body size and temperature sensitivity in the cichlid fish, *Aequidens portalegrensis* (Hensel). *Am. Nat.*, 96:35–50.

Peiss, C. N., and J. Field. 1950. The respiratory metabolism of excised tissues of warm- and cold-adapted fishes. *Biol. Bull.*, 99:213–224.

Precht, H., J. Christophersen, and H. Hensel. 1955. *Temperatur und Leben.* Springer, Heidelberg.

Prosser, C. L. 1962. Some temperature adaptations of poikilothermic vertebrates. *In* J. P. Hannon, and E. Viereck (eds.), *Proc. Symp. on Arctic Biol. and Medicine. II. Comparative Physiology of Temperature Regulation, Part I,* pp. 1–34. Arctic Aeromedical Laboratory, Fort Wainwright, Alaska.

Rao, K. P. 1967. Some biochemical mechanisms of low temperature acclimation in tropical poikilotherms. *In* A. S. Troshin (ed.), *The Cell and Environmental Temperature, Proc. of Intern. Symp. Cytoecology, Leningrad, Intern. Series Monographs in Pure and Applied Science,* Vol. XXXIV, pp. 98–112. Pergamon, New York.

Read, K. R. H. 1963. Thermal inactivation of preparations of aspartic/glutamic transaminase from species of bivalved molluscs from the sublittoral and intertidal zones. *Comp. Biochem. Physiol.*, 9:161–180.

Reish, D. J., and J. L. Barnard. 1960. Field toxicity tests in marine waters utilizing the polychaetous annelid *Capitella capitata* (Fabricius). *Pacific Nat.*, 1:1–8.

Rose, A. H. (ed.). 1967. *Thermobiology.* Academic Press, New York.

Salt, R. W. 1964. Terrestrial animals in cold: arthropods. In *Handbook of Physiology, Section 4, Adaptation to the Environment,* pp. 349–355. Am. Physiol. Soc., Washington, D.C.

Schlieper, C., H. Flugel, and J. Rudolf. 1960. Temperature and salinity relationships in marine bottom invertebrates. *Experientia,* 16:1–8.

Schmidt-Nielsen, K., and P. J. Bentley. 1966. Desert tortoise *Gopherus agassizii:* cutaneous water loss. *Science,* 150:911.

Shelford, V. E. 1913. *Animal Communities in Temperate America.* University of Chicago Press, Chicago.

Smith, J. M. 1957. Temperature tolerance and acclimatization in *Drosophila subobscura. J. Exp. Biol.*, 34:85–96.

Smith, R. I. 1955. Comparison of the level of chloride regulation by *Nereis diversicolor* in different parts of its geographical range. *Biol. Bull.*, 109:453–474.

Somero, G. N., and A. L. DeVries, 1967. Temperature tolerances of some Antarctic fishes. *Science,* 156:257–258.

Southward, A. J. 1958. Note on the temperature tolerance of some intertidal animals in relation to environmental temperatures and geographic distributions. *J. Mar. Biol. Assn. U.K.*, 37:49–66.

Spoor, W. A. 1955. Loss and gain of heat-tolerance by the crayfish. *Biol. Bull.*, 108:77–87.

Theede, H., A. Ponat, K. Hiroki, and C. Schlieper. 1969. Studies on the resistance of marine bottom invertebrates to oxygen-deficiency and hydrogen sulphide. *Mar. Biol.*, 2:325–337.

Thorson, T. B. 1955. The relationship of water economy to terrestrialism in amphibians. *Ecology,* 36:100–116.

Thorson, T. B. 1956. Adjustment of water loss in response to desiccation in amphibians. *Copeia,* 4:230–237.

Todd, M. E., and P. A. Dehnel. 1960. Effect of temperature and salinity on heat tolerance in two grapsoid crabs, *Hemigrapsus nudus* and *Hemigrapsus oregonensis. Biol. Bull.,* 118:150–172.

Tsukuda, H., and W. Ohsawa. 1958. The heat and cold coma temperatures in *Hebistes reticulatus* as determined by raising and lowering water temperature at constant rates. *J. Inst. Polytech., Osaka City, Univ. Series D,* 9:69–76.

Tulkki, P. 1965. Disappearance of the benthic fauna from the Basin of Bornholm (southern Baltic) due to oxygen deficiency. *Cahiers de Biologie Marine,* 6:455–463.

Vernberg, F. J. 1962. Comparative physiology: latitudinal effects on physiological properties of animal populations. *Ann. Rev. Physiol.,* 24:517–546.

Vernberg, F. J. 1970. Dissolved gases—animals. *In* O. Kinne (ed.), *Marine Ecology—Environmental Factors,* Book 1, Vol. III. Wiley, New York.

Vernberg, F. J., C. Schlieper, and D. E. Schneider. 1963. The influence of temperature and salinity on ciliary activity of excised cell tissue of molluscs from North Carolina. *Comp. Biochem. Physiol.,* 8:271–285.

Vernberg, F. J., and R. E. Tashian. 1959. Studies on the physiological variation between tropical and temperate zone fiddler crabs of the genus *Uca.* I. Thermal death limits. *Ecology,* 40:589–593.

Vernberg, F. J., and W. B. Vernberg. 1967. Thermal lethal limits of southern hemisphere *Uca* crabs. Studies on the physiological variation between tropical and temperate zone fiddler crabs of the genus *Uca.* IX. *Oikos,* 18:118–123.

Vernberg, W. B. 1961. Studies on oxygen consumption in digenetic trematodes. VI. The influence of temperature on larval trematodes. *Exp. Parasitol.,* 11:270–275.

Vernberg, W. B., and F. J. Vernberg. 1963. Influence of parasitism on thermal resistance of the mud-flat snail *Nassarius obsoleta* Say. *Exp. Parasitol.,* 14:330–332.

Warburg, M. R. 1967. On thermal and water balance of three central Australian frogs. *Comp. Biochem. Physiol.,* 20:27–43.

Wells, H. W., and I. E. Gray. 1960. The seasonal occurrence of *Mytilus edulis* on the Carolina coast as a result of transport around Cape Hatteras. *Biol. Bull.,* 119:550–559.

Wells, H. W. 1961. The fauna of oyster beds, with special reference to the salinity factor. *Ecol. Monographs,* 31:239–266.

Wohlschlag, D. E. 1964. Respiratory metabolism and growth of some Antarctic fishes. *Biologie Antarctique, Actualites Scientifiqués et Industrielles,* No. 1312.

4

Respiration and Habitat

INTRODUCTION

All organisms, regardless of the habitat they occupy, have at least one characteristic in common: they require some method of generating energy to survive. Although the ultimate source of energy is solar radiation, which enters the biosphere and is bound into compounds by photosynthetic activity, the immediate source of energy is food, and the energy-liberating system of all animals depends on oxidation, a process whereby electrons are unbound from compounds. In most animals, but not all, oxidation eventually involves oxygen, but unlike most molecules, oxygen cannot be stored in great quantities and therefore the demand for this substance is constant.

Organisms vary widely in the ecologic-physiologic strategies they have evolved so that they can survive successfully in their respective habitats. The importance of reduced or excessive concentrations of oxygen acting in the zone of resistance was emphasized in Chapter 3. Equally important to our understanding of the organism and its environment, however, is the study of the diversity of oxygen utilization by various animals within their biokinetic zone. This chapter stresses not only the similarity of respiratory adaptations that are common to all animals, but also the species variation that can be correlated with diverse ecological requirements.

RESPIRATORY ADAPTATIONS

In the course of evolution, respiratory mechanisms have undergone many physiological and morphological adaptations to environmental stress. These adaptations enable organisms to occupy extremely diverse habitats that vary widely in the availability of oxygen. Most terrestrial organisms, of course, live in an environment where a relatively constant amount of oxygen is available. In aquatic habitats the oxygen content of the water varies from a supersaturated region of high photosynthetic activity to oxygen-deficient areas, such as deeper lakes and bottom muds, while some wet marsh soils may be essentially anaerobic. In loose, well-aerated soils, the oxygen concentration may be similar to that in air, but rainfall may reduce the partial pressure to one-fourth of its original value. Only small amounts of oxygen are present in the habitats of many endoparasites. At higher altitudes the composition of air is unchanged, but the partial pressure of each component gas is reduced. For example, at 16,000 ft the oxygen pressure is only 88 mm Hg, in contrast to a value of about 159 mm Hg at sea level, while the percentage of oxygen at each altitude is essentially similar. Conversely, as one descends beneath the surface of the water, pressure is greatly increased: for every 10 meters of depth, the pressure is increased by 1 atm.

Despite the diversity of physiological and morphological adaptations in respiratory mechanisms, all have the same basic characteristics. A thin, moist membranc that is permeable to respiratory gases, chiefly oxygen and CO_2, separates the organism from its external environment. These gases pass from a high to a low concentration. Typically, the concentration gradient is a high oxygen, low CO_2 concentration on the outside of the organism (or cell) and a low oxygen, high CO_2 level on the inside of the animal (or cell) (Fig. 4-1).

In the protista and simpler multicellular animals like sponges and cnidarians, this exchange of respiratory gases takes place directly between the aqueous external environment and the cell. In the more complex metazoa, however, many modifications have evolved to enable oxygen to be supplied to the innermost cells. With increased internal organismic complexity and the exploitation of new habitats, certain regions of the body have become structurally adapted to assume a more important respiratory function. In general, this region of the body is increased markedly in surface area to allow a greater area for exchange of gases. Gills, branchial tufts, parapodia, and papulae represent a few of these morphological adaptations.

Increased movement of the fluids on one or both sides of the respiratory surface is another functional stratagem that ensures a rich supply of oxygen for the metabolic activities of the various cells. Without such movement, the oxygen tension gradient would be greatly reduced, and consequently little or no oxygen would be exchanged. The diverse mechanisms that create circulatory movements on either side of the respiratory membrane range from simple muscular activity to well-developed, closed

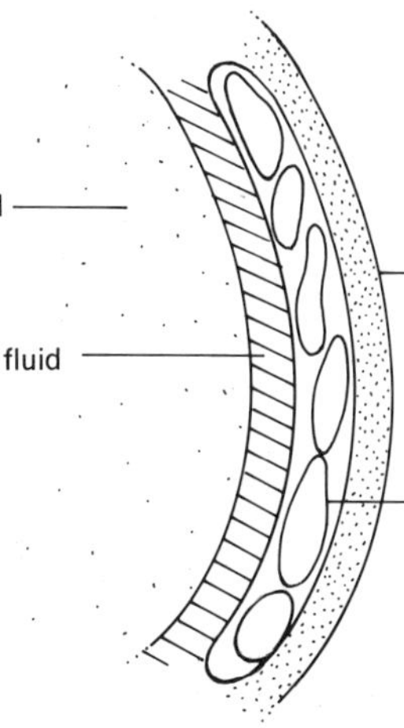

Figure 4-1. All body surfaces which serve for the exchange of respiratory gases have certain characteristics in common. As indicated in the diagram, this surface must be a thin, moist membrane with a high O_2, low CO_2 content on the outside of the organism or cell and a low O_2, high CO_2 content on the inside. Movement on either side of this membrane, aiding in maintaining an O_2 tension gradient may be accomplished either: (1) internally, by movement of the body fluid or cytoplasm; or (2) externally, by cilia or pumping mechanisms. In more complex animals, certain body surfaces may become enlarged and specialized for respiration. This adaptation may be (1) internal, such as lungs, or (2) external, such as parapodia or gills.

circulatory systems, in which a heart pumps blood through a system of blood vessels. The oxygen-carrying capacity of the body fluid is greatly increased in many animals by specific molecules in the fluid, called *respiratory pigments,* which have a high affinity for oxygen and thereby ensure that more oxygen is provided for cellular activity.

The movement of the ambient medium may be passively accomplished simply if the animal moves from a region of low oxygen tension to one of high tension. In other organisms, pumping devices have evolved (such as those associated with gills and lungs), which move water or air across respiratory surfaces. Certain animals such as bivalve molluscs depend on ciliary currents rather than pumping devices to move water. In yet other species, water currents are created by the limited fanning movement of respiratory surfaces, particularly of structures such as parapodia and external gills that extend from the body into the surrounding water.

If an animal is to be successful on land, its moist respiratory surface must be protected from the desiccating effects of a terrestrial environment. This has been accomplished by morphological changes and/or by behavioral responses. Morphological adaptations have taken many forms. In some animals the respiratory surface is covered with a protective structure and the amount of gill surface is decreased; in others, the respiratory surface is completely withdrawn into the body, as demonstrated by the tracheal system of insects and the lungs of terrestrial vertebrates. Some vertebrate animals such as certain salamanders, however, have neither gills nor lungs, and their skin acts as a respiratory membrane. In these animals the maintenance of a moist respiratory surface is in part accomplished by means of a behavioral adaptation, such as the timing of maximum locomotor activity to coincide with periods of relatively high moisture levels.

The type of adaptation to the stresses of a specific habitat does not necessarily follow systematic lines, but in many cases reflects physiological convergence by different animal groups. For example, some aquatic insects have gills, as do larval frogs. On the other hand, closely related species or different stages of the same species occupying different habitats may have different respiratory mechanisms to cope with their diverse ecological requirements. For example, many aquatic salamanders have gills, but the terrestrial species do not, a trend observed in the ontogeny of frogs. Various textbooks and reviews dealing with invertebrates and vertebrates may be consulted for details of structural diversity (Barnes, 1963; Romer, 1949; and Van der Klaauw, 1948).

TERMINOLOGY

Although the terms respiration and metabolism have been used interchangeably, the latter is broader in scope because it encompasses all vital processes of living matter, such as excretion and digestion, whereas respiration denotes one specific metabolic activity, that is, the utilization of oxygen and the production and release of CO_2. At times, however, the rate of respiration is used as an indication of metabolic rate, since all vital processes require energy and energy-producing reactions require oxidative processes.

Several additional terms used in respiratory studies are defined below.

1. Respiration, oxygen uptake, or oxidative metabolism generally refer to the amount of oxygen used per unit time.

2. Oxygen-consumption rate, oxygen-uptake rate, oxygen consumption (weight-specific), Q_{O_2}, all refer to the amount of oxygen consumed per

unit time and per unit weight. The unit weight may be expressed as wet weight, dry weight, or some other unit of measure. Under certain conditions, results are based on the number of animals or number of cells.

3. Basal metabolism denotes the oxygen-consumption rate of an organism under defined conditions of inactivity and a specific time period after ingestions of food. This value is the energy spent to maintain a minimal level of such vital physiological functions as circulation, excretion, muscle tone, and enzymatic activity. In humans, these conditions are rigidly defined; however, it is impossible to keep most other animals inactive during an experimental period without the use of disruptive restraints such as anaesthetics.

4. Standard metabolism is the oxygen-consumption rate determined while the animal is maintained under relatively constant and well-defined conditions of activity. This term differs from basal metabolism in that animals exhibit locomotor activity and the amount of locomotor activity varies with species and stage of development. However, by describing the standard experimental conditions, meaningful comparative data can be obtained.

5. Activity metabolism is the measurement of the rate of oxygen uptake during a period of controlled heightened activity under prescribed environmental conditions such as the rate during a period of swimming at a fixed speed at a given temperature.

The process of respiration in metazoa may be subdivided into a number of steps: (1) external phase—exchange between external environment and some body fluid, usually blood, across a respiratory membrane; (2) internal phase—exchange of gases between body fluid and tissues; (3) cellular phase—utilization and exchange between cell and its environment. Breathing, ventilation, pumping rate, and other terms are overt expressions of mechanical aspects of the external respiratory phase.

TECHNIQUES

Various techniques are used for determining the respiration rate of either the intact organism or its isolated parts. Changes in the oxygen content of water resulting from the respiratory activity of animals can be determined chemically by means of the standard Winkler technique (Welch, 1948). Recently developed oxygen electrode systems with recording devices are commercially available and permit continuous monitoring of the utilization of dissolved oxygen. Numerous manometric and volumetric techniques have been described for experimentation with either aquatic or terrestrial animals (Umbreit *et al.*, 1964). For a very small organism such as a protozoan, or for a small tissue sample, the Cartesian diver technique is excellent (Holter, 1943). Newly developed and highly sensitive electrical techniques, such as the Pauling paramagnetic oxygen sensor, measure continuously the amount of oxygen in air or dissolved in liquids. In addition to the rate of oxygen uptake, CO_2 production can be estimated by standard chemical methods or by newer physical techniques such as the infrared spectrophotometer. Valuable information on metabolism may be also obtained by measuring directly the heat production of a respiring system by means of direct calorimetry. Despite the available variety of instruments, their common purpose is to measure the amount of heat loss in calories by an organism over a specific time period.

VARIABLES INFLUENCING RESPIRATION

The rate of oxygen consumption determined under various experimental conditions is a good index of the metabolic capacity of an organism to face environmental stresses. Hence, a valid correlation exists between respiration and habitat. The metabolic rate, however, is subject to many variables of both intrinsic and extrinsic origin. Before reproducible data can be obtained to furnish a firm basis for ecological interpretation, the experimenter must adequately control or assess the importance of these variables. Since the variables discussed below generally influence all animals, irrespective of their habitat, they will be examined before considering specific metabolic adaptations for distinctive habitats.

Body Size

The relationship between body size and respiration of both tissues and intact organisms has received considerable attention from many workers (see review papers of Hemmingsen, 1960; Kleiber, 1947; and Zeuthen, 1953). Although there are numerous exceptions, in general the younger and/or smaller animals have a higher rate of oxygen consumption on a weight-specific basis than older and/or larger ones. This relationship has been reported for both inter- and intraspecific comparisons.

Data from these studies are commonly expressed as a power function of body size. The results are written in two general ways (Figs. 4-2 and 4-3): For total metabolism, the equation is

$$O_2 = aW^b \qquad \text{or} \qquad \log O_2 = \log a + b \log W$$

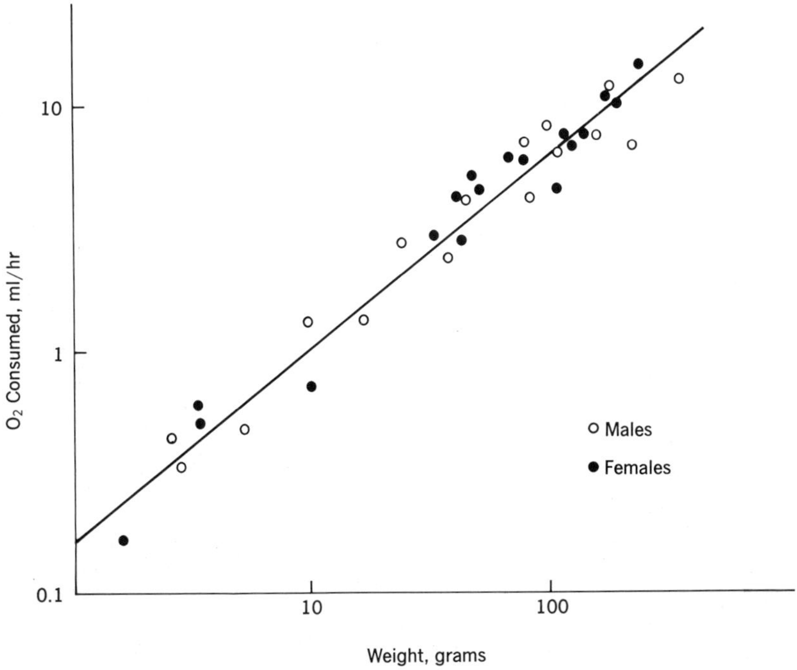

Figure 4-2. Total oxygen consumption in the kelp crab, *Pugettia producta*, plotted against body weight on a double logarithmic scale. (After Weymouth *et al.*, 1944.)

For weight-specific, the formula is

$$\frac{O_2}{W} = aW^{(b-1)} \qquad \text{or} \qquad \log \frac{O_2}{W} = \log a + (b-1) \log W$$

In these equations,

O_2 = oxygen consumed per unit time
W = weight of animal, variously expressed as wet weight, dry weight, nitrogen, etc.
a and b = coefficients, in which a represents the intercept of the y-axis and b represents the slope of the function in the logarithmic plot

The relationship of oxygen consumption to body size can be determined from the slope of the line (the constant b). If b is 1, then the rate of oxygen utilization is significantly proportional to weight. Often this con-

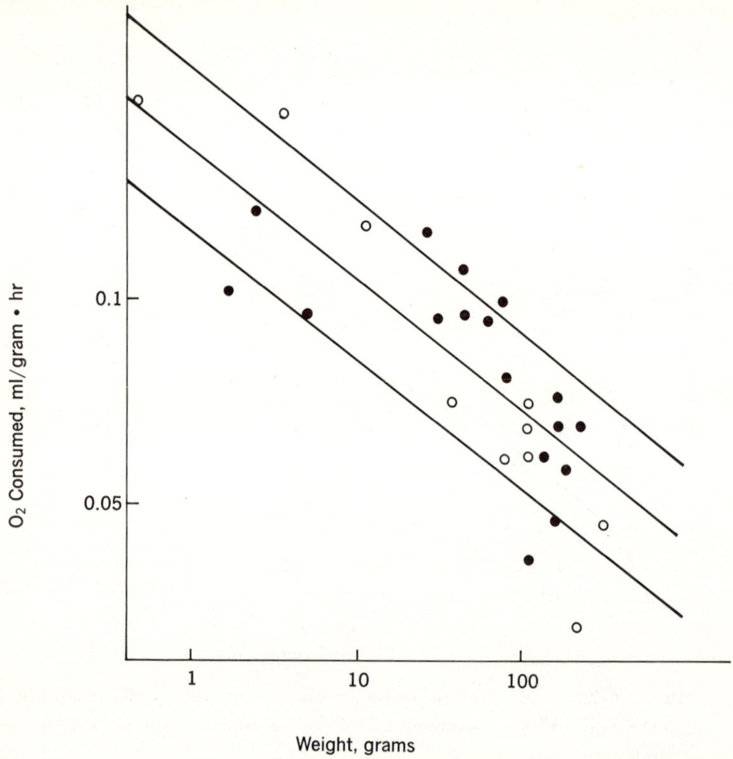

Figure 4-3. The oxygen consumption in the kelp crab, *Pugettia producta.* The weight-specific oxygen consumption rate is plotted against body weight. The central line represents the regression line. The lines on either side are at a distance of one standard error of estimate. The area between the two outer lines represents 68 per cent of the cases. (After Weymouth *et al.,* 1944.)

stant is less than 1, and then the correlation between size and oxygen consumption decreases as the *b* value decreases. As seen in Table 4-1, *b* values may vary, dependent on different groups of animals, the stage of development within a species, and environmental parameters such as temperature.

According to Zeuthen (1953), the value of *b* is approximately 0.76 for poikilothermic animals containing more than 1 mg N, while for animals containing less than 1 mg N, *b* is about 0.95. These values, however, represent generalizations, and care must be exercised if they are applied to a restricted number of closely related species. For example, in adult fiddler crabs (crustaceans), *b* values vary significantly and unpredictably from the value of 0.80, which is characteristic of many Crustacea (Vernberg,

Table 4-1. *b* Values as Determined under Different Conditions.

Conditions	*b Value*
Various Environmental Factors	
Temperature	
Gillichthys mirabilis (Gobiid fish)[a]	
10°C	0.814
17°C	0.862
24°C	0.423
31°C	0.474
Uca rapax (fiddler crab)[d]	
12°C	0.623
15°C	0.764
17°C	0.429
28°C	0.657
36°C	0.626
Salinity	
Metapenaeus monoceros (marine prawn)[b]	
Marine population from 33.5 ‰	
100% sea water	0.93
50% sea water	0.50
25% sea water	0.60
Brackish water population from 20 ‰	
100% sea water	0.73
50% sea water	0.97
25% sea water	1.05
Latitudinally Separated Populations of Same Species	
Gallichthys mirabilis (Gobiid fish)[a]	
San Francisco (Lat 37° N)	0.423
San Carlos (Lat 28° N)	0.549
Uca rapax (fiddler crab)[d,e]	
Florida (Lat 30° N)	0.790
Jamaica (Lat 19° N)	0.657
Brazil (Lat 24° S)	0.665
Brazil (Lat 13° S)	0.560
Ontogenetic Development	
Mytilus edulis (marine clam)[h]	
Larval stage (1 mg N)	0.7–0.8
Immature adult stage	0.9–1.0
Mature adult stage	0.7–0.8
Variation within One Taxonomic Group	
Average of 54 crustaceans[f]	0.826
Pachygrapsus crassipes (23.5°C)[c]	0.730
Porcellio scober (23°C)[g]	0.830

(continued on page 112)

Table 4-1. *(continued)*

Conditions	*b Value*
Oniscus oscellus (23°C)[g]	0.720
Asellus aquaticus (23°C)[g]	0.650
Uca uruguayensis (25°C)[e]	0.811
Uca rapax (25°C)[e]	0.660
Uca salsisitus (25°C)[e]	0.780
Uca thayeri (25°C)[e]	0.840
Variation with Different Size Groups[i]	
Protozoa	0.55
Metazoa (containing less than 1 mg N)	0.95
Metazoa (poikilotherms containing more than 1 mg N)	0.76

[a] Barlow, G. W., 1961.
[b] Rao, K. P., 1958.
[c] Roberts, J. L., 1957.
[d] Vernberg, F. J., 1959b.
[e] Vernberg, F. J., and Vernberg, W. B., 1966.
[f] Weymouth *et al.*, 1944.
[g] Will, A., 1952.
[h] Zeuthen, E., 1947.
[i] Zeuthen, E., 1953.

1959a; Weymouth *et al.*, 1944). Moreover, biological factors such as the locomotor behavior of the species, appear to alter the slope of this curve. In nine species of marine crabs, the active species had a higher rate of oxygen consumption than similar sized, less active species, as seen when comparing the rate of *Callinectes* and *Libinia*. These consumption-rate differences apparently are not due entirely to locomotor movements but are also influenced by basic cellular metabolic differences, since the oxygen consumption of isolated sections of gill tissue of these species followed a similar relationship (Fig. 4-4).

A number of theories have been advanced to explain the variations observed in size/respiration experiments. These include differences in the rate of enzyme production with increased size, variations in the cell surface to volume relationship (for example, as cell size increases, surface area is increased by the square while volume is increased by the cube), and in larger or older animals a greater production of tissues that have low metabolic rates, such as fat and connective tissues.

The relationship between metabolic rate and body size has fundamental significance to ecological bioenergetics (see Chapters 8 and 10). Energy utilized by organisms must come from the environment. In order

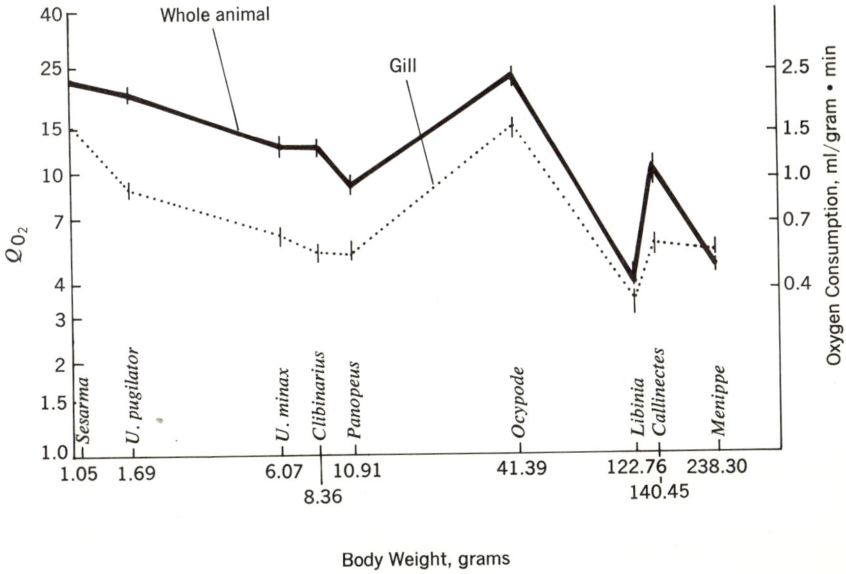

Figure 4-4. Relation of oxygen consumption of whole animal and gill tissue to body size. (From Vernberg, 1956.)

to estimate the energy requirements of a group of organisms, their metabolic rate must be determined under conditions approximating that of their environment. Since body-size/metabolic-rate relationships are complex and no one respiratory pattern is common to all species, a detailed analysis must be made for each species.

Activity

Animals engaged in locomotor activity consume oxygen at a far greater rate than when at rest (activity metabolism versus standard metabolism). Measurements of the increased energy requirements during activity have been made on various animals either in flight or while swimming. The amount of increase in oxygen utilization varies with a number of factors, including the speed of flight or swimming, temperature, and oxygen tension. The percentile increase of the active rate over the standard rate is one important index of the metabolic capacity of an organism for activity under different environmental conditions. This value is termed the *scope of activity* by Fry (1947). When the scope of activity is low, the animal has little capacity for meeting any additional stressful component of the environment because most of its metabolic machinery is involved in standard metabolic activities. Conversely, when this value is large, the

Table 4-2. Scope of Activity of Various Species of Fish in Relation to Different Environmental Factors.

Factor	*Speckled Trout*[a]	*Bullhead*[b]	*Goldfish*[b]
Varying Oxygen Tension			
Temperature, °C	20	30	30
Tension % air sat of water			
100	220	125	340
50	63	90	245
25	—	30	100
Varying Temperature, °C			
10	276	100	270
15	291		
20	220	150	315
30	—	125	340
Varying CO_2 Tension			
Temperature, °C	20	30	30
Tension, mm Hg			
0	220	125	340
3	300	112	318
6	180	110	315

SOURCE: After Basu, 1959.
[a] Lives in cold, well-oxygenated waters.
[b] Lives in warm waters of variable oxygenation.

large metabolic reserve enables an organism to expend energy to meet environmental challenges. Various environmental parameters influence the scope of activity of various species of fish, as can be seen in Table 4-2.

Along with the influence of overt locomotor activity on respiration, inherent interspecific differences have been demonstrated. The metabolic rate of active surface-dwelling fish tends to be higher than that of sluggish bottom-dwelling species. These differences cannot be explained entirely by species differences in locomotor activity, since similar results were obtained at the tissue level (Vernberg and Gray, 1953). A similar correlation between activity of animals and respiration rates of whole animals and of tissues has been reported for marine crabs (Vernberg, 1956).

Sex

In some species the males consume oxygen faster than females, and in other species the females have higher rates. In still others, no discernible difference between males and females can be observed. Moreover, differences in respiration correlated with the sex of an animal may appear only at certain temperatures; for example, male *Drosophila melanogaster* consumed oxygen at a faster rate than females only at elevated temperatures (Vernberg and Meriney, 1957). The metabolic rate also may vary with the stage of the reproductive cycle (Sparck, 1936; Thorson, 1936).

At the tissue level, variations in metabolic response of males and females also have been reported. Muscles from male cockroaches consume oxygen five times faster than those from females (Barron and Tahmisian, 1948). In contrast, the metabolic rates of tissues from male and female crustaceans were similar (Belding *et al.,* 1942; Vernberg, 1956).

Photoperiod

That photoperiod influences respiration was shown when crabs maintained on a short-day regime (8 hr of light, 16 hr of darkness) consumed oxygen faster than animals on a long day (Dehnel, 1958). Photoperiod effects on fish are closely related to reproductive activity. During the spring, increasing photoperiod length triggers the elevation of metabolism. Once gonadal development begins, however, physiological sensitivity to photoperiod is lost until the completion of the breeding cycle (Roberts, 1960). Metabolic changes in response to photoperiod have also been demonstrated at the tissue level (Evans *et al.,* 1962).

Cyclic Changes

Since the habitat in which an organism lives is characterized by rhythmic changes in environmental factors, it is not unexpected that physiological functions such as respiration rate will vary cyclically, even when measured under relatively constant conditions in the laboratory. These inbuilt rhythms must be considered in the design of experiments to demonstrate the influence of various environmental parameters. Rhythms may occur at different time intervals. For example, the cyclic release of CO_2 by certain insects occurs on a short-term basis, whereas metabolic rhythms of both whole animals and tissues have been demonstrated to occur on a diurnal, tidal, lunar, or seasonal frequency.

Two examples will illustrate the adaptive significance of seasonal metabolic changes. At low temperatures, the metabolic rate of *Emerita talpoida,* an intertidal-zone crustacean, was about four times higher during the winter than in the summer. This species was active and continued to grow at lower winter temperatures. In contrast, the beach flea *Talorchestia megalopthalma* did not show any seasonal adjustment in metabolism, and entered an apparent state of hibernation during the winter (Edwards and Irving, 1943a,b). Seasonal cycles may vary with latitudinal distribution; for example, intertidal crabs from the temperate zone, which has wide fluctuations in temperature, show seasonal metabolic cycles, whereas closely related species in the tropics, where there is scant seasonal variation, do not show these cycles. Seasonal cycles, however, may not reflect inherent seasonal differences but rather may result from thermal acclimation (Roberts, 1957; Vernberg, 1959a,b; Dehnel, 1958).

Starvation

Food is obviously a necessary source of energy for an organism; as might be expected, the absence of food influences the metabolic rate. In some cases, a marked decrease in respiration rate occurs within 24 hr of death from starvation (Berg and Ockelmann, 1959). When the respiration rate of fiddler crabs was measured over a long period of time, an initial drop in rate was observed, followed by a rather long period when the rate did not change significantly (Fig. 4-5). When animals are placed in either

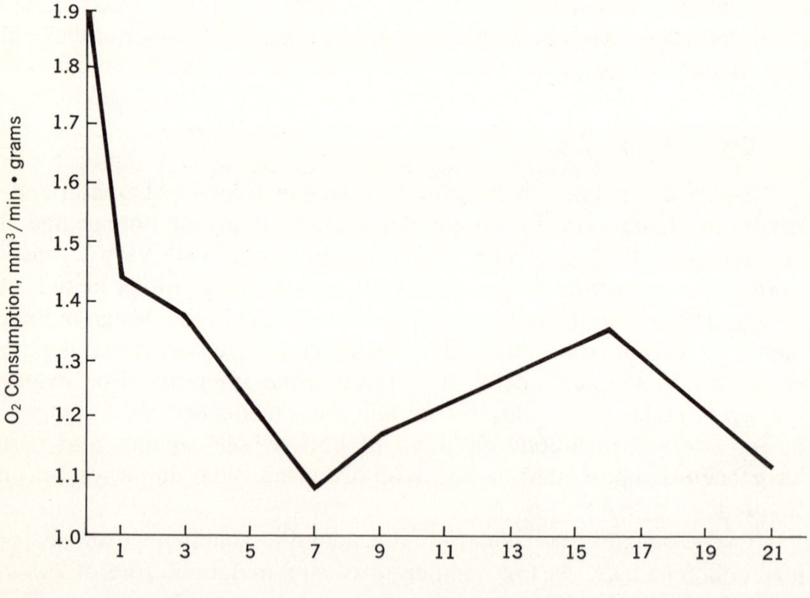

Figure 4-5. The influence of starvation on the rate of oxygen consumption of *Uca pugnax*. (From Vernberg, 1959a.)

high or low temperature, the shape of the metabolic-rate/starvation curve may be altered. Animals with large food-storage depots may withstand starvation for long periods of time without marked metabolic changes, whereas the metabolic rate of animals with small reserves can be greatly influenced in a relatively short time.

When animals are deprived of their "normal" diet and must subsist on stored foodstuffs, a different RQ (respiratory quotient) value often results as a consequence of the shift in type of food being utilized. The RQ value is the ratio obtained when the amount of CO_2 produced is divided by the quantity of oxygen consumed. The type of food material being metabolized is indicated by the RQ value. Some representative values are: carbohydrate, 1.0; fat, 0.71; and protein, 0.80. At low oxygen tensions the RQ for salt-marsh crabs rises sharply above 1. A value of 10 was reported for *Uca pugnax*. This marked increase in CO_2 release probably is caused by dissolution of carbonates of shell and tissue as a result of the action of metabolic acid produced through fermentation at low oxygen pressures (Teal and Carey, 1967).

Temperature

It is well known from the physical sciences that the rate of physiochemical reactions is controlled by temperature. Although biological entities are much more complex than simple physical and chemical reacting systems, the rate of many physiological functions increases with temperature up to a maximum point; then any further increase causes a decline in rate of activity. Respiration, especially in poikilotherms, is an example of a physiological function that is greatly influenced by temperature.

Many attempts by means of mathematical formulas have been made to characterize the effects of temperature on animals. The resultant constant is termed a *temperature coefficient* or characteristic, and each coefficient is represented by a specific symbol. One of the most commonly used coefficients is Q_{10}, which originally was formulated for chemists and is known as van't Hoff's law. It states that for every 10°C change in temperature, the rate of activity changes two to three times; in this case, the temperature characteristic, or Q_{10}, is 2–3. If the activity, expressed as Q_{10}, is increased or decreased more than this, the rate of reaction is due not merely to simple chemical changes. Generally, the Q_{10} values for physical processes vary from this 2–3 range; for example, the Q_{10} value for diffusion is approximately 1.1. Values of Q_{10} can be calculated on reactions at thermal points that are more or less than 10°C apart. The formula for determining Q_{10} is

$$Q_{10} = \left(\frac{k_1}{k_2}\right)^{\frac{10}{t_1 - t_2}}$$

or

$$\log Q_{10} = \frac{10(\log k_1 - \log k_2)}{t_1 - t_2}$$

where k_1 = rate at one temperature (t_1)
k_2 = rate at second temperature (t_2)

Another well known temperature coefficient, the μ value, is calculated by the Arrhenius equation. This temperature coefficient is used to express the amount of energy necessary for the activation of a process, for example, the number of calories necessary to transform 1 gram-equivalent of inactive molecules into an active state. If the rate of a thermochemical reaction is plotted according to the Arrhenius equation, theoretically a straight line is obtained. A break in this line indicates a new μ value and a new reaction. Specific μ values have been reported for certain enzymatic reactions, and it has been suggested that a comparison of μ values of different catalysts with the μ values of a complex multistep reaction gives an indication of the one reaction that is limiting the entire biological system. Unless further kinetic studies are made, however, this apparent relationship between μ values gives only a hint of the identity of the controlling mechanism.

When the rate of oxygen consumption of poikilothermic and homeothermic animals is determined at different temperatures and then plotted, the resultant representation is called a metabolic-rate/temperature (or M–T) curve. Typically, the shape of the curve is different for each of these groups. In general, the metabolic rate for homeotherms tends to be relatively independent of ambient temperatures, whereas that of poikilotherms varies with temperature. Throughout biology, exceptions to the general rule exist, and the subject of metabolic-rate/temperature reaction is no exception. Some homeothermic animals lose their ability to regulate their body temperature effectively, and as the ambient temperature drops, their metabolic rate drops. This may happen seasonally (as seen in hibernators) or on a short-term basis, as reported for certain birds that periodically become extremely sluggish (torpid). For example, the metabolic rate of a torpid poorwill with a body temperature of 5°C was only 3 percent of its rate at a normal body temperature (Bartholomew *et al.*, 1957). Conversely, rather than varying with temperature, the metabolic rate of some poikilothermic animals is relatively stable over a restricted thermal range. This plateau is at lower temperatures for cold-

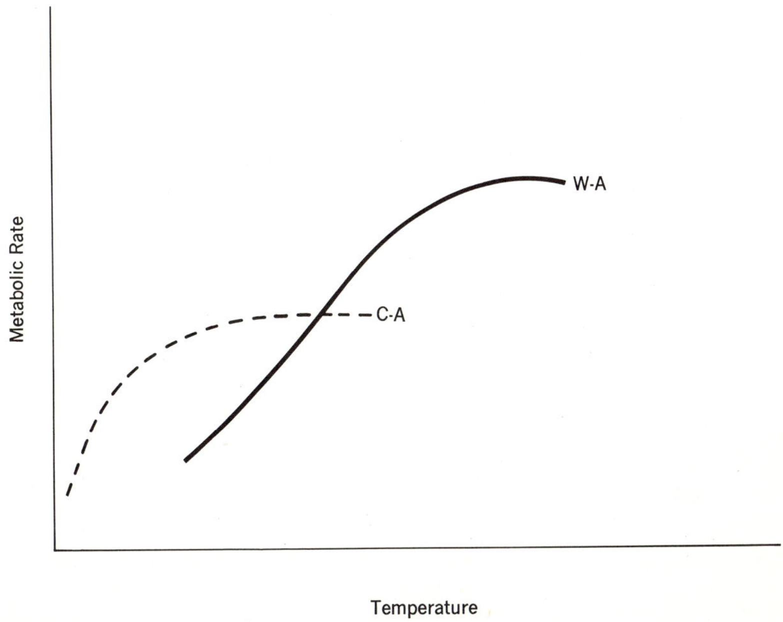

Figure 4-6. The relative relationship of metabolic-temperature (M–T) curves of cold-adapted and warm-adapted animals.

adapted species and higher for warm-adapted species (Fig. 4-6). The net effect is that, between the extreme metabolic responses of complete independence and complete dependence on temperature, organisms show a series of responses that appear to have ecological significance. Specific examples will be presented in a later section of this chapter dealing with adaptations to particular environments.

As a general response, when animals are exposed to a new thermal level, the metabolic rate undergoes an initial change. Although this response is more pronounced in poikilotherms, it is observed in some homeotherms, especially at extreme temperatures. Following this initial deflection the metabolic rate may oscillate before establishing a new steady level. These positive and negative oscillations are called over- and undershoots (Fig. 4-7). Generally, this new rate of oxygen uptake is reached within minutes and may persist for hours. However, if the animal is maintained at this new temperature for a prolonged period of time, the metabolic rate may show additional compensatory changes that establish a new metabolic rate. This animal is said to be *acclimated* to this new temperature. If the animal is returned to its former acclimation temperature, typically the metabolic rate will be different from the original value, but

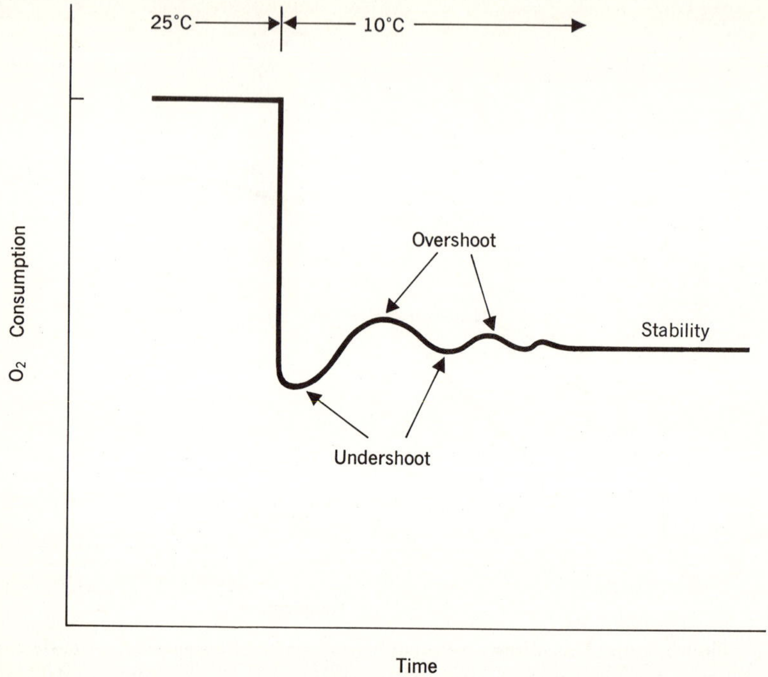

Figure 4-7. The overshoot-undershoot effect. When exposed to a new environment, the organismic response may show oscillations above (overshoot) and below (undershoot) the resultant level of stability.

after a period of time the metabolic rate will return to the original value. When the metabolic rate of an animal is determined at different temperatures, using animals acclimated to each of these temperatures, the resultant M–T curve is called an acclimation curve. In contrast, an M–T curve based on the initial rate of the oxygen consumption determined at temperatures to which the animals have not been acclimated is called the acutely determined curve. Figure 4-8 is a generalized representation of both acclimated and acutely determined M–T curves of a species that is capable of acclimating. If animals acclimate only to certain temperatures, the position of these two curves with respect to each other will be different. Both curves will be the same in those species that do not acclimate.

The speed at which animals metabolically acclimate to temperature varies interspecifically and with environmental parameters. Fiddler crabs (genus *Uca*) from the temperate zone show significant metabolic compensation after seven days of exposure to low temperature. If the animals continue to be starved, however, this adaptive response is lost, in contrast

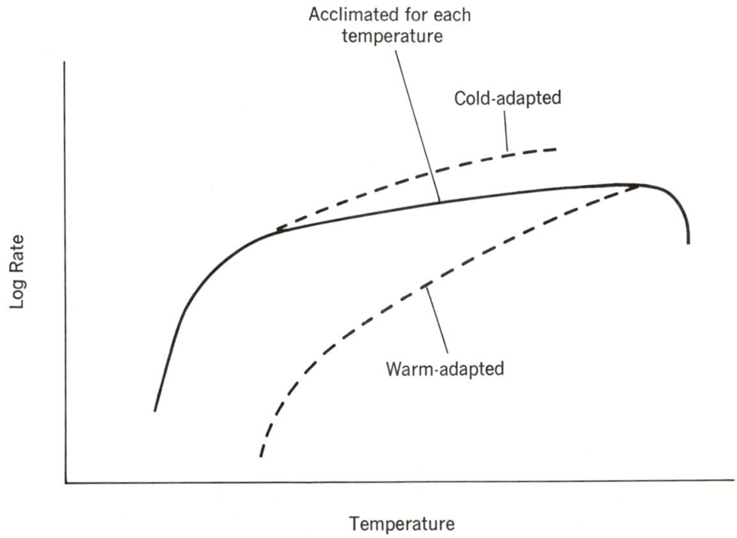

Figure 4-8. Generalized metabolic-temperature curves representing the relationship between acutely-determined curves of warm- and cold-adapted organisms and an acclimated curve. (From Bullock, 1955.)

to the response of fed animals who reach a steady rate after 14 days (Vernberg, 1959b). Populations of *Uca rapax* from the tropics show an immediate adaptive metabolic response to low temperature, but after three days no acclimation is observed (Vernberg and Vernberg, 1966).

Two systems of classifying the patterns of compensatory changes to temperature have been proposed in an effort to give a comparative basis for dealing with the diverse responses reported in the scientific literature. The use of these two systems is not restricted to oxygen-consumption rate versus temperature reaction, but can be used to express thermal acclimation patterns for any physiological function. For example, the rate of heart beat, the rate of enzyme activity, or swimming rates may show an acclimatory response to temperature.

Precht's Classification

When an animal completely acclimated to one temperature (T_1) is moved to an elevated temperature (T_2), various response patterns may occur (Fig. 4-9a). Typically, the initial steady rate of metabolism would be higher, at least in most poikilotherms. After a period of continuous exposure to T_2, the metabolic rate is remeasured. If the metabolic rate is unchanged from the initial acutely determined value, this response is classified as type 4, no acclimation. If the metabolic rate is now the same

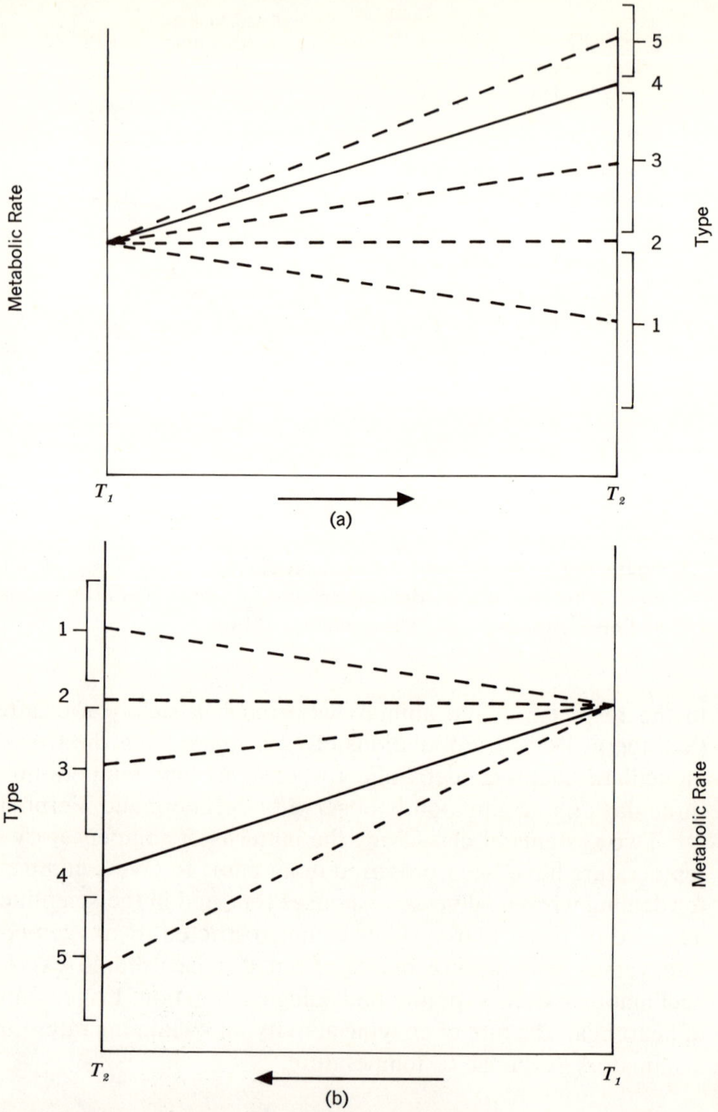

Figure 4-9. Precht's classification scheme of response patterns to temperature change. See text for explanation. (After Precht, 1958.)

as it was at T_1, this is type 2, complete acclimation. Any value between type 4 and type 2 is classified as type 3, partial acclimation. If one assumes that a type 4 pattern is zero percent acclimation and type 2 is 100 percent acclimation, the relative position of the metabolic rate of a type 3 response between these two points will give a percentile estimate of the acclimatory ability of a species. For example, if the metabolic rate is midway between type 2 and type 4 responses, the species has a 50 percent capacity to acclimate. Any value less than a type 2 response is classified as type 1, supraoptimal acclimation, and any value more than type 4 is type 5, inverse acclimation. The same terminology is used when T_2 is a lower temperature than T_1 (Fig. 4-9b).

Prosser's Classification

Since Precht's classification is based on only two temperatures, it should be borne in mind that the pattern of response might be different if other temperatures were selected. The classification of Prosser (1958) takes into account determinations made over a wide temperature range of the steady metabolic rate of groups of animals acclimated to two temperatures. The following patterns have been proposed (Fig. 4-10):

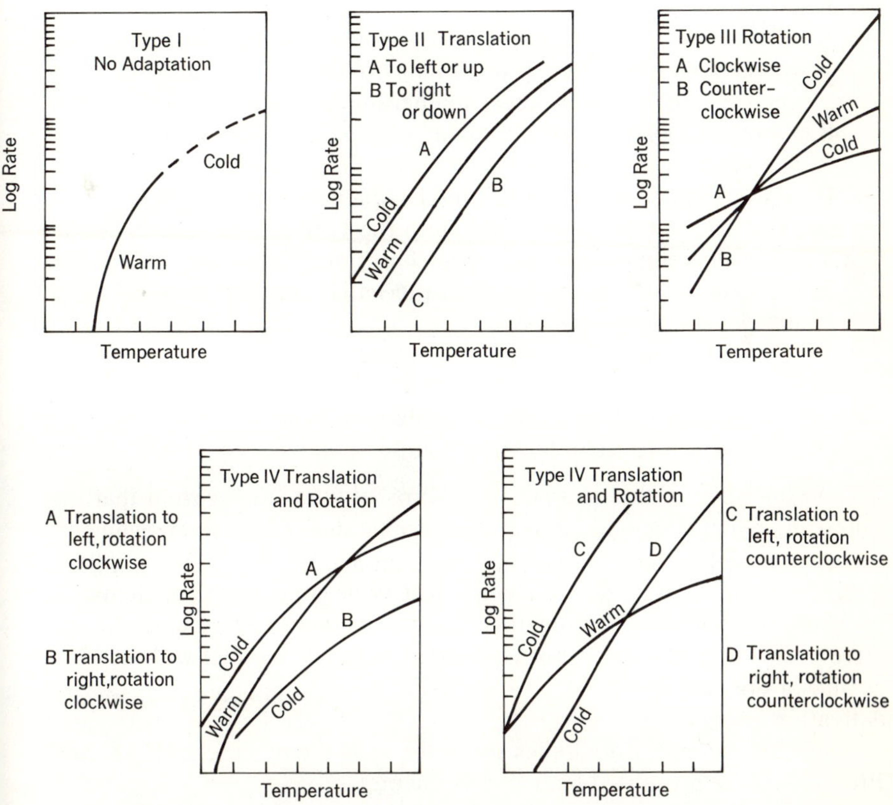

Figure 4-10. Prosser's classification scheme of response patterns to temperature change. See text for explanation. (After Prosser, 1958.)

Type I. No Acclimation. The M–T curves of both cold- and warm-acclimated animals overlap.

Type II. Translation

A. Translation to the left: The M–T curve of the cold-acclimated animals is similar in shape to that of warm-acclimated animals, but it is translated to the left; that is, at a common temperature the metabolic rate of cold-acclimated animals is higher than that of warm-acclimated animals.

B. Translation to the right: Now the M–T curve of cold-acclimated animals is positioned to the right of that of the warm-acclimated animals.

Type III. Rotation. The slope of the M–T curve for cold-acclimated animals may be different from that of warm-acclimated animals, with the result that they cross at some midpoint on the thermal gradient. Changes in Q_{10} values will alter the slope of the M–T curves. Two kinds of patterns are observed:

A. Clockwise rotation: The cold-acclimated animals consume less oxygen at higher temperatures than do warm-acclimated ones, but the reverse occurs at lower temperatures.

B. Counterclockwise rotation: The positioning of the M–T curves is the reverse of Type III-A.

Type IV. Translation and Rotation. Frequently a combination of both rotation and translation is observed, with the two curves intersecting at either end of the temperature spectrum. This intersection may take place within the zone of compatibility or in the lethal zone by extrapolation. Four subtypes have been proposed:

A. Translation to left, rotation clockwise

B. Translation to the right, rotation clockwise

C. Translation to left, rotation counterclockwise

D. Translation to right, rotation counterclockwise

On the basis of this classification, Prosser (1958) suggested that the following conclusions be drawn concerning biochemical mechanisms involved in acclimation: (1) a change in position of the M–T curve implies altered enzyme activity due to change in enzyme concentration, change in its medium (such as ionic strength, pH, or water content), or change in relative proportion to some other enzyme in series or parallel with it; (2) a change in slope of the M–T curve due to Q_{10} changes indicates a change in activation energy.

Specific examples of studies using the systems of Precht (1958) and Prosser will be presented later in this chapter.

Oxygen Tension

The metabolic response of an organism to fluctuations in ambient oxygen tension appears to be correlated with the animal's habitat preference. When an organism is exposed to a low oxygen tension, it may exhibit an avoidance response and seek regions of higher oxygen content or it may remain, either to adjust or to die. Although there is a wide spectrum of respiratory responses exhibited by animals, two main categories have been proposed: conformers and regulators. If the rate of respiration varies directly with oxygen tension—that is, with decreasing tension, a decreasing metabolic rate, and vice versa—the organism is a conformer. Regulators, in contrast, maintain a relatively constant consumption level over a wide range of oxygen tensions. But in this group of organisms there is a transition point at the lower end of the oxygen-tension gradient where the organism becomes a conformer. This value, called either the critical pressure (P_c) or the critical oxygen tension (T_c), is valid only for the particular experimental conditions under which it was determined, since it may change with various external and internal factors. Typically, the P_c values of regulators are below air saturation, whereas those of conformers are above.

In general, animals from oxygen-rich environments have higher P_c values than related species from oxygen-poor environments. However, a correlation between the ecology of a species and its respiratory response to oxygen tension is not always apparent, since a critical comparative evaluation based on the results of various workers is difficult because of differences in experimental procedures and techniques. For example, the rate at which animals are subjected to a reduced oxygen tension must be controlled. If too much time elapses, compensatory adjustments to reduced tension may occur, resulting in data that are not comparable to data for animals exposed to rapidly decreasing oxygen tensions.

Numerous variables have been demonstrated to influence the level of critical concentration of oxygen within a species. With an increase in temperature, the P_c value increases (Fry and Hart, 1948). At 5°C this value is 100 mm Hg for the fresh-water speckled trout, but at 24.5°C it is 150 mm Hg. An animal in an active state (activity metabolism) has a higher P_c value than at rest (standard metabolism) (Graham, 1949). The P_c value, along with the respiratory response to oxygen tensions, may vary with the developmental stage of the life cycle (Hayes *et al.*, 1951; Vernberg, 1963; Marshall *et al.*, 1935). Small-sized trout and marine prawns have lower P_c values than large-sized specimens (Job, 1955; Subrahmanyan, 1962). This respiratory characteristic may be further altered by acclimation to different oxygen tensions, in much the same manner as the lethal limits of animals may be altered (see Chapter 3).

Measurements of standard metabolism of fishes (brook trout, *Salvelinus fontinalis;* carp, *Cyprinus carpio;* and goldfish, *Carassius auratus*) that were acclimated to different levels of oxygen tension showed that the fish respired at the same rate as in ambient oxygen down to an oxygen tension of 80 mm Hg, regardless of the acclimation oxygen tensions. However, below 80 mm Hg, consistently lower standard Q_{O_2} values of fish acclimated to a low oxygen level indicated that acclimation to low oxygen enhances the anaerobic fraction of standard metabolism (Beamish, 1964). Two effects on the metabolism of goldfish, *Carassius auratus,* acclimated to low oxygen were reported: (1) the standard oxygen consumption rate is reduced; (2) the critical oxygen pressure is shifted to lower values (from about 3.1 to 1.5 ml O_2 per liter). Brain and muscle tissue from goldfish acclimated to low oxygen consume oxygen at a slower rate than similar tissues from fish acclimated to high oxygen levels, whereas liver tissue is not influenced (Prosser *et al.*, 1957).

A regulating organism that finds itself in an environment of reduced oxygen concentration must have some mechanism(s) for obtaining more ambient oxygen; otherwise its metabolic rate will drop. This may be done through increased ventilation obtained by forcing more of the external media over the respiratory epithelia, or through increased efficiency in removing oxygen from the media. In the puffer, a marine fish, the removal of oxygen remains constant over a wide range of oxygen concentration, while the ventilation rate increases (Hall, 1931). On the other hand, the octopus increases its ventilation rate and increases the efficiency of O_2 withdrawal from water when placed under low oxygen stress (Wintersstein, 1925).

The pattern of ventilation in animals is usually associated with respiratory stress. In filter-feeding animals, however, it is not always possible to delineate between the necessity of ventilation for food gathering and for respiration. Environmental factors greatly influence ventilation rhythms but not always in a predictable way, especially in invertebrates. For example, the tube-dwelling marine polychaete *Arenicola* pumps water in a rhythmical manner, and although decreased oxygen concentrations did not affect the timing of the rhythm, they did decrease the strength of pumping (Wells, 1949). Other burrowing or tube-dwelling worms reacted differently to lack of oxygen. Some decreased or stopped pumping, while in another the rate increased (Wells and Dales, 1951). Similar species differences in response to oxygen deficiency have been reported in crustaceans (Walshe-Maetz, 1952).

High Oxygen Tensions

Organisms that normally live in regions of low oxygen tension are more sensitive to high oxygen tensions than are animals from air-saturated habitats. For example, the metabolic rate of the adult stage of the endoparasite *Gynaecotlya adunca* was greatly depressed when measured in 100 percent oxygen (Hunter and Vernberg, 1955). A temporary metabolic increase was noted when certain free-living invertebrates were placed at high oxygen concentration, whereas the metabolic rate was depressed in the larval and pupal stages of the insect *Prodenia*.

Anaerobiosis

There are times not only when the environmental oxygen concentration varies greatly, but also when, under certain conditions, the animal may be faced with anaerobic conditions. Some animals (obligatory anaerobes) live permanently in oxygen-depleted regions, other species may tolerate anoxia for varying periods of time (facultative anaerobes), and still others cannot survive even short periods of anoxia. These various responses can be observed when species are compared, and are exhibited by an individual animal at different seasons of the year or at different stages of its life cycle.

Most animals do not store oxygen, but must provide for its utilization on a continuing basis. Fortunately, some energy for locomotion and nervous integration can be derived from anaerobic metabolism for a limited period of time through the incomplete biochemical oxidation of compounds, which are further degraded in the presence of oxygen at a future time. As a result of the formation and storage of these compounds, the organism incurs an oxygen debt. The repayment and the amount of the debt may be indicated by a transitory increase in the rate of oxygen utilization when oxygen becomes available to the organism. Since much variation can be observed in repayment of an oxygen debt, however, the significance of this factor to the physiological ecology of an organism must be interpreted in general terms. Typically, a large oxygen debt is incurred by species that are sensitive to oxygen lack, whereas a resistant species has an insignificant oxygen debt. Some animals, however, such as diving mammals and birds, withstand anaerobiosis well and yet exhibit a pronounced oxygen debt. Obligatory anaerobes generally excrete the products of nonoxidative energy production and therefore do not require oxygen.

Diving mammals and birds have well-developed physiological mechanisms, which enable them to survive long periods of anaerobiosis (Scholander, 1964). During a dive, decreased body temperatures, even when the animal is maintained in a thermoneutral environment, suggest that aerobic processes are greatly reduced.

Although other environmental parameters also influence oxygen consumption of animals, they will be discussed in the next section of this chapter, which will emphasize metabolic adaptations to the ecological stresses of animals from different habitats.

AQUATIC HABITATS

Animals that live in an aquatic environment do not have the problem of drying out of respiratory surfaces as do terrestrial and semiterrestrial animals. In many ways the environment of aquatic habitats is much more constant than that found on land. Environmental temperatures are less varied, and when changes occur, they are apt to be slower. In the deeper parts of the ocean, the oxygen content is relatively constant, although in fresh-water lakes and in shallow marine waters, oxygen concentrations and other factors may vary considerably. Salinity, ionic composition, and dissolved substances are, of course, very important to animals in aquatic environments. Thus, the four environmental parameters that have the greatest influence on respiration in aquatic animals are temperature, oxygen tension, salinity, and dissolved substances.

Temperature

As animals extend their distributional limits, they are confronted with new environmental stresses. Through the process of natural selection, distinct populations of animals arise, which may show differential degrees of physiological adaptations to the environmental parameters characteristic of these geographically separated regions. Early in this century it was suggested that respiratory responses of animals could be correlated with habitat temperatures, since animals from polar regions living at low temperatures have metabolic rates comparable to those of tropical animals experiencing higher temperatures (Krogh, 1916). Twenty years later, three papers were published (Fox, 1936; Sparck, 1936; and Thorson, 1936) which clearly demonstrated that this hypothesis was correct. From comparisons of a metabolic-rate/temperature response of molluscs and crustaceans from the coastal waters of Greenland, Denmark, Sweden, England, and the Mediterranean Sea, it is possible to make several generalizations. Species with a northerly distribution have a higher metabolic rate than southerly distributed species of the same genus at the same temperature. The M–T curves of the cold-water species are displaced to the left of those curves for warm-water species. Within one climatic region, epifaunal forms like *Pecten* and *Modiolaria* have a high rate of oxygen consumption, particularly at higher temperatures, whereas digging species and level-bottom forms have not only a relatively lower rate of oxygen uptake, but also a slower rate of increase with temperature. Thus, the rate of oxygen consumption seems to be correlated with the mode of life of the individual species within a given geographical area; also, this rate represents a form of metabolic adaptation to differences in temperature in southern and northern waters. Deep-water species from areas where temperature is typically below 0°C are more temperature-sensitive than shallow-water forms. If a northern species moves to warm waters, the increased energy demands resulting from higher metabolic rate cause the animal literally to starve to death, and thereby limit its southern distribution. On the other hand, the northern distribution of southern forms appears to be limited by thermal influences on larval development and reproduction.

Additional data on M–T curves of a number of unrelated species of arctic and tropical poikilotherms prompted Scholander *et al.* (1953) to conclude that, on an interspecific basis, there is considerable but incomplete metabolic adaptation in aquatic arctic forms relative to aquatic tropical species, while terrestrial insects exhibit slight adaptation, if any. Studies of the metabolic rate-temperature response of closely related species of fiddler crabs (genus *Uca*) found along the east coast of the Americas and Caribbean islands provide further insight into the problem of climatic adaptation. The adult stage of *Uca,* which is a cosmopolitan genus, inhabits marshes and protected beaches in regions adjacent to estuarine and marine waters, and the larval stages, being planktonic, are found in estuarine and oceanic waters. Using M–T curves for different populations ranging from Massachusetts to southern Brazil, certain generalizations can be made.

Adults

At temperatures below 12°C, species from the temperate zone respire at a higher rate than tropical species (Fig. 4-11). In contrast, at higher temperatures, no systematic difference could be correlated with geography, although interspecific differences did exist (Vernberg, 1959a; Vernberg and Vernberg, 1966). Variations in the shape of the M–T curves suggest that temperature does not uniformly influence respiratory rate, but certain temperatures are more critical than others. For example, over the temperature range of 12–15°C, Q_{10} values less than 2 were typical for temperature-zone species (Table 4-3). Values are high for temperate-zone species over the lower thermal range of 7–12°C. At intermediate tempera-

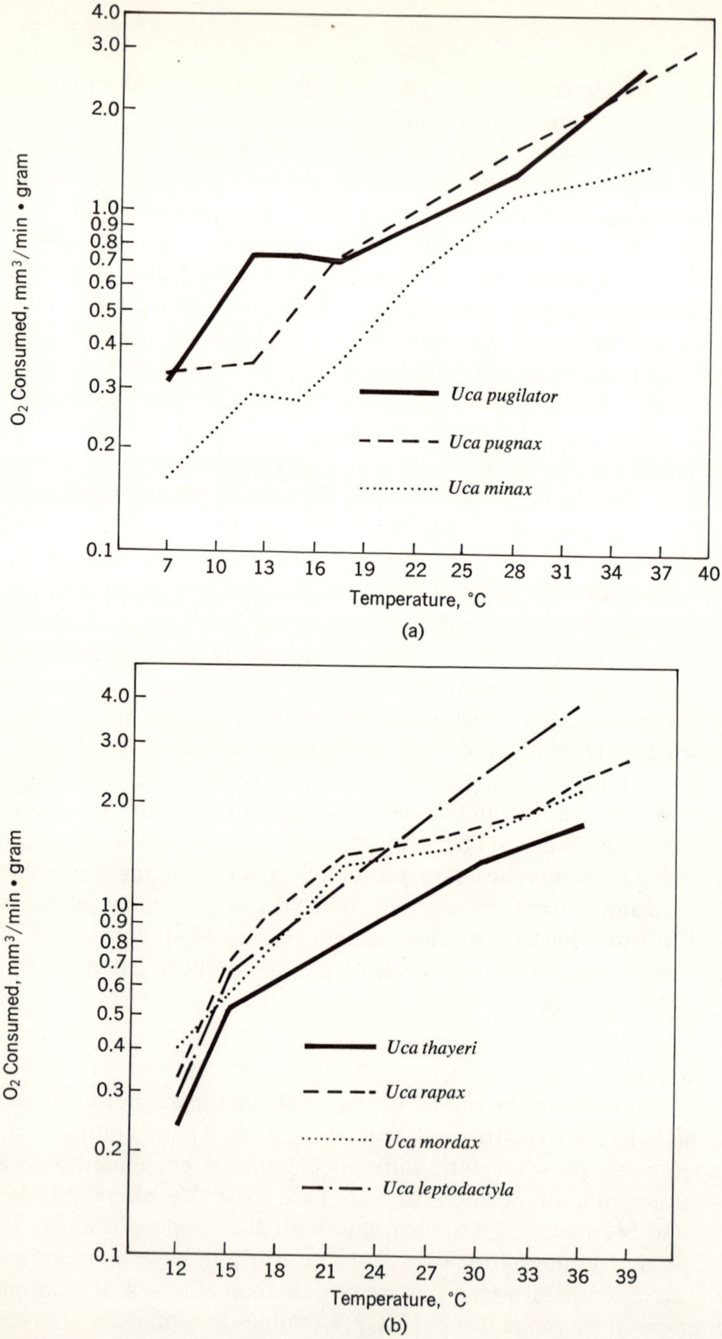

Figure 4-11. M–T curves of fiddler crabs (genus *Uca*) from different climatic regions: (*a*) temperature zone species and (*b*) tropical zone species. (From Vernberg, 1959a.)

Table 4-3. Q_{10} of Oxygen Consumption of Seven Species of *Uca* from Tropical and Temperate Zones.

Temperature Range, °C	*U. minax*	*U. pugnax*	*U. pugilator*	*U. rapax (Florida)*	*U. rapax (Jamaica)*	*U. mordax*	*U. lepto-dactyla*	*U. thayeri*
7–12	3.06	1.20	5.83	4.36	—	—	—	—
7–17	2.04	2.24	2.32	2.89	—	—	—	—
7–28*	2.54	2.10	2.00	2.45	—	—	—	—
7–36	2.10	—	2.11	2.07	—	—	—	—
12–15	1.00	—	1.00	—	8.58	3.73	15.9	13.7
12–17	1.37	4.23	1.00	1.93	—	—	—	—
12–22	—	—	—	—	4.20	3.66	—	—
12–28*	2.40	2.51	1.43	2.02	2.65	2.41	3.28	2.67
12–36	1.95	—	1.71	1.76	2.23	2.27	2.92	2.33
15–17	2.67	—	1.00	—	6.20	—	—	—
15–22	—	—	—	—	3.10	3.49	—	—
15–28*	2.51	—	1.55	—	2.02	2.18	2.39	1.92
15–36	2.18	—	1.84	—	1.84	2.12	2.29	1.68
17–22	4.04	—	—	—	2.35	—	—	—
17–28*	3.09	1.97	1.75	1.86	1.63	—	—	—
17–36	2.45	—	2.01	1.71	1.62	—	—	—
22–28*	2.57	—	—	—	1.23	1.32	—	—
22–33	1.90	—	—	—	1.29	—	—	—
28–33*	1.22	1.90	—	1.42	1.38	—	—	—
28–36*	1.28	—	2.46	1.33	1.57	2.03	2.05	1.55
33–36	2.49	—	—	2.22	1.98	—	—	—

SOURCE: Vernberg, 1959a.

* Determinations were made at 27°C for *U. rapax* (Florida) and 30°C for *U. leptodactyla* and *thayeri*.

tures, Q_{10} values are low, indicating that there is a thermal range within which these animals are relatively thermal-insensitive. This temperature range of thermal insensitivity appears to coincide with the thermal regime encountered by the animals in their "normal" habitat. *U. pugnax* from Jamaica had a Q_{10} of 1.29 over the temperature range of 22-33°C, while the temperate-zone *U. pugilator* exhibited a low Q_{10} (1.55) over the range of 15–28°C.

Another expression of climatic adaptation difference between tropical- and temperate-zone species is seen in their response to thermal acclimation. *Uca pugnax,* a temperate-zone species, is more thermally labile at low temperatures than is *U. rapax,* a tropical species. But at high temperatures, the reverse responses are observed (Fig. 4-12). This reponse may have adaptive significance, since temperate-zone animals are exposed to wide yearly fluctuations in temperature. In North

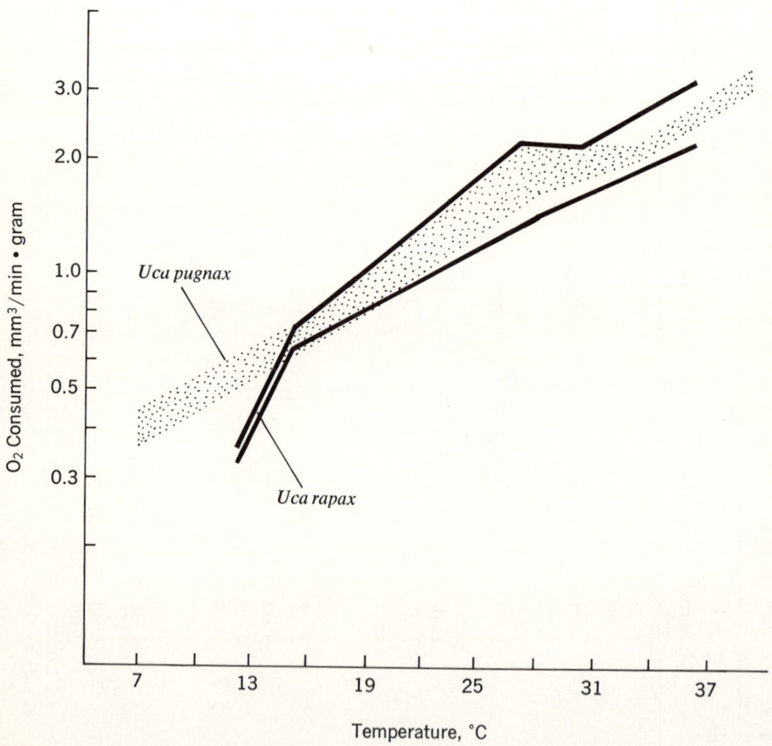

Figure 4-12. Genotypic limits of oxygen consumption of adult fiddler crabs. *Uca pugnax* is a temperature zone species; *Uca rapax* from the tropics. At low temperature the cold-adapted species shows greater lability; at high temperature the warm-adapted species is more labile. (Based on data of Vernberg, 1959b.)

Carolina, *U. pugnax* may experience a range of about 2–35°C, whereas *U. rapax* from Jamaica lives in a restricted thermal environment having a range of about 18–40°C. This acclimatory ability is further demonstrated by the seasonal respiratory cycle of *U. pugnax,* which is absent in *U. rapax* (Vernberg, 1959a,b).

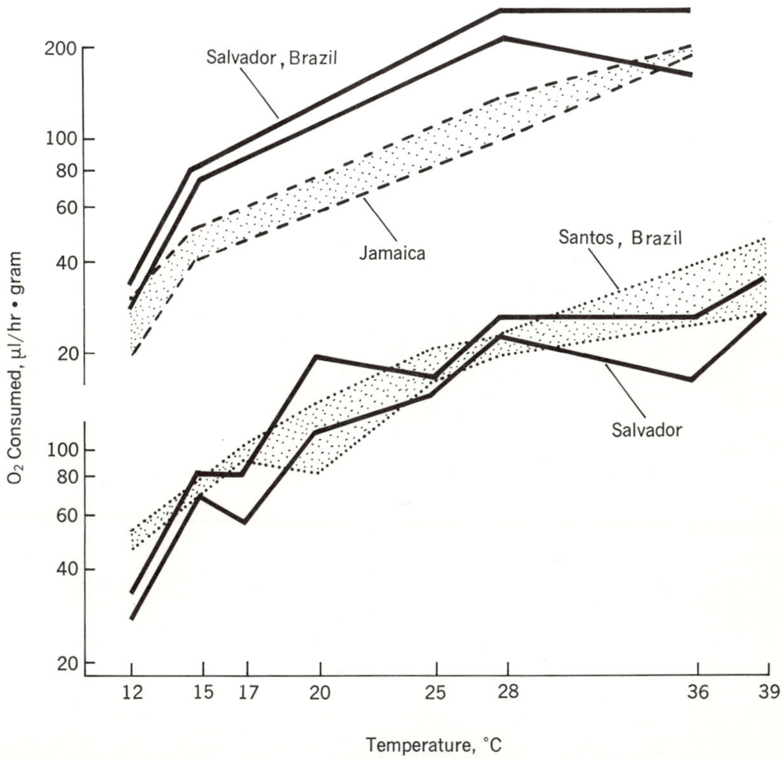

Figure 4-13. Differences in the "genotypic" limits of geographically separated populations of fiddler crabs (*Uca rapax*). These limits are determined using metabolic data from cold- and warm-acclimated animals. (From Vernberg and Vernberg, 1966.)

There are marked differences not only between tropical- and temperate-zone species, but also between geographically distinct populations of the same species, *U. rapax* (Fig. 4-13). Populations at the coldest parts of the range are metabolically better adapted to low temperature. These differences in metabolic adaptation are further illustrated when their patterns of acclimation are classified according to the system of Prosser (Table 4-4) and Precht (Table 4-5).

Table 4-4. Comparisons of Patterns of Acclimation in *Uca*.

Species	*Stage of Life Cycle or Type of Tissue*	*Pattern Number (according to Prosser, 1958)*	*Source of Populations or Individuals*	*Reference*
U. pugilator	Adults	II A	Mass. and Fla.	Demeusy (1957)
U. pugilator	Young adults	II A	Mass., N.C., and Fla.	Vernberg and Costlow (1966)
U. rapax	Adult	I or III B	Jamaica	Vernberg (1959)[a]
U. rapax	Adult	III A	Fla.	Vernberg (1959)[a]
U. pugnax	Adult	IV A	N.C.	Vernberg (1959)[a]
U. pugnax	Adult	II A	N.Y. and N.C.	Tashian (1956)
U. rapax	Adult	II A	Fla. and Trinidad	Tashian (1956)
U. pugilator	1st stage zoea	IV D	Fla. and N.C. and Mass.	Vernberg and Costlow (1966)
U. pugilator	1st stage zoea	I	N.C. and Mass.	Vernberg and Costlow (1966)
U. pugnax	1st stage zoea	IV A	N.Y. and N.C.	Vernberg and Costlow (1966)
Five species of *Uca*	1st stage zoea	I	Three temperate zone species from N.C. and two tropical species	Vernberg and Costlow (1966)
U. pugnax	Megalops	III B	N.C. and Mass.	Vernberg and Costlow (1966)
U. pugilator	Megalops	No match of any described patterns	Fla. and N.C. and Mass.	Vernberg and Costlow (1966)
U. pugnax	Gill	IV A	N.C.	Vernberg and Vernberg (1964)
U. pugnax	Midgut gland	III A	N.C.	Vernberg and Vernberg (1964)
U. rapax	Gill	III A	Jamaica	Vernberg and Vernberg (1964)
U. rapax	Midgut gland	I	Jamaica	Vernberg and Vernberg (1964)
U. pugilator	Brain	II B	N.C.	Vernberg and Vernberg (1964)
U. pugilator	Muscle	III B	N.C.	Vernberg and Vernberg (1964)
U. pugilator	Heart	IV C	N.C.	Vernberg and Vernberg (1964)

Table 4-5. Types of Thermal Acclimation by Geographically Isolated Populations of *Uca* (after the classification of Precht).

Species and Location of Population	*Original Acclimation Temperature*	
	High	*Low*
Uca rapax		
Santos	IV	III
Salvador	IV	III
Jamaica	IV	IV
Florida	III	V
Uca pugnax		
North Carolina	III	III
Uca uruguayensis		
Torres	III	IV
Santos	V	III

SOURCE: Vernberg and Vernberg, 1966.

Larvae

No distinctive difference of zoogeographical significance has been noted between the M–T curves of tropical- and temperate-zone fiddler crab larvae (Vernberg and Costlow, 1966). Nonetheless, there is intraspecific divergence, and northern populations of *U. pugnax* tend to have higher metabolic rates than southern populations, particularly at the lower temperatures; in contrast, the southern population of *U. pugilator* exhibits higher rates than the more northern populations (Fig. 4-14). However, when the aquatic larvae of *U. pugilator* develop to the semiterrestrial, young crab stage, their metabolic rate is then similar to the pattern of the adult crabs (Fig. 4-15).

Oxygen Tension

The physiological ecologist's viewpoint on oxygen tension and respiration was clearly expressed by Fry (1957): "Any reduction of oxygen content below the level where active metabolic rate begins to be restricted is probably unfavorable to the species concerned. From the ecological point of view this incipient limiting level can be taken as the point where oxygen content begins to be unstable." An obvious example of the correlation of oxygen tension and habitat preference is seen in the work on seven species of crabs representing different habitats (Van Weel *et al.*, 1954). Four species from oxygen-rich habitats consumed oxygen in proportion to the ambient tensions; in two species living on sandy flats in shallow water, the rate decreased very rapidly with a declining oxygen

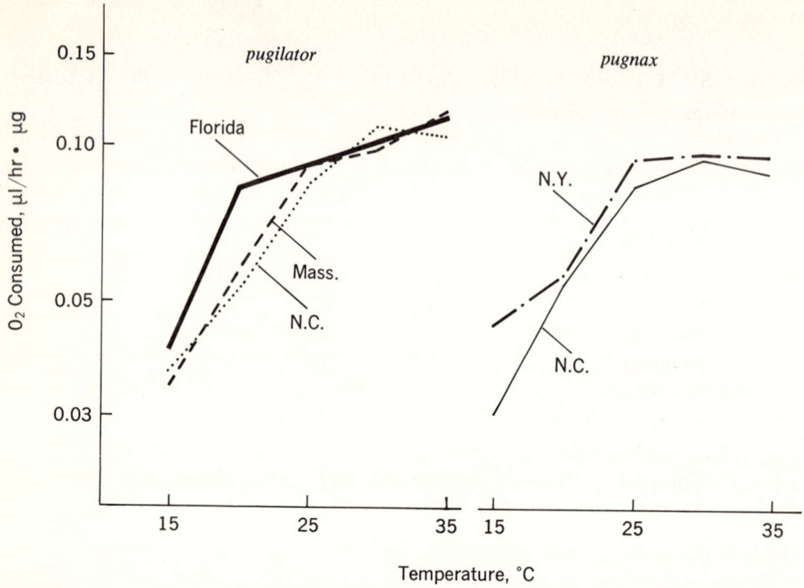

Figure 4-14. Oxygen consumption of first-stage zoeae of latitudinally separated populations of fiddler crabs (*genus Uca*). (From Vernberg and Costlow, 1966.)

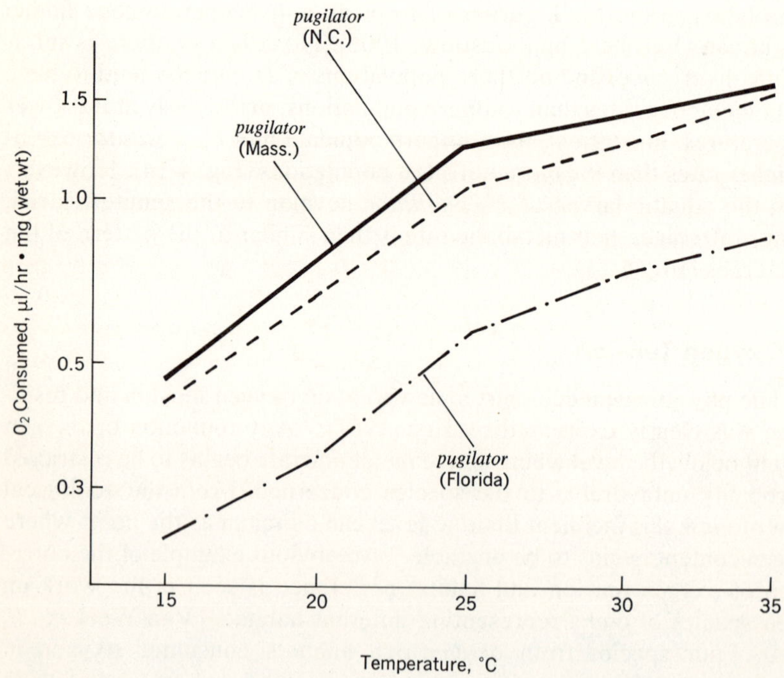

Figure 4-15. Oxygen consumption of young adult laboratory-raised fiddler crabs (*Uca pugilator*) from different geographical areas. (From Vernberg and Costlow, 1966.)

tension, while the mud-dwelling species consumed oxygen independently of ambient oxygen tension until a low external level was reached. At an O_2 value below 2.5 ml/liter, all seven species sharply increased the percentage of oxygen removed from sea water. This response demonstrated that at a low critical oxygen tension, these animals begin to regulate their metabolism in order to ensure a minimum amount of cellular oxidation. The P_c is lower in brackish and fresh-water species of isopods than in semiterrestrial species that live in an environment with a higher oxygen concentration (Walshe-Maetz, 1956). In salt marshes, crabs exposed to waters of low oxygen content are able to continue consuming oxygen down to a level of 0.4 percent of 1 atm, while their critical oxygen tension is about 1–3 percent of 1 atm (Teal and Carey, 1967). An actively swimming fish like the scup is oxygen-independent over a wide range of tensions (40–120 mm Hg), but the metabolic rate of the sluggish toadfish is oxygen-dependent (Hall, 1929).

Different stages in the life history of an animal may occupy diverse habitats, which vary with respect to the availability of oxygen. The adaptive responses of these different stages is well illustrated in parasitic flatworms. The metabolic rate of the adult stage of most trematodes is oxygen-dependent. Although one marine species, *Gynaecotyla adunca,* exhibited this response over the range of 21-1.5 percent oxygen, at low tensions of 1.5–0.5 percent, which is characteristic of the gut in which it lives, its metabolic rate was independent of oxygen tension. In contrast, another parasitic stage, the redia, of another species, *Himasthla quissetensis,* was metabolically independent of oxygen tension over the range of 5–1.5 percent. But the metabolic rate of the free-living cercarial stage of this species is oxygen-dependent, and it becomes inactive at a low tension of 0.5 percent (Vernberg, 1963).

Animals differ greatly in their ability to withdraw oxygen from their external environment. In general, filter feeders like sponges, lamellibranchs, and ascidians have low efficiency values for extracting oxygen (about 13 percent of the oxygen is removed from the incoming water). Crustaceans and annelids tend to have higher values (about 50 percent). Certain gastropods, cephalopods, and fishes have still higher values (Hazelhoff, 1938). Presumably, circulatory adjustments account for this greater efficiency. It is impossible, however, to generalize on the response of a group of organisms, since animals in the same genus may show more diversity in response than species belonging to different phyla. For example, *Octopus vulgaris* removes 70 percent of the oxygen from sea water (Winterstein, 1925), but *O. dofleini* extracts only 27 percent of the oxygen (Johansen and Lenfant, 1966). In fishes there is a distinct positive correlation between their ability to remove dissolved oxygen from water and their habitat. The sluggish bottom-dwelling toadfish can remove oxygen until the concentration of oxygen is almost 0 mm Hg; at the other extreme, the actively swimming, surface-dwelling mackerel is unable to withdraw oxygen when the tension drops below 70 mm Hg (Hall, 1930).

The behavior of aquatic organisms changes with decreases in ambient oxygen levels. When the oxygen level drops, the ability of salmon to maintain a sustained maximum swimming rate is reduced (Davis *et al.*, 1963), and bivalve molluscs differentially alter the rhythmic closure of their valves (Salanki, 1966). Other organisms move toward a source of oxygen from an oxygen-poor region (*Holothuria,* Newell and Courtney, 1965; *Gammarus,* Cook and Boyd, 1965), but this avoidance response may vary with the stage of development; for example, salmonid alevins do not respond to abnormally low oxygen concentrations, young fry are very sensitive and avoid hypoxic waters, and older animals are insensitive (Bishai, 1962). Many mobile animals respond to a decrease in oxygen content by an increase in movement of respiratory apparatus, such as the beating of pleopods or scaphognathites in crustaceans, or opercular movements in fish. In this manner a greater amount of water passes over the respiratory surface. However, other species, particularly sessile animals, show a decrease in respiratory movements when the oxygen tension decreases; for example, ciliary activity in some clams and in *Branchiostoma* amphioxus.

Salinity

No simple generalization concerning the influence of salinity on respiration is apparent. Reduced salinity depresses the respiratory rate of certain species and increases it in others, while some respire at the same rate over a wide salinity gradient. Furthermore, the interaction of temperature and salinity complicates the response patterns. For example, in two closely related species of crabs, very different response patterns were reported: *Hemigrapsus oregonensis,* when acclimated to low salinity, consumed oxygen at a higher rate at all acclimation temperatures than when acclimated to 75 percent sea water; however, at high acclimation temperature, the second species, *H. nudus,* had higher metabolic rates only at high salinity, but when acclimated to low temperature, the higher metabolic response occurred at low salinity (Dehnel, 1960).

Salinity conditions, either in nature or in the laboratory, appear to influence the metabolic response of animals. The mollusc *Mytilus edulis* from a high-salinity region (35 ‰), consumes oxygen faster at high salinities than at low salinities, whereas this same species of mollusc from the low saline waters of Finland consumes oxygen at a greater rate at low salinities.

The change in metabolic rate with salinity may be related to the degree of hydration of tissues; that is, any alteration of the water content may influence the hormonal and enzymatic balance in the cell. In general, however, the energy requirements for ionic regulation are not sufficient to account for all the change in metabolic rate when animals are subjected to a new salinity regime.

Carbon Dioxide

Excess CO_2 has been reported to increase, decrease, or have no effect on the ventilation rate of invertebrates. Generally, respiratory centers of aquatic animals are less sensitive to CO_2 stimulation than are those of terrestrial animals. Since CO_2 is highly soluble in water and is readily altered chemically to form H_2CO_3 by the presence of carbonic anhydrase and carbonate and bicarbonate salts, CO_2 concentration in the organism never reaches a high level. Thus, low oxygen concentration is a better indicator of respiratory stress for aquatic species. However, increased complexity of the respiratory system with the evolution of terrestrial animals has resulted in a shift from oxygen to CO_2 as the principal chemoregulatory agent.

SEMITERRESTRIAL AND TERRESTRIAL HABITATS

Modifications of Respiratory Systems for Terrestrialism

Both structural and functional modifications of the respiratory system have evolved with increased terrestrialism. When species of crabs from wholly aquatic habitats, from the intertidal zone, or from land were compared, the following changes were reported. The number and volume of gills and the gill area tended to be reduced per unit of weight as the different species approached land habitats (Gray, 1957). Any reduction in gill area probably reflects the fact that in a habitat with abundant oxygen, less respiratory surface is needed than in an aquatic environment, although in some land crabs this trend is negated, since the regions of the wall of the gill chambers are highly vascularized, presumably for the purpose of increased gaseous exchange. In both habitats, however, the exchange of respiratory gases in the crab must be through an aqueous layer on the respiratory surface. The structure of the gill has also undergone change. Some species have strongly sclerotized gills; in others, the gills are held at right angles to the gill bar; and yet in others, the gill leaves are not easily closed (Raben, 1934). The land crabs, however, have not evolved structural modifications that will completely prevent desiccation. Instead, these animals are nocturnally active, at a time when the danger of desiccation is diminished. The metabolic rate of the intact crab and its gill tissue is increased in species adapted to a more terrestrial habitat (Vernberg, 1956).

The same tendency also was observed at the molecular level of organization with the activity of cytochrome *c* oxidase in gill tissue from three species of crabs: *Libinia emarginata* (the spider crab), *Callinectes sapidus* (the blue crab), and *Ocypode quadrata* (the ghost crab). Two of these species, *L. emarginata* and *C. sapidus,* are aquatic, although the range of *L. emarginata* extends into much deeper water than that of *C. sapidus*. The ghost crab is the most terrestrial of these three species of decapod Crustacea (Fig. 4-16). Cytochrome *c* oxidase activity as well as metabolic rate is significantly greater in the whole organism and in gill tissue from the terrestrial ghost crab than in the other two species.

Metabolic adaptations have also been noted for intertidal-zone barnacles (Augenfeld, 1967). The whole-animal oxygen uptake and cytochrome oxidase activity were higher in barnacles that live in the higher regions (where they are out of water and exposed to high oxygen tensions) than in those from the lower intertidal-zone regions.

Salamanders represent another group of animals in transition from an aquatic (in this case, fresh water) to a purely terrestrial environment. Unlike marine crustaceans, the more aquatic species of Plethodontid salamanders have a higher metabolic rate than more terrestrial species (Evans, 1939; Vernberg, 1952). However, the tendency of aquatic plethodontids to be more active than terrestrial species is not the general rule for most animal groups. When closely related salamander species occupying similar habitats were compared, the more active species had the higher metabolic rate (Vernberg, 1955).

In addition to studies on salamanders, the metabolic rate-temperature responses of anurans from different geographical ranges indicate respiratory adaptation. At their normal habitat temperature, the more northern species had a higher rate of respiration than the tropical species (Tashian and Ray, 1957). Geographical differences in the percentage of the total oxygen uptake supplied by the pulmonary surface and the cutaneous surface were noted in temperate and tropical species of anurans. The lungs accounted for a greater percentage of oxygen uptake with increasing temperature. Although some species variation was reported, the increased contribution of the lungs to respiration occurred at lower temperatures for northern species (Hutchison *et al.,* 1968).

In the higher taxonomic categories, a causal relationship between physiology and distribution becomes progressively more obscure as homeostatic mechanisms become more complex. Corroborative detailed

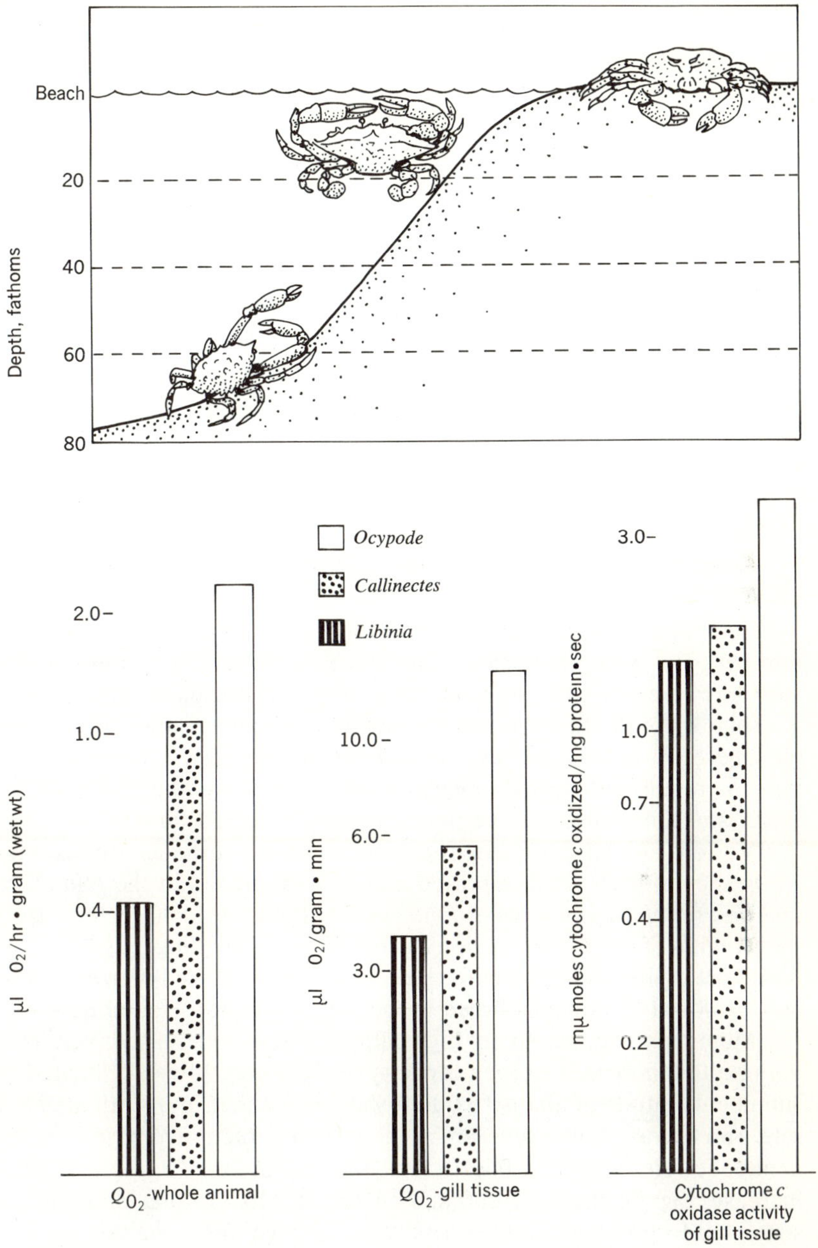

Figure 4-16. Whole-animal metabolism, gill-tissue metabolism, and cytochrome c oxidase activity of gill tissue can be correlated with zonal distribution in three species of Crustacea: *Ocypode quadrata,* the ghost crab, is terrestrial and found along open beaches; *Callinectes sapidus,* the blue crab, is found in shallow waters; and *Libinia emarginata,* the spider crab, is found from water near shore out to about 68 fathoms. (From Vernberg and Vernberg, 1968.)

ecological analysis, including behavioral patterns and microhabitat selection, are needed. The dynamics of these homeostatic mechanisms is a complex physiological problem which is of great importance in understanding the physiological ecology of land animals. A discussion of these mechanisms is not within the scope of this book and therefore the student is referred to various references for details. Many pertinent papers are found in the following books: *Handbook of Physiology* (edited by Dill, Adolph, and Wilber, 1964), *Comparative Physiology of Thermoregulation* (edited by Whittow, 1970), *Comparative Physiology of Temperature Regulation* (edited by Hannon and Viereck, 1962), and *Thermobiology* (edited by Rose, 1967).

High Altitude

Since barometric pressure is reduced at high altitudes, animals living here are subjected to an oxygen-poor atmosphere. At sea level, the average barometric pressure is 760 mm Hg (range from 664 mm to 805 mm Hg). For every increase in altitude of 18,000 ft, the barometric pressure is reduced by half; for example, at 18,000 ft, the barometric pressure is about 380 mm Hg. Although the amount of oxygen and CO_2 in the atmosphere (based on percent by volume of dry air) is constant, most humans require additional oxygen in the air they breathe at 10,000 ft or more, and they must breathe pure oxygen above 35,000 ft.

Various animal groups are represented at high altitudes. For example, 11 species of frogs were reported above 3700 meters in the Himalayas, and one reptile, *Leiolopisma ladacense,* is found at 5500 meters. In the puna region of the Andes, various insect groups, including may flies, stone flies, beetles, bee flies, and butterflies, have been reported between 10,000 and 13,000 ft. Flamingos breed in the salt depressions found here.

Mammals acclimated to high altitudes show increased ventilation volume and increased oxygen capacity of the blood. Mice acclimated to a simulated altitude of about 16,000 ft show no change in overall metabolic rate. In contrast, cattle and lambs show a 20 percent decrease in resting metabolic rate, while the basal metabolic rate of man from low altitudes increases significantly at 5200 and 14,150 ft. Most tissues of albino rats subjected to prolonged hypoxia show no change in Q_{O_2} values.

RESPIRATORY PIGMENTS

In many organisms the ability to provide oxygen for cellular metabolism is enhanced by respiratory pigments. These molecules, which are distributed in various parts of an organism, function to combine reversibly with oxygen. Normally, these pigments form a loose chemical combination with oxygen as they pass near a respiratory surface where the external oxygen tension is higher than that of the body fluids. When a region of low tension is reached, as found in tissues and cells, oxygen is released. Respiratory pigments may also serve in oxygen storage and as a buffer to control any harmful effects of excess oxygen.

Since the discovery of respiratory pigments, most studies have progressed along three main lines: (1) The characterization of these molecules in terms of their chemical and physical properties—many excellent studies using powerful modern research techniques have dealt with the structure and evolution of these molecules. (2) The identification and distribution of these various pigments throughout the biotic kingdoms—this includes descriptive information on their location in the body; for example, some are either in cells or free in various body fluids. (3) Although a respiratory pigment may be present, it is necessary to demonstrate that it is functional within the range of environmental conditions experienced by the organism. Only the third area of research will be emphasized here. Various reviews and textbooks are available which describe the first two major lines of investigation (Ghiretti, 1968; Manwell, 1960).

The capacity of respiratory pigments to combine with oxygen varies intra- and interspecifically, but there are certain phylogenetic trends. Sea water saturated with air at 20°C contains 0.538 vol % oxygen (vol % is the abbreviation for number of volumes of a substance in 100 volumes of a solution or air sample).

Invertebrates with respiratory pigments have concentrations ranging from 0.5–10 vol %, and in poikilothermic vertebrates the range is 5–12 vol %. Within one group of animals, a correlation between the oxygen capacity of blood and the habitat can be found. For example, among fish, the active surface-dwelling mackerel has a higher oxygen capacity than the bottom-dwelling sluggish toadfish. The amount of respiratory pigment present in the blood may vary on an intra- and interspecific basis. Generally, active species have more respiratory pigment per unit blood than the more sluggish species (Hall and Gray, 1929), and animals living at high altitudes have higher values than those from low altitudes (Hall, 1936). In aquatic environments the tendency is for the amount of blood pigment to increase with a decrease in ambient oxygen, although not all animals behave in the same way (Fox, 1955).

Although an organism possesses a respiratory pigment, its functional significance to the organism must be demonstrated. One useful measurement of function is to characterize the oxygen-equilibrium curve.

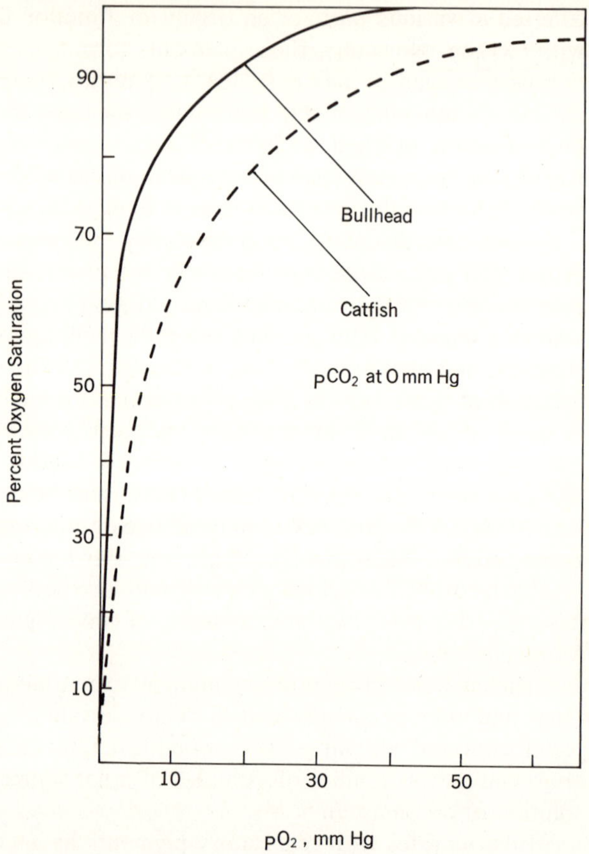

Figure 4-17. Oxygen equilibrium curves of two species of fish from different habitats. The bullhead is from low oxygen tension waters while catfish live in more oxygen-rich waters. (From Haws and Goodnight, 1962.)

Essentially, this is done by measuring and plotting the percent of saturation of the respiratory pigment at different oxygen pressures. The pressures required for 95 percent saturation (the loading tension, P_{sat}) and 50 percent saturation (the unloading tension, P_{50}) are of particular importance for comparative purposes. An illustration of this approach is the study involving two species of fish. The oxygen-equilibrium curve of the brown bullhead is to the left of that of the catfish, which indicates that the respiratory pigment of the brown bullhead is fully saturated at a lower ambient oxygen tension than is that of the catfish (Fig. 4-17). These laboratory findings agree with the habitats of these two species in that the brown bullhead lives in areas where the oxygen tension is lower than that normally encountered by the catfish (Haws and Goodnight, 1962). A similar pattern of response was noted when the free-swimming cod was compared with the plaice, which lives partially buried in the sea bottom. The P_{50} values were 18 mm Hg for cod and 10 mm Hg for plaice, while the P_{sat} values were 70 mm and 40 mm, respectively. An extremely low P_{50} value of 0.19 mm Hg was reported for a mud-dwelling bivalve clam, while the value for chitons living in well-aerated waters ranged from 20–26 mm Hg (Read, 1962). In general, the oxygen-equilibrium curves for animals living in oxygen-poor environments are farther to the left than for animals from well-aerated waters, which in turn tend to be to the left of terrestrial animals (Fig. 4-18). However, a noticeable exception to this generalization has been reported for the land crab *Cardisoma* which has a P_{50} value of 3.5–4 mm Hg; this is surprisingly low for a terrestrial animal (Redmond, 1968). Many terrestrial mammals and birds have P_{50} values ranging from 25–58 mm Hg (Prosser and Brown, 1961). Intraspecific differences also have been reported; for example, the position of the oxygen-equilibrium curve for the respiratory pigment of the fetal or larval stage may be different from that of the adult.

The position of the oxygen-equilibrium curve along an oxygen tension gradient is important in assessing the ecological role of respiratory pigments, but the influence of other environmental factors on the P_{50} and P_{sat} values must also be known.

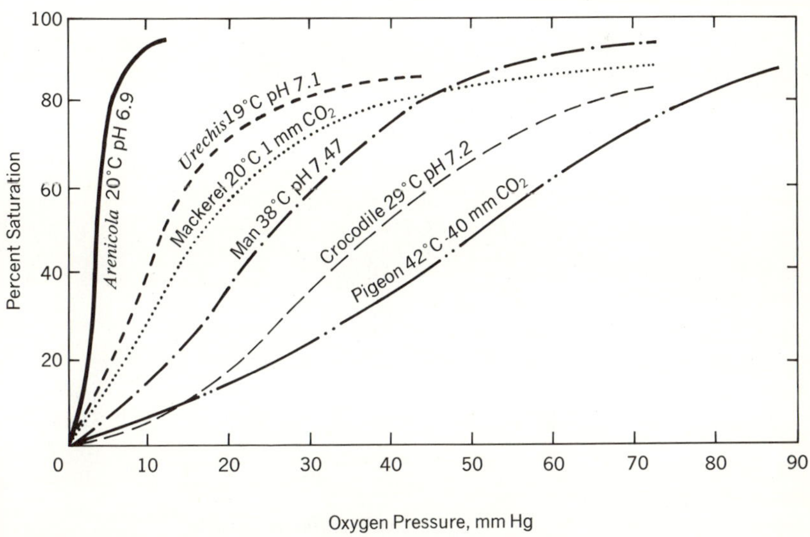

Figure 4-18. Oxygen equilibrium curves of hemoglobin from various species. (From Prosser and Brown, 1961.)

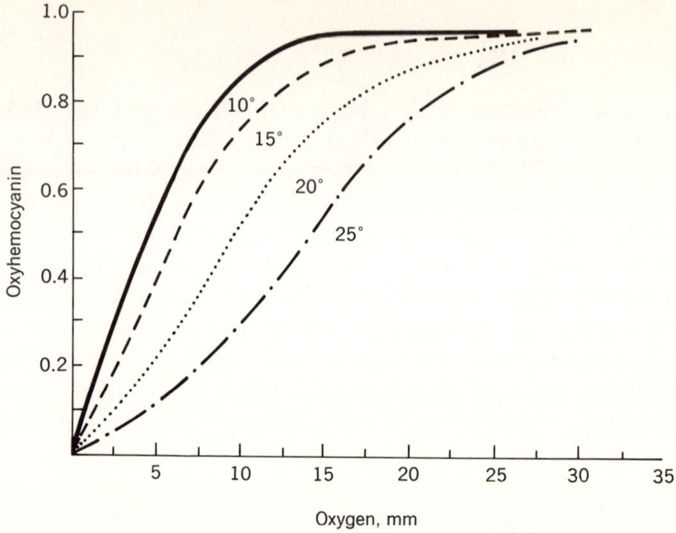

Figure 4-19. The effect of temperature on the oxygen equilibrium curves of hemocyanin from the lobster, *Panulirus interruptus*. (From Redmond, 1955.)

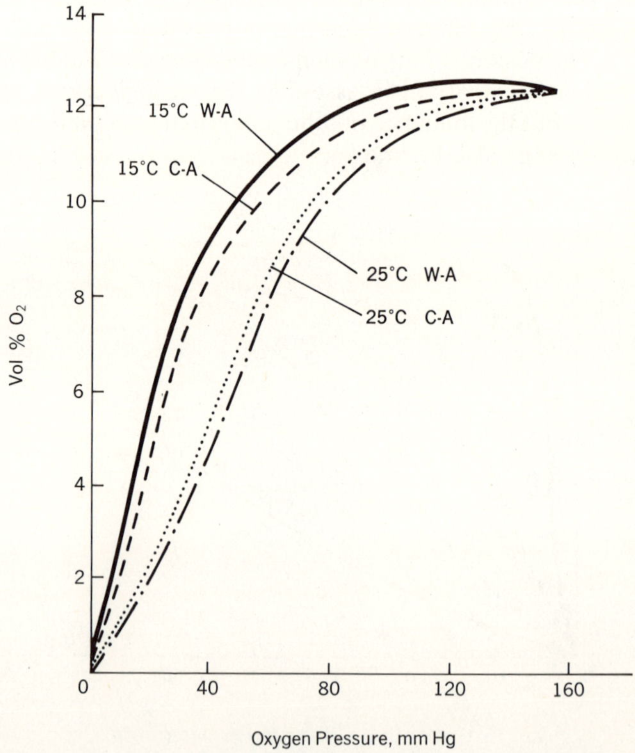

Figure 4-20. The influence of temperature and temperature acclimation on the oxygen equilibrium curves of hemoglobin from a frog (*Rana esculenta*). WA is warm-acclimated (15°C) and CA is cold-acclimated (25°C). Determinations made at 15° and 25°. (From Straub, 1957.)

Temperature

Reduced temperature causes a shift of the position of the oxygen-equilibrium curve of crabs to the left and alters its shape (Fig. 4-19). Not only was the same response observed in frogs, but the curve for cold-acclimated animals was to the left of that of warm-acclimated frogs, when both were determined at the same temperature (Fig. 4-20). The differential thermal sensitivity of hemocyanin of two species of crabs was correlated with their survival ability: at 16–17°C the P_{50} values were similar for both species, but at 25°C this value was lower for the species that survived best (Redmond, 1955). Interspecific differences exist between species of chitons and crabs from the tropics and those from the colder waters of the Pacific. At its habitat temperatures of 28°C, the P_{50} value for the tropical chiton was 30 mm Hg, while this value was 8 mm for the temperate-zone species at its habitat temperature of 10°C. At 25°C, however, both species had similar values (Redmond, 1962). In contrast, the oxygen-equilibrium curve of the spiny lobster from the tropics, determined at its environmental temperature (25°C), corresponds with the curve for the cold-water spiny lobster determined at its 15°C habitat temperature (Fig. 4-21).

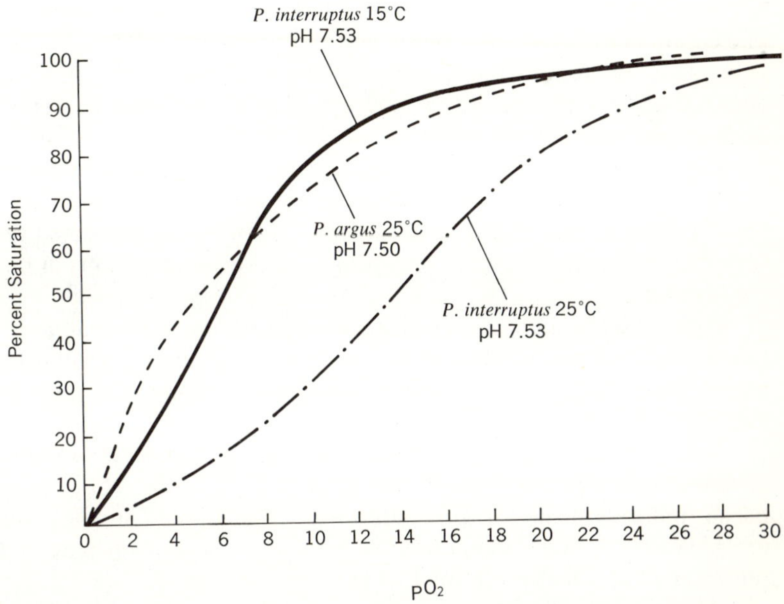

Figure 4-21. Oxygen equilibrium curves of hemocyanins from spiny lobsters. *Panulirus argus* is from tropical waters; *P. interruptus* from temperate waters. (From Redmond, 1968.)

pH

An increase in acidity of CO_2 may shift the pH curve to the right (the Bohr effect), to the left (the reverse Bohr effect), or may produce little change. The result of the Bohr effect is to facilitate the unloading of oxygen from the respiratory pigment in regions of increased acidity, such as in tissues. If the ambient pH is lowered, however, the organism would be at a disadvantage, since a higher oxygen tension would be required for the pigment to be saturated. The reverse Bohr effect would facilitate the uptake of oxygen from habitats with low oxygen content and high CO_2.

SUMMARY

All organisms must generate energy if they are to sustain biological integrity in a demanding environment. This energy results from biochemical transformations and eventually involves oxidation of oxygen and hydrogen. During the course of evolution, numerous physiological and morphological adaptations have enabled animals to exploit habitats that differed widely in the availability of oxygen. These habitats vary from anaerobic muds, to aquatic environments with varying oxygen concentration, to high altitude with decreased oxygen pressures. Different organisms have become metabolically adapted not only to these various levels of oxygen, but also to other biotic and abiotic factors influencing metabolism, such as temperature, salinity, stage of life cycle, and oxygen acclimation. To understand the energetics of an organism in terms of its ecology, the influence of these factors on respiration must be studied.

Certain basic characteristics are common to respiration of animals in either water or air, although different mechanisms evolved to meet specific habitat restrictions—for example, an internally protected respiratory membrane in terrestrial vertebrates or elongated siphons in burrowing bivalves.

In contrast to animals in terrestrial habitats, those from aquatic habitats do not face the problem of having their respiratory surfaces dry out, and in general they live in a more constant environment, except for the varying concentration of oxygen.

When geographically separated organisms are compared, metabolic adaptation to different climatic regions is evident at the whole-organism, tissue, and enzyme levels. High latitude or cold-adapted animals consume oxygen faster at lower temperatures, while tropical animals reach a respiratory peak at a higher thermal level.

Although the respiratory performance of an animal is determined by its genotype, the rate of oxygen uptake is variable and depends in part on the external environment. In addition, acclimation to various factors, such as oxygen and/or temperature, will affect the intensity of respiration.

The exciting challenge to present-day students is to determine the metabolic characteristics of organisms under different ecological conditions and to understand the mechanisms of adaptations. These are not simple tasks, since respiration is complex and is influenced by so many factors.

References

Augenfeld, J. M. 1967. Respiratory metabolism and glycogen storage in barnacles occupying different levels of the intertidal zone. *Physiol. Zool.,* 40:92–96.

Barlow, G. W. 1961. Intra- and interspecific differences in rate of oxygen consumption in gobiid fishes of the genus *Gillichthys. Biol. Bull.,* 121:209–229.

Barnes, R. D. 1963. *Invertebrate Zoology.* Saunders, Philadelphia.

Barron, E. S. G., and T. N. Tahmisian. 1948. The metabolism of cockroach muscle (*Periplaneta americana*). *J. Cell. Comp. Physiol.,* 32:57–76.

Bartholomew, G. A., T. R. Howell, and T. J. Cade. 1957. Torpidity in the white-throated swift, Anna hummingbird, and poor-will. *Condor,* 59:145–155.

Basu, S. P. 1959. Active respiration of fish in relation to ambient concentrations of oxygen and carbon dioxide. *J. Fish. Res. Bd. Canada,* 16:175–212.

Beamish, F. W. H., 1964. Respiration of fishes with special emphasis on standard oxygen consumption. III. Influence of oxygen. *Can. J. Zool.,* 42:355–366.

Belding, H. S., J. Field, II, F. W. Weymouth, and S. C. Allen. 1942. Studies on the metabolism of marine invertebrate tissue. I. Respiration of the midgut gland of the kelp crab (*Pugettia producta*). *Physiol. Zool.,* 15:75–88.

Berg, K., and K. W. Ockelmann. 1959. The respiration of freshwater snails. *J. Exptl. Biol.,* 36:690–790.

Bishai, H. M. 1962. Reactions of larval and young salmonids to water of low oxygen concentration. *J. Cons. Perm. Int. Explor. Mer.,* 27:167–180.

Bullock, T. H. 1955. Compensation for temperature in the metabolism and activity of poikilotherms. *Biol. Rev.,* 30:311–342.

Cook, R. H., and C. M. Boyd. 1965. The avoidance of *Gammarus oceanicus* Segerstole (Amphipoda, Crustacea) of anoxic regions. *Can. J. Zool.,* 43:971–975.

Davis, G. E., J. Foster, C. E. Warren, and P. Doudoroff. 1963. The influence of oxygen concentration on the swimming performance of juvenile Pacific salmon at various temperatures. *Trans. Am. Fish. Soc.,* 92:111–124.

Dehnel, P. A. 1958. Effect of photoperiod on the oxygen consumption of two species of intertidal crabs. *Nature,* 181:1415–1417.

Dehnel, P. A. 1960. Effect of temperature and salinity on the oxygen consumption of two intertidal crabs. *Biol. Bull.,* 118:215–249.

Demeusy, N. 1957. Respiratory metabolism of the fiddler crab, *Uca pugilator* from two different latitudinal populations. *Biol. Bull.,* 113:245–253.

Dill, D. B.; E. F. Adolph, and C. G. Wilber (eds.). 1964. Adaptation to the environment. In *Handbook of Physiology, Section 4.* Am. Physiol. Soc., Washington, D.C.

Edwards, G. A., and L. Irving. 1943a. The influence of temperature and season upon the oxygen consumption of the sand crab, *Emerita talpoida* Say. *J. Cell. Comp. Physiol.,* 21:169–182.

Edwards, G. A., and L. Irving. 1943b. The influence of season and temperature upon the oxygen consumption of the beach flea, *Talorchestia megalophthalma. J. Cell. Comp. Physiol.,* 21:183–189.

Evans, G. 1939. Factors influencing the oxygen consumption of several species of Plethodontid salamanders in aerial and aquatic media. *Ecology,* 20:74–95.

Evans, R. M., F. C. Purdie, and C. P. Hickman, Jr. 1962. The effect of temperature and photoperiod on the respiratory metabolism of rainbow trout (*Salmo gairdneri*). *Can. J. Zool.,* 40:107–118.

Fox, H. M. 1936. The activity and metabolism of poikilothermal animals in different latitudes. I. *Proc. Zool. Soc. London,* 1936:945–955.

Fox, H. M. 1955. The effect of oxygen on concentration of haem in invertebrates. *Proc. Roy. Soc., Ser. B,* 143:203–214.

Fry, F. E. J. 1947. Effects of the environment on animal activity. *University of Toronto Stud. Biol.,* 55:1–62.

Fry, F. E. J. 1957. The aquatic respiration of fish. *In* M. E. Brown (ed.), *Physiology of Fishes,* Vol. I, pp. 1–63. Academic Press, New York.

Fry, F. E. J., and J. S. Hart. 1948. Cruising speed of goldfish in relation to water temperature. *J. Fish. Res. Bd. Canada,* 7:169–175.

Ghiretti, F. (ed.). 1968. *Biochemistry and Physiology of Haemocyanins.* Academic Press, New York.

Graham, J. M. 1949. Some effects of temperature and oxygen pressure on the metabolism and activity of the speckled trout, *Salvelinus fontinalis. Can. J. Res., Ser. D,* 27:270–288.

Gray, I. E. 1957. A comparative study of the gill area of crabs. *Biol. Bull.,* 112:34–42.

Hall, F. G. 1929. The influence of varying oxygen tensions upon the rate of oxygen consumption in marine fishes. *Am. J. Physiol.,* 88:212–218.

Hall, F. G. 1930. The ability of the common mackerel and certain other marine fishes to remove dissolved oxygen from sea water. *Am. J. Physiol.,* 93:412–421.

Hall, F. G. 1931. The respiration of puffer fish. *Biol. Bull. Mar. Biol. Lab., Woods Hole,* 61:457–467.

Hall, F. G. 1936. The effect of altitude on the affinity of hemoglobin for oxygen. *J. Biol. Chem.,* 115:485–490.

Hall, F. G., and I. E. Gray. 1929. The hemoglobin concentration of the blood of marine fishes. *J. Biol. Chem.,* 81:589–594.

Hannon, J. P., and E. Viereck (eds.). 1962. *Comparative Physiology of Tempera-*

ture Regulation—Parts I, II, III. Arctic Aeromedical Laboratory, Fort Wainwright, Alaska.

Haws, T. G., and C. J. Goodnight. 1962. Some aspects of the hematology of two species of catfish in relation to their habitats. *Physiol. Zool.,* 35:8–17.

Hayes, F. R., I. R. Wilmot, and D. A. Livingstone. 1951. The oxygen consumption of the salmon egg in relation to development and activity. *J. Exptl. Zool.,* 116:377–395.

Hazelhoff, E. H. 1938. Über die Ausnutzung des Sauerstoffs bei verschiedenen Wassertieren. *Z. vergl. Physiol.,* 26:306–327.

Hemmingsen, A. M. 1960. Energy metabolism as related to body size and respiratory surfaces, and its evolution. *Rept. Steno Memorial Hospital, Nordisk Insulinlaboratorium,* Vol. IX, Part II. Copenhagen.

Holter, H. 1943. Technique of the Cartesian diver. *Compt. Rend. Trav. Lab. Carlsberg, Ser. Chim.,* 24:399–478.

Hunter, W. S., and W. B. Vernberg. 1955. Studies on oxygen consumption in digenetic trematodes. II. Effects of two extremes in oxygen tension. *Exptl. Parasitol.,* 4:427–434.

Hutchison, V. H., W. G. Whitford, and M. Kohl. 1968. Relation of body size and surface area to gas exchange in anurans. *Physiol. Zool.,* 41:65–86.

Job, S. V. 1955. The oxygen consumption of *Salvelinus fontinalis. University of Toronto Stud. Biol. Ser. 61,* 73:1–39.

Johansen, K., and C. Lenfant. 1966. Gas exchange in the cephalopod, *Octopus dofleini. Am. J. Physiol.,* 210:910–918.

Kleiber, M. 1947. Body size and metabolic rate. *Physiol. Rev.,* 27:511–541.

Krogh, A. 1916. *Respiratory Exchange of Animals and Man.* Longmans, Green, London.

Manwell, C. J. 1960. Comparative physiology: blood pigments. *Ann. Rev. Physiol.,* 22:191–244.

Marshall, S. M., A. G. Nicholls, and A. P. Orr. 1935. On the biology of *Calanus finmarchicus.* VI. Oxygen consumption in relation to environmental conditions. *J. Mar. Biol. Assoc. U.K.,* 20:1–28.

Newell, R. C., and W. A. M. Courtney. 1965. Respiratory movements in *Holothuria forskali* Delle Chiaje. *J. Exptl. Biol.,* 42:45–57.

Precht, H. 1958. Concepts of the temperature adaptation of unchanging reaction systems of cold-blooded animals. *In* C. L. Prosser (ed.), *Physiological Adaptation,* pp. 50–77. Am. Physiol. Soc., Washington, D.C.

Prosser, C. L. 1958. General summary: the nature of physiological adaptation. *In* C. L. Prosser (ed.), *Physiological Adaptation,* pp. 167–180. Am. Physiol. Soc., Washington, D.C.

Prosser, C. L. 1962. Acclimation of poikilothermic vertebrates to low temperatures. *In* J. P. Hannon and E. Viereck (eds.), *Comparative Physiology of Temperature Regulation,* pp. 1–34. Arctic Aeromedical Lab., Fort Wainwright, Alaska.

Prosser, C. L., and F. A. Brown, Jr. 1961. *Comparative Animal Physiology.* Saunders, Philadelphia.

Prosser, C. L., L. M. Barr, R. D. Pine, and C. Y. Lauer. 1957. Acclimation of goldfish to low concentrations of oxygen. *Physiol. Zool.,* 30:137–141.

Raben, K. van. 1934. Veränderungen in dem Kiemendeckel und in den Kiemen einiger Brachyuren (Decapoden) im Verlauf der Anpassung an die Feuchtluftatmung. *Z. wiss. Zool. A,* 145:425–461.

Rao, K. P. 1958. Oxygen consumption as a function of size and salinity in *Metapenaeus monoceros* Fab. from marine and brackish-water environments. *J. Exptl. Biol.,* 35:307–313.

Read, K. R. H. 1962. The hemoglobin of the bivalved mollusc, *Phacoides pectinatus. Biol. Bull. Mar. Biol. Lab., Woods Hole,* 123:605–617.

Redmond, J. R. 1955. The respiratory function of hemocyanin in Crustacea. *J. Cell. Comp. Physiol.,* 46:209–247.

Redmond, J. R. 1962. The respiratory characteristics of chiton hemocyanins. *Physiol. Zool.,* 35:304–313.

Redmond, J. R. 1968. The respiratory function of hemocyanin. *In* F. Ghiretti (ed.), *Biochemistry and Physiology of Haemocyanins,* pp. 5–23. Academic Press, New York.

Roberts, J. L. 1957. Thermal acclimation of metabolism in the crab *Pachygrapsus crassipes* Randall. I. The influence of body size, starvation, and molting. *Physiol. Zool.,* 30:232–242.

Roberts, J. L. 1960. The influence of photoperiod upon thermal acclimation by the crucian carp, *Carassius carassius* L. *Ver. der Deutschen Zool. Ges. in Bonn/Rhein.* 73–78.

Romer, A. S. 1949. *The Vertebrate Body.* Saunders, Philadelphia.

Rose, A. H. (ed.). 1967. *Thermobiology.* Academic Press, New York.

Salanki, J. 1966. Comparative studies on the regulation of the periodic activity in marine lamellibranchs. *Comp. Biochem. Physiol.,* 18:829–843.

Scholander, P. F. 1964. Animals in aquatic environments: diving mammals and birds. In *Handbook of Physiology, Section 4, Adaptation to the Environment*, Am. Physiol. Soc., Washington, D.C.

Scholander, P. F., W. Flagg, V. Walters, and L. Irving. 1953. Climatic adaptation in arctic and tropical poikilotherms. *Physiol. Zool.,* 26:67–92.

Somero, G. N., and A. L. DeVries. 1967. Temperature tolerances of some Antarctic fishes. *Science,* 156:257–258.

Sparck, R. 1936. On the relation between metabolism and temperature in some marine lamellibranchs, and its zoogeographical significance. *Biol. Medd.,* 13:No. 5.

Straub, M. 1957. Weitere untersuchungen zur temperaturadaptation der Sauerstoffbindung des Blutes von *Rana esculenta* L. *Z. vergl. Physiol.,* 39:507–523.

Subrahmanyan, C. B. 1962. Oxygen consumption in relation to body weight and oxygen tension in the prawn *Penaeus indicus* (Milne Edwards). *Proc. Indian Acad. Sci.,* 55:152–161.

Tashian, R. E. 1956. Geographic variation in the respiratory temperature coefficient in tropical and temperate forms of the fiddler crab, *Uca pugnax. Zoologica,* 41:39–47.

Tashian, R. E., and C. Ray. 1957. The relation of oxygen consumption to temperature in some tropical, temperate and boreal anuran amphibians. *Zoologica,* 42:63–68.

Teal, J. M., and F. G. Carey. 1967. The metabolism of marsh crabs under conditions of reduced oxygen pressure. *Physiol. Zool.,* 40:83–91.

Thorson, G. 1936. The larval development, growth, and metabolism of arctic marine bottom invertebrates compared with those of other seas. *Medd. Gronland,* 100: No. 6.

Umbreit, W. W., R. H. Burris, and J. F. Stauffer. 1964. *Manometric Techniques.* Burgess, Minneapolis.

Van der Klaauw, C. J. 1948. Ecological studies and review. IV. Ecological morphology. *Series D, Bibliotheca Biotheoretica,* 4:27–111.

Van Weel, P. B., J. E. Randall, and M. Takata. 1954. Observations on the oxygen consumption of certain marine Crustacea. *Pacific Science,* 8:209–218.

Vernberg, F. J. 1952. The oxygen consumption of two species of salamanders at different seasons of the year. *Physiol. Zool.,* 25:243–249.

Vernberg, F. J. 1955. Correlation of physiological and behavior indexes of activity in the study of *Plethodon cinereus* (Green) and *Plethodon glutinosus* (Green). *Am. Midland Naturalist,* 54:382-393.

Vernberg, F. J. 1956. Study of the oxygen consumption of excised tissues of certain marine decapod Crustacea in relation to habitat. *Physiol. Zool.,* 29:227–234.

Vernberg, F. J. 1959a. Studies on the physiological variation between tropical and temperate zone fiddler crabs of the genus *Uca.* Oxygen consumption of whole organisms. *Biol. Bull.,* 117:163–184.

Vernberg, F. J. 1959b. Studies on the physiological variation between tropical and temperate zone fiddler crabs of the genus *Uca.* IV. The influence of temperature acclimation on oxygen consumption of whole organisms. *Biol. Bull.,* 117:582–593.

Vernberg, F. J., and J. D. Costlow, Jr. 1966. Studies on the physiological variation between tropical and temperate zone fiddler crabs of the genus *Uca.* IV. Oxygen consumption of larval and young crabs reared in the laboratory. *Physiol. Zool.,* 39:36–52.

Vernberg, F. J., and I. E. Gray. 1953. A comparative study of the respiratory metabolism of excised brain tissue of marine teleosts. *Biol. Bull.,* 104:445–449.

Vernberg, F. J., and D. K. Meriney. 1957. The influence of temperature and humidity on the metabolism of melanistic strains of *Drosophila melanogaster. J. Elisha Mitchell Sci. Soc.,* 73:351–362.

Vernberg, F. J., and W. B. Vernberg. 1964. Metabolic adaptation of animals from different latitudes. *Helg. wiss. Meeres.,* 9:476–487.

Vernberg, F. J., and W. B. Vernberg. 1966. Studies on physiological variation between tropical and temperate zone fiddler crabs of the genus *Uca.* VII. Metabolic-temperature acclimation responses in southern hemisphere crabs. *Comp. Biochem. Physiol.,* 19:489–524.

Vernberg, F. J., C. Schlieper, and D. E. Schneider. 1963. The influence of temperature and salinity on ciliary activity of excised gill tissue of molluscs from

North Carolina. *Comp. Biochem. Physiol.*, 8:271–285.

Vernberg, W. B. 1963. Respiration of digenetic trematodes. *Ann. N.Y. Acad. Sci.*, 113:261–271.

Vernberg, W. B., and F. J. Vernberg. 1968. Physiological diversity in metabolism in marine and terrestrial Crustacea. *Am. Zool.*, 8:449–458.

Walshe-Maetz, B. M. 1952. Environment and respiratory control in certain Crustacea. *Nature,* 169:750–751.

Walshe-Maetz, B. M. 1956. Controle respiratoire et métabolisme chez les Crustacés. *Vie Milieu,* 7:523–543.

Welch, P. S. 1948. *Limnological Methods.* Blakiston (McGraw-Hill) New York.

Wells, G. P. 1949. Respiratory movements of *Arenicola marina* L.: intermittent irrigation of the tube, and intermittent aerial respiration. *J. Mar. Biol. Assoc. U.K.,* 28:447–464.

Wells, G. P., and R. P. Dales. 1951. Spontaneous activity patterns in animal behavior; the irrigation of the burrow in the polychaetes *Chaetopterus variopedatus* and *Nereis diversicolor. J. Mar. Biol. Assoc. U.K.*, 29:661–680.

Weymouth, F. W., J. M. Crimson, V. E. Hall, H. S. Belding, and J. Field. 1944. Total and tissue respiration in relation to body weight: a comparison of the kelp crab with other crustaceans and with mammals. *Physiol. Zool.,* 17:50–71.

Whittow, G. C. (ed.). 1970. *Comparative Physiology of Thermoregulation.* Academic Press, New York.

Will, A. 1952. Körpergrösse, Körperzeiten und Energielbilanz. VI. Mitteilung Körpergrösse und O_2-Konsum bei Schaben und Asseln (Isopoden). *Z. vergl. Physiol.,* 34:20–25.

Winterstein, H. 1925. Über die chemische regulierung der atmung bei den Cephalopoden. *Z. vergl. Physiol.,* 2:315–328.

Zeuthen, E. 1947. Body size and metabolic rate in the animal kingdom with special regard to the marine micro-fauna. *Compt. Rend. Trav. Lab. Carlsberg, Ser. Chim.,* 26:17–161.

Zeuthen, E. 1953. Oxygen uptake as related to body size in organisms. *Quart. Rev. Biol.,* 28:1–12.

5

Water and Ions

As a dynamic metabolizing system, an organism must maintain an internal chemical environment that will sustain protoplasmic integrity independent of external variations. Numerous physiological mechanisms have evolved throughout the vastly diverse animal kingdom which permit different organisms successfully to exploit habitats that vary markedly in chemical composition. All life is thought to have originated in the sea and subsequently to have invaded brackish and fresh waters and terrestrial regions. As indirect evidence of this supposition, many biologists have cited the fact that the compositions of sea water and the body fluids of many animals are similar. When organisms invaded new habitats, however, they had to contend with an external environment which often differed drastically from that of sea water. Consequently, they had to overcome two basic problems: (1) The cells had to withstand wide fluctuations in osmoconcentration; and/or (2) some functional method was required to control the internal chemical environment. Therefore it is obvious that the ecology of an organism is intricately interwoven with the availability and method of control of water and concentration of ions. The limiting nature of these environmental parameters was highlighted in Chapter 3; the purposes of this chapter are to emphasize physiological diversity with respect to comparative ecology of species and to review the different control mechanisms operative in the zone of compatibility.

THE NATURE OF THE PROBLEM

A specific range of water concentration must be maintained in the body of an organism or it will perish. This concentration of water, however, may vary at different stages in the ontogeny of an animal or at any one stage under different physiological-ecological states. Further, this value varies, from those species that are about 95 percent water (such as jellyfish) to other species that have low levels of water. For example, gastrotrichs have dormant stages in which there is a greatly reduced level of body water after storage in a dry state for two years (Brunson, 1949), and seeds of many plants contain only 5–10 percent water (Fuller and Tippo, 1954). Most protoplasm in an active state, however, has a water content of 70–90 percent, whereas many active animals have values above 50 percent (Hall, 1922).

During the long period of evolutionary time since marine acellular organisms originated and gave rise to multicellular animals occupying diverse habitats, complex adaptations have taken place. Whereas acellular marine organisms and the cells of certain multicellular organisms are bathed directly in sea water, the cells of complex organisms are surrounded by blood or some other extracellular fluid. In the latter case the basic problem still remains—the water content of the individual cell must be maintained.

If a cell is immersed in a solution that does not cause the cell to shrink or to increase in size, the solution is said to be *isotonic*. If the cell volume is increased by the solution, it is called *hypotonic;* if volume decreases in size the solution is *hypertonic*. These changes essentially represent a net movement of water molecules across a concentration gradient from a region of high salt concentration to low concentration. The water is the solvent containing various kinds and numbers of particles; the total number of particles per liter of solvent is referred to as the *osmotic concentration* of a solution and is commonly expressed as an osmole, osmole per liter (osm/liter) or milliosmole (mOsm/liter). One osmole of a given solution is equal to the concentration (number of particles) of an ideal nonelectrolyte solution that has the same physical characteristics. Since many solutes do not act as ideal nonelectrolytes, the osmolal concentration of a solution is not equal to the total chemical concentration. Although molal and molar concentrations are not the same, a difference of only a few percent exists at the concentration at which biologically important solutes normally are found. The concentration of a solution can be determined by measuring any one of its four colligative physical properties: the freezing-point depression, osmotic pressure, vapor pressure, and boiling point. These four properties are directly related to one another, and the measurement of one will give an estimate of another. Only the number of particles in the solute are measured—not the kind of particle. Probably the most common property to be measured as an index of osmotic pressure is the freezing point of a solution: an osmolal aqueous solution freezes at $-1.86°C$.

The problem of maintaining a functional water balance probably was not critical for marine protista, but arose only when animals invaded brackish and fresh waters, and land. On the other hand, evidence seems to indicate that ionic differences on opposite sides of a membrane have always been present in cells. Cells of marine organisms have a greater internal concentration of potassium than sea water, but the reverse is observed for sodium.

EVOLUTIONARY PATTERNS

A number of different morphological-physiological adaptations that have occurred in the animal kingdom allow organisms to exploit and colonize different habitats. Some animal groups have been very successful and have invaded fresh-water and terrestrial regions, whereas others have remained virtually restricted to oceanic waters. Along with this evolutionary sequence whereby organisms faced a new external environmental complex, the cells were also exposed to a new internal environment. With the increasing complexity of multicellular organisms, the individual cell became more distantly removed from the external environment, and the cell became surrounded by an extracellular fluid that had a composition different from that of the environmental fluid bathing the organism.

Generally, animals may be placed into one of two major categories, depending on their response to fluctuations in environmental osmotic concentrations: (1) osmoconformers and (2) osmoregulators. An osmoconforming (*poikilosmotic*) animal does not regulate the osmotic concentration of its extracellular body fluid; rather this fluid conforms to the concentration of the external environment. In contrast, the osmoconcentration of the extracellular fluid of an osmoregulating (*homoiosmotic*) animal remains relatively constant when the external environmental fluid fluctuates in osmotic properties. As will be discussed in detail shortly, there are many intergrades between these general artificial categories, for species not only differ from one another, but also vary in osmoregulatory ability even during the life history of an individual. It should be remembered that the preceding terms refer to the total osmotic concentration, and do not indicate the presence or absence of any capacity to regulate differentially the kinds and amounts of ions.

Two terms used to describe the influence of salt concentration on organisms are *stenohaline* and *euryhaline*. If an organism can survive only a small change in external salt concentration, it is said to be stenohaline. Although this term usually is applied to organisms restricted to oceanic waters, it can also denote organisms restricted to other parts of a salinity

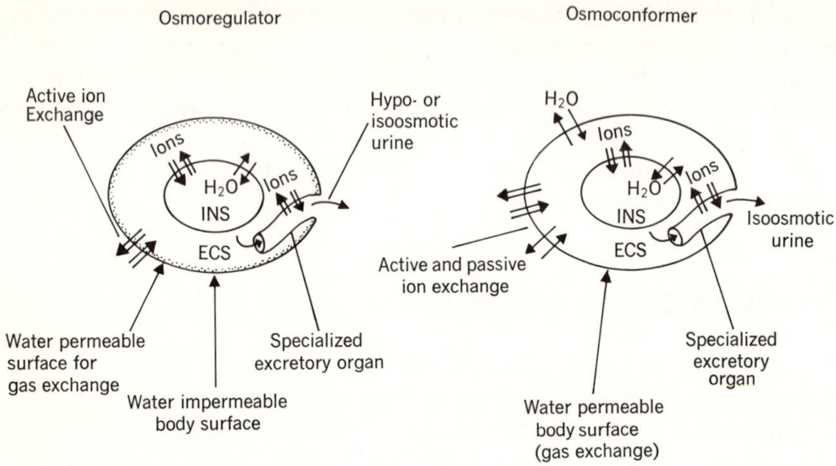

Figure 5-1. Generalized comparison of principal adaptations of osmoregulating and osmoconforming organisms.

gradient, such as fresh-water animals. Euryhalinic organisms tolerate wide fluctuations in salt concentrations. Euryhalinic and stenohalinic organisms may be either osmoconformers or osmoregulators, and therefore it should not be assumed that a stenohalinic animal is necessarily an osmoconformer.

Throughout the animal kingdom various strategies have been employed to enable organisms to maintain an optimum osmoconcentration of their body fluids. In general, these mechanisms function by walling off the organism from the external environment, thus preventing excess water and ion exchange, and/or hyper- or hyporegulating the osmoconcentration of the body fluids by means of specialized structures (Fig. 5-1). Even though elaborate devices have evolved to control the osmoconcentration of the extracellular fluid, the cell actively regulates the specific ionic concentration of the intracellular fluid. Detailed discussions of this subject can be found in the publications of Krogh (1939), Lockwood (1964), Parry (1966), Potts and Parry (1964). The following principal mechanisms are utilized by organisms invading osmotically difficult habitats, and two or more may be employed by an individual species.

Reduced Permeability of the Body Covering to Loss of Uptake of Salts and/or Water

Oceanic invertebrates tend to have body surfaces that are permeable to ions and water. With the invasion of brackish or fresh water and land, a marked decrease in permeability has occurred (Fig. 5-2). Although much

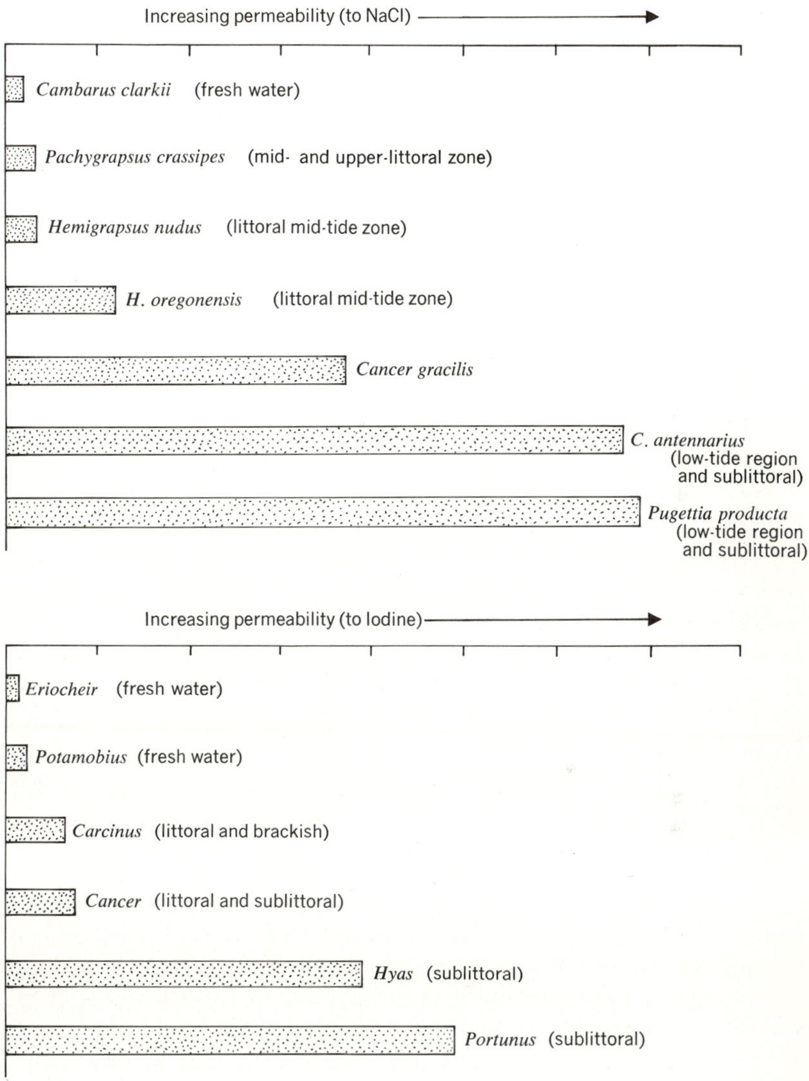

Figure 5-2. The correlation of the cuticular permeability of crustaceans from different environments to NaCl (*a*) and to iodine (*b*). (From Lockwood, 1964.)

of the outer body covering may be rendered almost impermeable, certain surfaces, such as the gill, remain quite permeable to gases, ions, and water. The amount of gill surface exposed varies interspecifically and may represent an extensive area. As an example of animals of comparable body size (Gray, 1954), the total gill area of an active surface-dwelling fish, the menhaden, is about ten times greater than that of the sluggish bottom-dwelling toadfish. Various structural devices have decreased permeability: these include increased calcium deposits in the exoskeleton; waxy outer covering, as in the insect exoskeleton; thickened skin; increased number of mucous glands; and scales. There are also morphological adaptations such as opercula to protect and isolate respiratory surfaces from adverse environmental conditions, and other adaptations such as internal lungs or a tracheal system to cover the gills; moreover, many of these surfaces have been placed in a more central location in the body.

Maintenance of Water and Salt Balance by Renal and Extrarenal Organs

A hyperregulating organism (one living in a hypo-osmotic environment) must cope with continuous osmotic inflow of water, whereas a hyporegulator must contend with the opposite effect. Animals living in a hyperosmotic medium have been observed to drink the surrounding water and actively excrete salts, so that ultimately the water balance is maintained. Certain marine teleosts drink about 0.2–25 percent of their body weight per hour (Conte, 1969), and small fish drink proportionately more than large fish. The monovalent ions are absorbed from the gut along with the water, leaving the divalent ions behind in the rectal fluid. In addition, the volume of urine produced is very low, and the urine is either typically iso-osmotic or hypo-osmotic to the blood. The regulating methods used by various marine vertebrates to maintain salt and water balance is represented in Fig. 5-3.

Because animals living in a medium of low osmoconcentration must constantly dispose of excess body water, reduced permeability to water uptake and the production of a hypotonic urine are important to these organisms. Active uptake of salts and the utilization of salts from food compensate for the loss of salts by diffusion and in the urine. The kidney or renal organ of an animal living in low-salinity or fresh water is of great value in volume regulation and salt retention, as indicated by the comparative anatomy of these organs from fresh-water and marine animals. Despite some exceptions, renal structures of animals from regions of low osmoconcentrations tend to have more renal units, a greater total surface of glomerular filtration, and longer and more highly differentiated renal tubules that function efficiently for salt resorption (Hickman and Trump, 1969).

Along with the renal organ, extrarenal pathways aid in maintaining the salt balance in organisms. The gills are especially important in salt excretion and/or uptake, but other specialized structures have also been

	Liability to Osmotic Water Loss	Liability to Evaporative Water Loss	Blood Conc. Relative to Medium Conc.	Urine Conc. Relative to Blood Conc.
Elasmo-branchs	○	○	Isotonic	Isotonic
Teleosts	●	○	Hypotonic	Isotonic
Reptiles	○	●	Hypotonic	Isotonic
Mammals	○	●	Hypotonic	Hypertonic
Birds	○	●●●	Hypotonic	Hypertonic

○ = No liability
● = Liability

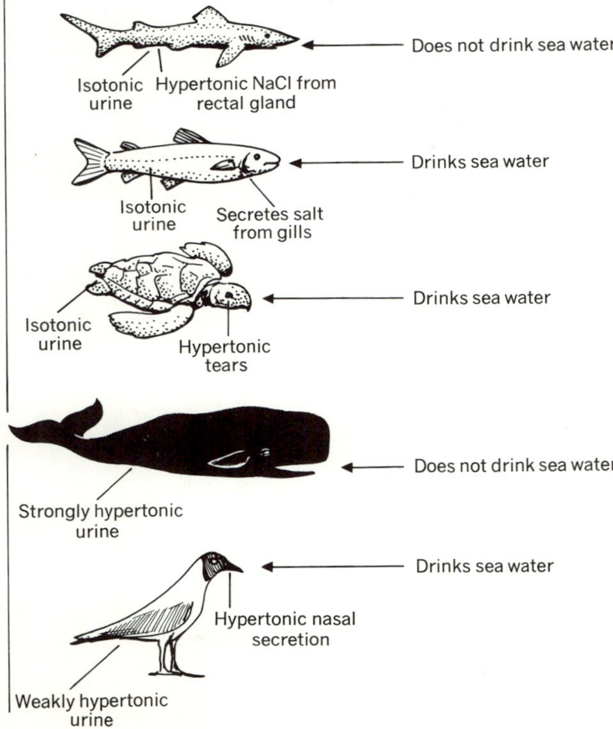

Figure 5-3. Salt and water balance regulating mechanisms of marine vertebrates. (From Lockwood, 1964.)

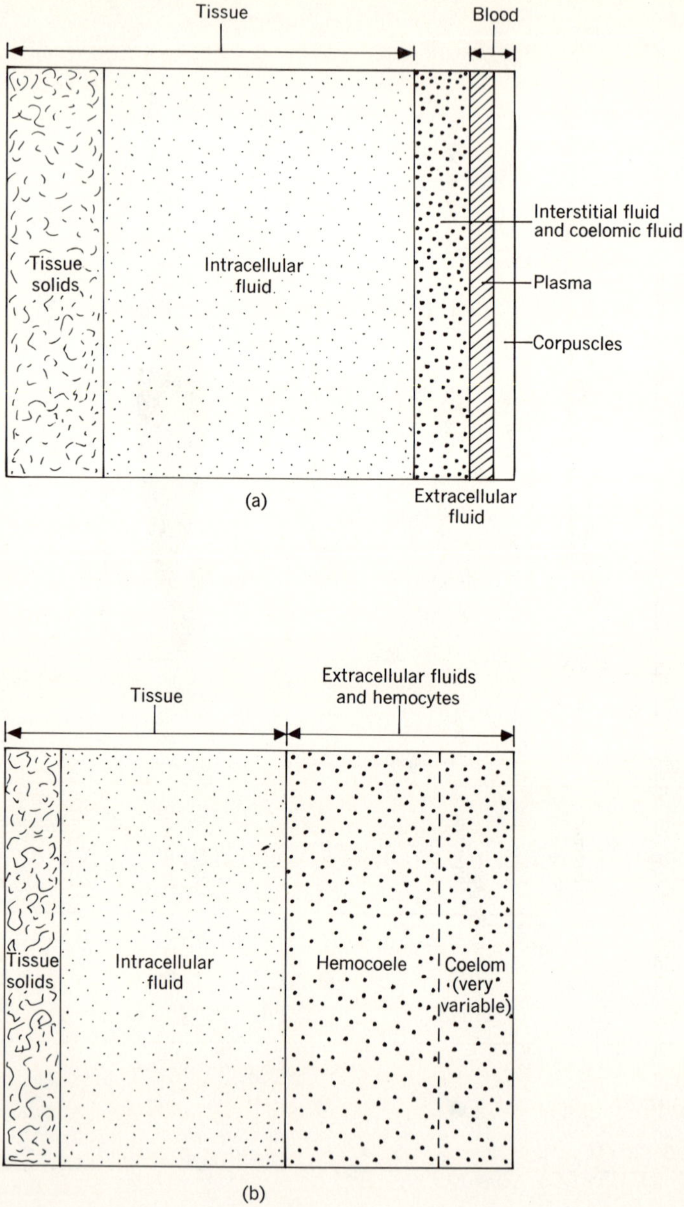

Figure 5-4. The comparative distribution of water in a mammal (*a*) and a crustacean (*b*). (From Potts and Parry, 1964.)

described, such as the nasal or lachrymal gland present in certain birds and the rectal gland of elasmobranchs. In addition, the body surface is important in this function. The "salt secretion" activity of gills has been attributed to special cells (Copeland, 1968), although the unequivocal demonstration of their function has been questioned (Parry, 1966; Conte, 1969).

BODY FLUIDS

The body fluid of most multicellular organisms is compartmentalized, with the two main compartments being the extracellular fluid and the intracellular fluid (Fig. 5-4). The chemical composition of each is

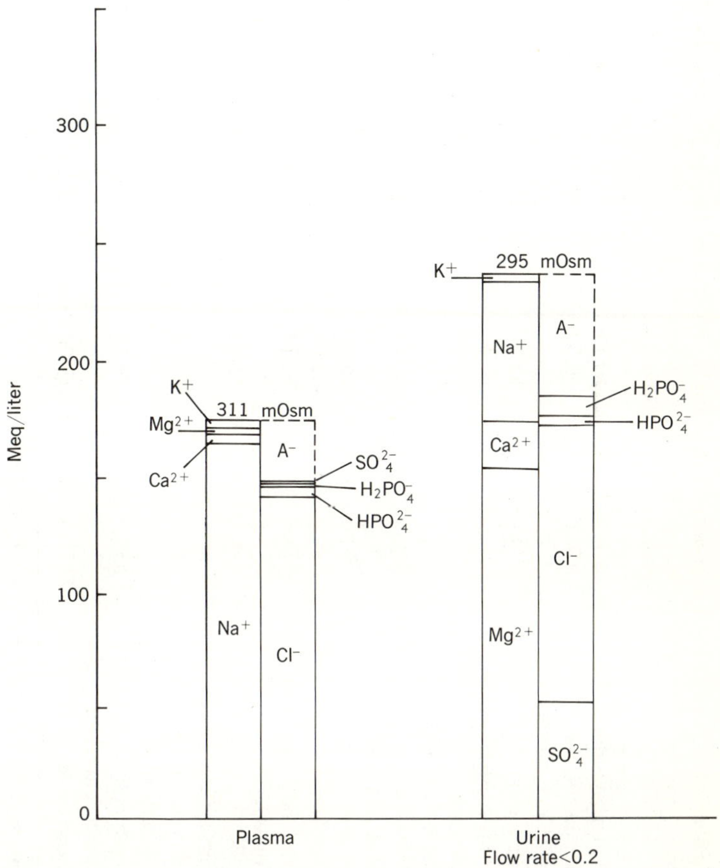

Figure 5-5. Ionic composition of plasma and urine from the southern flounder (From Hickman, 1968.)

Table 5-1. Ionic Concentration in Extracellular Fluid and Tissues in Some Marine Animals (mM/Kg water).

Medium	*Sodium*	*Potassium*	*Calcium*	*Magnesium*	*Chloride*	*Sulfate*
Sea Water Standard	478.3	10.13	10.48	54.5	558.4	28.77
Aurelia mesogleal fluid (Coelenterata)	474.0	10.72	10.03	53.0	580.0	15.77
Mytilus blood (Bivalve, Molluscs)	474.0	12.00	11.90	52.6	553.0	28.90
Mytilus fast adductor muscle	79.0	152.00	7.30	34.0	94.0	8.80
Loligo blood (squid, molluscs)	456.0	22.20	10.60	55.4	578.0	8.14
Loligo nerve	50.0	400.00	0.40	10.0	40.0	—
Carcinus nerve	41.0	422.00	—	—	27.0	—
Echinus coelomic fluid (echinoderm)	474.0	10.13	10.62	53.5	557.0	28.70
Maia blood (crab, arthropods)	488.0	12.37	13.56	44.1	554.0	14.50
Carcinus blood (crab, arthropods)	531.0	12.26	13.32	19.5	557.0	16.46
Myxine blood (hagfish, chordata)	537.0	9.12	5.87	18.0	542.0	6.33
Myxine muscle	137.0	113.00	2.20	10.8	121.0	3.70

SOURCE: Adapted from Potts and Parry, 1964.

different, and that of the extracellular fluid more closely resembles sea water (Table 5-1). High concentrations of potassium are present in tissues relative to that in extracellular fluids, but low ionic concentrations of sodium, chloride, magnesium, and sulfate are present. This chemical difference between tissue fluid and extracellular fluid is graphically represented by ionograms of fish fluids (Fig. 5-5).

The chemical composition of the extracellular fluid may vary directly with changes in the chemical composition of the ambient fluid (ionic conformers). This type of response is well illustrated in certain molluscs and annelids (Potts and Parry, 1964). Other animals are ionic regulators and maintain a relatively constant concentration of ions in the extracellular fluid over a wide range of changes in the ambient medium (Lockwood, 1964). The ability to regulate ionic concentrations, however, may vary seasonally or with temperature (Dehnel and Carefoot, 1965).

When animals are moved from one salinity to another, changes in the intracellular ionic composition have been reported even in brackish-water animals that can osmoregulate. Sodium, chloride, and amino acid levels are particularly decreased, and potassium is lost to a lesser degree (Potts and Parry, 1964).

OSMOREGULATION

Types of Osmoregulators

Many groups of organisms show varying degrees of ability to osmoregulate. In general, only those invertebrates living in low salinity waters or fresh water have evolved physiological mechanisms for the control of osmotic concentrations, whereas all fishes but the jawless hagfishes can osmoregulate. In land animals the necessity for water conservation is a dominant aspect of osmoregulation.

An organism that exhibits perfect osmoregulation is able to maintain a constant level of osmoconcentration in its extracellular fluids regardless of changes in the external environment. This relationship is represented graphically when the freezing-point depression (Δ°C) of the blood is plotted against the freezing-point depression of the external fluid (Fig. 5-6, type I). If the internal fluid is iso-osmotic with the external fluid (type II), as in the case of a perfect osmoconformer, this reponse is graphically represented as a single line of iso-osmoticity. Between these extremes, various types of responses are found. Some organisms are hyperosmotic to their external environment, but only when it is dilute (type III). Others are not only hyperosmotic to media of low osmotic concentration, but

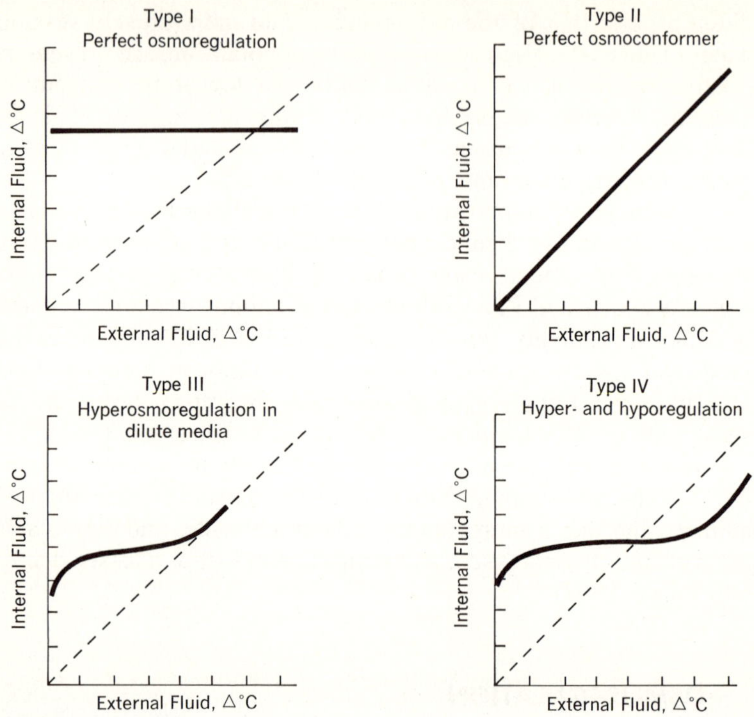

Figure 5-6. Four generalized types of osmoregulatory responses. The dotted line is the line of iso-osmoticity.

also are hypo-osmotic to highly concentrated waters, even though the internal osmoconcentration is not constant throughout the total osmotic gradient (type IV).

Although an organism generally may be placed in one of these categories, species differ in their intensity of response. For example, within a number of crabs classified as type IV, some animals are better hyporegulators (*Artemia* and *Uca*) than others (*Pachygrapsus* and *Sphaeroma*), whereas some are better hyperregulators (*Paleomonetes* and *Uca*) than others (*Artemia* and *Leander*) (Fig. 5-7). In addition, not only various factors may influence the response of one species, but also each stage of development may respond differently, since each stage may face a different set of environmental factors.

Variables

The osmoregulatory ability of an organism is influenced by many factors, some of which originate from the external environment and

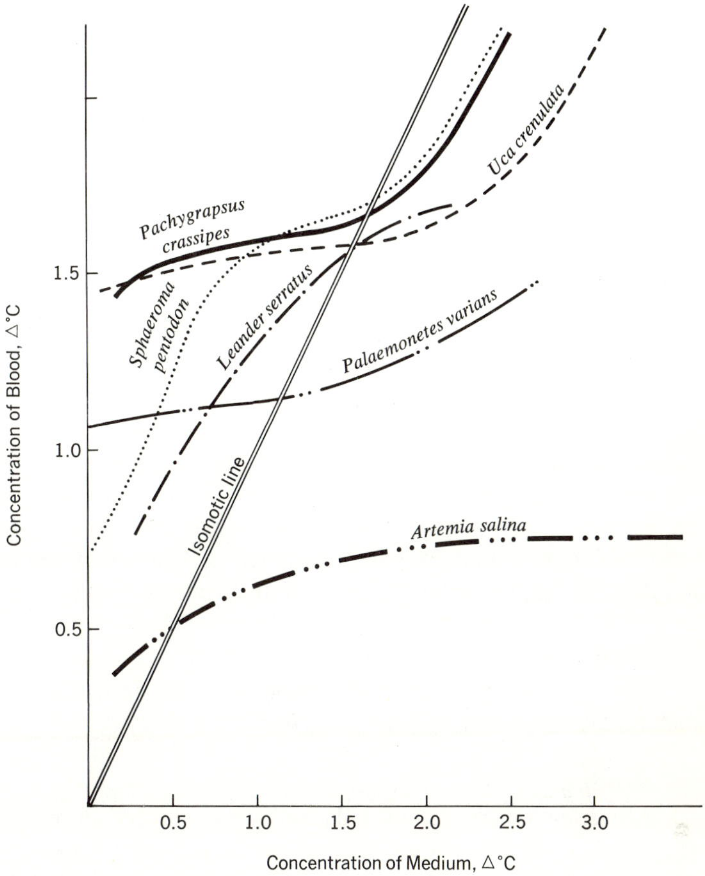

Figure 5-7. The osmoregulatory response of arthropods inhabiting various habitats. (Adapted from Lockwood, 1962.)

others of which are intraorganismic phenomena. The following examples represent some of these variables, which must be considered before valid generalizations can be drawn for correlating the physiological activity of an organism with its habitat.

Stage of Life Cycle

Adult and larval stages of the mud crab, *Rhithropanopeus harrisii,* are able to hyperregulate in dilute sea water (Jones, 1941; Kalber and Costlow, 1966). In contrast, adult fiddler crabs are excellent hyperregulators, but the larvae (first-stage zoeae) are very sensitive to reduced salinity and fail to complete development when exposed to 20‰ (Vernberg, 1969). These ontogenetic changes in osmoregulative ability influence the spatial distribution of these organisms. Adult mud crabs and fiddler crabs normally live in estuaries in regions of reduced salinity, whereas the larvae of fiddler crabs occur in greatest abundance in regions of high salinity. Fish also exemplify the adaptiveness of organisms to salinity change. The mode of osmoregulation of migratory species changes during their development, particularly with respect to electrolytes and organic substances (Parry, 1966).

Size

Some evidence indicates that body size may affect osmoregulative ability, for the small-sized flounder is more sensitive to osmotic disturbances than is a large-sized individual (Hickman, 1959). In the invertebrates, Smith (1964) noted that the weight-specific rate of salt loss in smaller-sized *Nereis limnicola* was greater than in larger-sized worms, but this tendency was not so pronounced in another polychaete species, *N. succinea*. Although the mechanisms for this type of response are unknown, it is significant that in many species the amount of body surface is relatively higher in young or small individuals than in larger ones.

Food

The ionic composition of the blood of a terrestrial crab is significantly dependent upon food (Gross and Holland, 1960).

Molting

Abnormalities in osmoregulation were reported in molting crabs by Prosser *et al.* (1955). Some species can hyperosmoregulate to sea water prior to molting, and presumably this response aids in the rapid inward diffusion of water during molting (Kalber and Costlow, 1966). The role of antidiuretic and diuretic hormones during molting is important in assuring that water and ionic balance of crabs is not critically disturbed (Bliss *et al.*, 1966).

Symbionts

The osmoregulatory characteristics of different species living in some type of symbiotic relationship show interesting adaptations both to host species and to the host's environment. When the parasitic copepod *Lernaeocera branchialis* is attached to the codfish, its hemolymph is hypo-osmotic to sea water, as is the blood of the fish. When separated from its host, however, the hemolymph of the copepod is iso-osmotic with the surrounding sea water (Panikkar and Sproston, 1941). Unlike their host, cestodes living in the gut of the dogfish shark are poikilosmotic (Read *et al.*,

1959). Pinnotherid crabs live as commensals in or with many invertebrate hosts, and the osmoregulatory ability of two species of this crab can be directly correlated with the ability of the host animal to survive salinity extremes. *Pinnotheres maculatus* is a stenohalinic osmoconformer that is unable to withstand salinities below 15‰; these characteristics parallel those of its host species, the scallop *Aequipecten irradians*. In contrast, the closely related crab *Pinnotheres ostreum*, which lives in oysters, can hyperregulate at salinities as low as 5‰ for at least one month. Oysters, *Crassostrea virginica*, typically live in low-salinity water, whereas the scallop is restricted to more saline conditions (Read, 1968).

Multiple Factors

As pointed out previously, the environmental complex consists of numerous parameters, each of which exerts an influence on an organism. The interaction of multiple factors may produce an organismic response different from that when only one factor is varied, as seen when organisms are subjected to fluctuations in both temperature and salinity. At their summer habitat temperature, two species of shrimp, *Penaeus duorarum* and *P. aztecus*, can maintain their blood osmotic concentration as hypoosmotic to sea water and hyperosmotic to dilutions of sea water below 30‰. At reduced temperatures (8.8°C), however, the hyperosmoregulating mechanism is impaired and the blood tends to become iso-osmotic with the surrounding media. These species can also survive low temperature better when in high-salinity water. These data have ecological significance in that the species *P. duorarum*, which regulates better at low temperature, is also the species that overwinters in the colder waters of North Carolina (Williams, 1960). A similar response was reported for a species of shrimp from the colder water offshore of Germany, where a greater osmoregulatory capacity was exhibited at 7°C than at 20–21°C.

However, the long-standing observation that marine animals penetrate farther into reduced saline water in the tropics and subtropics than in temperate-zone estuaries has prompted the hypothesis that the optimum osmotic concentration of body fluid may be lower at higher temperatures. Since ionic regulation is thermally influenced, when the rate of ion uptake as compared with ion loss increases faster with increasing temperature, an organism may be expected to penetrate a dilute media more successfully (Lockwood, 1962).

Seasonal differences in the performance of an osmoregulatory mechanism were demonstrated in the southern flounder, *Paralichthys lethostigma*. During the winter when this fish migrates to a fully marine habitat on the continental shelf, the glomerular filtration rate is low, but a high rate is reported during the warmer months when this species returns to low-salinity inshore waters. Glomerular filtration is essential to volume regulation in a dilute medium (Hickman, 1968). Shallow-water Arctic fishes show a unique adaptation (Fig. 5-8) to thermal stress by seasonal alteration in plasma osmoconcentration (Scholander *et al.*, 1957). In the summer when the water temperature is +5°C, the freezing-point depression of their plasma is −0.8°C. During the winter when the water temperature is −1.73°C, the plasma osmoconcentration is almost doubled. Deep-

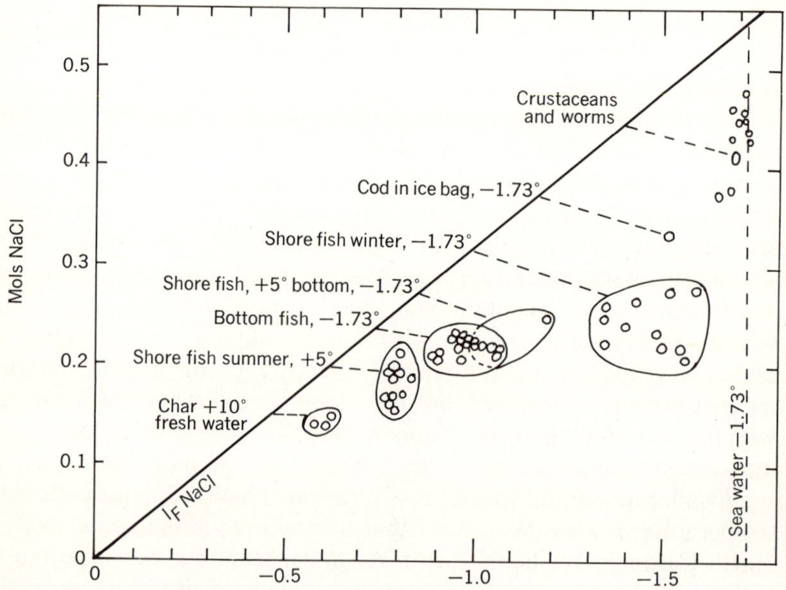

Figure 5-8. Freezing point versus chlorides in fishes and invertebrates from Arctic waters. (From Scholander *et al.*, 1957.)

water species living at −1.73°C throughout the year have plasma freezing points ranging from about −0.9 to −1.0°C, which probably do not change with season. It was suggested that these deep-water species are supercooled and, very importantly, are in little danger of having their plasma frozen by being seeded with an ice crystal. In contrast, during the winter the plasma of the surface fish faces the risk of seeding by direct contact with ice in the sea. This process of seeding could cause a supercooled fluid to freeze. An increase in the freezing-point depression of the plasma of these shallow-water fish reduces the likelihood of this fluid turning to ice.

ACCLIMATION

As stressed in earlier chapters, the response of an organism to a given environmental parameter may depend on the previous conditions of its exposure to this parameter. Whether the observed responses are environmentally influenced (phenotypic) or whether they reflect genetic differences may be determined through acclimation studies. When animals are placed in water with a new osmoconcentration, they must adjust to the stress. In osmoconformers that lack specific organs for osmoregulation, it is assumed that the individual cells must adapt to different levels of osmoconcentration. Obviously, if the change in osmoconcentration exceeds genetically determined limits, these cells will die. Within these limits, however, the cells may show various environmentally induced (nongenetic) responses.

Although most oceanic stenohalinic osmoconforming crabs, echinoderms, and tunicates cannot be acclimated to reduced salinity much below 75 percent sea water, a few species can acclimate to 50 percent sea water. For example, the starfish *Asterias rubens,* living at 30‰ in the North Sea, can be acclimated to live at 15‰. Natural populations living at this reduced salinity in the Baltic Sea are smaller, have a higher water content and a lower ash content than the high-salinity North Sea populations (Schlieper, 1957). Colonies of the hydroid *Cordylophora,* a euryhalinic osmoconformer, can be acclimated to both fresh water and sea water (Kinne, 1958).

Osmoregulators, on the other hand, have discrete organs that enable the body-fluid osmoconcentration to be controlled over a wide range of external change. The activity of these organs also represents a nongenetic form of adjustment to changes in external osmoconcentration.

Animals may show an immediate response to a sudden change in salinity, as illustrated by the following examples. Within minutes after the transfer of blennies (fish) to a new salinity, sodium efflux occurred, but osmostability of the blood took longer (House, 1963). When suddenly moved from sea water to a low-saline medium (5.45‰), the serum osmolarity of the starry flounder was lower at 5 hr than at 12 hr (an undershoot effect); after an increase at 12 hr (an overshoot response), the concentration dropped slowly until a steady state was reached after about 24 hr (Hickman, 1959). When the oyster crab *Pinnotheres ostreum* is transferred from 30‰ sea water to a reduced salinity, most of the change in osmoconcentration occurs within 3 hr, and no further significant change occurs between 24 and 48 hr (Read, 1968).

At the opposite end of the salinity gradient, some species acclimate to conditions present in hypersaline lagoons and salt flats. The mud crab

Hemigrapsus oregonensis, however, is unable to hyporegulate immediately when taken from normal sea water and placed in concentrated sea water (Gross, 1963). Although detectable acclimation to high salinity was observed after an exposure of 5 days, excellent hyporegulatory ability is exhibited only after 22 days immersion in concentrated sea water.

Following the immediate response to changes in salinity, some organisms exhibit a period of stabilization during which the composition of body fluid does not change markedly. Generally, this stabilized peak is reached only if the salinity change is neither too extreme nor too rapid. The ability of organisms to reach a new stabilized state varies from species to species, and may vary with the age of the organism involved. The rate of acclimation to salinity change may have survival value in that animals occupying inshore waters, in which the salinity may quickly vary over a wide range, adapt more rapidly than species from relatively constant saline environments. For example, the steelhead trout requires 80–170 hr for the osmoconcentration of the blood to stabilize after it is transferred to sea water (Houston, 1959), but salmon fry and young adults require 200 hr (Parry, 1960). In contrast, estuarine fishes such as blennies and flounders show a relatively rapid response, ranging from minutes to 24 hr.

If young fish (for example, *Cyprinodon macularius*) are exposed to different salinities during embryonic development or immediately on hatching, they exhibit irreversible but nongenetic adaptation to that particular salinity. These marked differences persist to sexual maturity in certain individuals (Kinne, 1962). In addition to changes in physiological functions as a result of salinity acclimation, structural changes also occur, as shown in the hydroid *Cordylophora.*

ECOLOGICAL CONSIDERATIONS

Osmoconformers

Typically, type II animals (Fig. 5-6) are restricted to marine waters and, with the exception of some intertidal burrowing animals, occur in the sublittoral zone. In general, these species experience only small changes in salinity. Not all osmoconformers have restricted distributions with respect to a salinity gradient, for some are euryhaline. For example, echinoderms are restricted to high-salinity waters, whereas certain osmoconforming molluscs tolerate low-salinity waters. Marked species differences exist in the amount of intracellular fluid dilution that animals can withstand before protoplasmic integrity is disrupted with a resultant metabolic breakdown. The ability to survive may depend more on the activity of specific ions than on osmoconcentration. The ability of isolated gill pieces of the mollusc *Mytilus edulis* to survive various temperature-salinity combinations is dependent on the calcium content of the medium (Schlieper and Kowalski, 1956). Tissue from animals living in high-salinity water (30‰) is more thermally resistant than tissue from animals from low-salinity conditions. When animals are acclimated to low-salinity water that has been calcium-enriched to equal that of the high-salinity water, the cellular thermal resistance is increased and equal to that of high-salinity acclimated animals.

Osmoconforming organisms exposed to waters of reduced salinity may behave as osmometers in that they swell and increase in body weight, owing to the influx of water. If the animal is not returned to normal sea water, the volume change persists and it may die. This response has been observed in certain stenohalinic marine crabs, annelids, molluscs, echinoderms, sipunculids, protozoans, and the eggs of many marine organisms. With time, however, certain species either show varying degrees of volume regulation by losing weight or adjust to more protoplasmic dilution. These responses may have some adaptive significance to species living in an estuary subjected to rhythmic tidal changes in salinity.

An excellent example of physiological adaptation to different salinity regimes has been reported in an ecologically arranged series of closely related species of nereid worms whose habitats extend from the open ocean to fresh water (Fig. 5-9). Nereids restricted to high-salinity water (*Perinereis* and *Nereis vexillosa*) will gain weight rapidly when placed in low-salinity waters. In contrast, the estuarine species *Nereis diversicolor* demonstrated a degree of volume control and with time lost weight. Moreover, *Perinereis* is three times as permeable to sodium as is *N. diversicolor* (Fretter, 1955). Furthermore, high-salinity species are iso-osmotic in all those salinities in which they can survive, whereas estuarine forms show varying degrees of osmoregulation at salinities below sea water of 36‰, (Oglesby, 1965). This limited osmoregulatory ability of *N. diversicolor* may have little survival value to the organism, since isolated muscles of this species continue to function at the same low salinities experienced by the intact organism (Wells and Ledingham, 1940). Species successfully living in fresh water have reduced permeability to salts and do not swell as much as the purely estuarine species when placed in distilled water (Table 5-2). Further reduction in the permeability to salts can take place in *N. limnicola* acclimated to low salinities, with complete acclimation to a new salinity taking more than two days (Oglesby, 1968). Although the influx of water, as measured by D^2O, is similar in *N. succinea* and *N. limnicola*, only the latter species can live in fresh water.

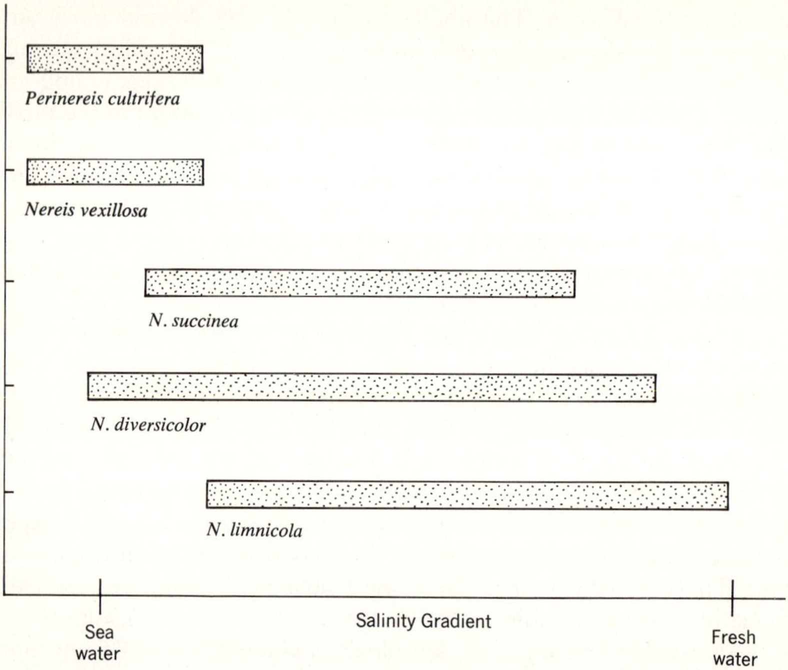

Figure 5-9. The relative distribution of nereid worms along a salinity gradient.

Apparently the difference in survival of these species in fresh water is related to the ability of *N. limnicola* to produce more urine and/or to differences in the architecture of the integument (Smith, 1964).

A classical example of adaptation to an estuarine environment by an osmoconformer is that of the intertidal-zone triclad flatworm, *Procerodes ulvae* (formerly *Gunda*). When these animals are exposed to soft tap water, they rapidly increase in volume by about 110 percent and die after

Table 5-2. Average Percent Swelling in Nereid Worms when Exposed to Distilled Water for One Hour.

Species	*Percent Increase*
Nereis succinea	50–60
Nereis diversicolor	24–25
Nereis limnicola	10

SOURCE: From Smith 1963.

about 48 hr. If maintained in calcium-enriched water or natural stream water, their weight rapidly increases to 70 percent and then decreases, probably because of reduced membrane permeability (Weil and Pantin, 1931). The gut epithelium removes excess water by forming vacuoles, and thereby reduces the dilution of the extracellular fluid bathing the other cells (Beadle, 1934).

A degree of internal osmoconstancy is imparted to certain species by their behavioral pattern rather than by purely physiological mechanisms. The terrestrial hermit crab *Coenobita perlatus,* for example, has poorly developed regulatory mechanisms for maintaining inorganic ion levels in the blood. When the blood ionic levels are grossly modified experimentally, the crab can control the concentration of ions by selecting water of the appropriate salinity from a number of available choices (Gross and Holland, 1960).

Many other organisms such as molluscs and hydroids exhibit another type of behavioral response; when first confronted with a change in salinity, they undergo marked body contraction, which causes a reduction of surface exposed to an adverse environment. Similarly, the false limpet *Siphonaria pectinata,* which lives attached to rocks, has no physiological capability to osmoregulate, but instead relies on a mecho-behavioral response. Over the salinity range of 5–60 ‰, the hemolymph is iso-osmotic with the external medium, but the hemolymph of animals that remain firmly attached to a substrate is hyperosmotic at salinities below 20 ‰ and is hypo-osmotic at salinities above 40 ‰. By clamping itself tightly to the substrate to form a firm bond with the margin of the shell, the animal is able to reduce exposure to an osmotic gradient (McAlister and Fisher, 1968). Similar behavioral responses have been noted in bivalves and barnacles. Still another variation was reported for the limpet *Acmaea limulata,* in which extravisceral water trapped between the shell and the soft parts may serve an osmoregulatory function when the animal is under osmotic stress (Segal and Dehnel, 1962).

Although some organisms live in estuaries that undergo rapid and radical changes in salinity, they may escape this unstable condition by burrowing in the substratum, since the salinity of interstitial water of the substratum does not change rapidly (Reid, 1932).

Because the composition of the extracellular fluid of euryhalinic osmoconformers is not regulated, reduced osmotic and ionic concentrations may dramatically influence the growth and structure of their cells. The influence of salinity on the structure and function of cells is well illustrated in the studies on the brackish-water hydroid *Cordylophora* (Kinne, 1958). Colonies or sections of a colony of this species can be grown in the laboratory at various salinities, ranging from almost full-

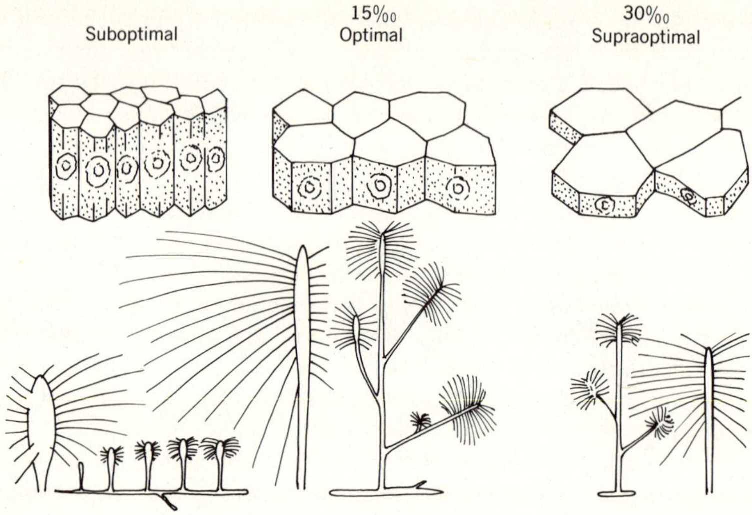

Figure 5-10. Structural alterations of the hydroid, *Cordylophora,* with identical genotypes adapted to different salinities. The entire colony and cells are modified. (From Kinne, 1958.)

strength sea water to fresh water. Because one colony of this organism is considered to be an individual, all sections from one colony have the same genotype, and therefore any morphological differences between sections from one colony grown in different salinities reflect individual nongenetic (phenotypic) variation and not population or genotypic variation. The major structural changes are represented in Fig. 5-10. The gross form of the entire colony is a function of salinity; colonies grown in fresh water have short and unramified hydrocauli, and 76 percent of the length of the colony consists of stolons. For colonies grown at 15 ‰, this figure is 16 percent stolons, whereas at 30 % it is only 7 percent. The hydrocauli are greatly ramified at 15 ‰ and 30 ‰. Hydranths of colonies grown at the optimum salinity (15 ‰) are longest and have a greater number of tentacles that tend to be longer. The net effect is that the surface area is maximal in 15 ‰ and minimal in 30 ‰. In fresh water, cells are longer and narrower, with decreased free surface to act directly with the environment. With increasing salinity, the cells become flatter and broader, and have a greater free-surface area.

A similar decrease in nuclear size and total number of cells per hydranth is observed with increased salinity. These structural modifications have obvious adaptive significance to the functional qualities of cells: (1) Generally columnar cells with large nuclei indicate cells having a high level of physiological activity; and (2) a reduced free cell surface, an increased number of cells, and increased cell metabolism indicate adaptation to a hypotonic solution. This work demonstrates how the success of an organism in a changing environment is determined by the principle of interdependence of the morphological-physiological-ecological complex.

Animals belonging to type III (Fig. 5-6) typically occur in estuarine and coastal waters where salinity may vary considerably. Some species normally live in regions of low salinity; others are concentrated in that part of the estuary where the salinities tend to be higher, but where marked fluctuations may occur. Still others show affiinities for the open ocean or high-salinity regions of estuaries. Fresh-water crustaceans maintain their blood concentrations hyperosmotic to the surrounding medium in a manner similar to that of estuarine animals; in general, however, the concentration level is lower than that of estuarine species.

Animals capable of hyper- and hyporegulation (type IV, Fig. 5-6) generally live in habitats where salinity may vary markedly. Many crabs exhibiting this response are semiterrestrial; others live in salt marshes or mangrove swamps. Some migrate from fresh water to the sea to breed, and still other species, such as *Artemia,* live in inland concentrated-salt waters. It has been observed that crabs living in a hypersaline lagoon, where the salinity reached 53 ‰, are able to hyporegulate (Gross, 1961).

Ionic Composition

Although changes in total osmoconcentration appear to greatly influence the ecology of organisms, some work has indicated that the change in relative proportions of ions can exert a considerable effect on organisms. The ratio of one ion to another probably does not change in open marine water, but it can be critical in regions of reduced salinity, fresh water, or in hypersaline waters. The ionic composition of oceanic and coastal waters differs from that of lakes, rivers, and certain inland salt-water bodies such as the Dead Sea. In ocean waters, sodium and chloride ions are most abundant, whereas in fresh water, calcium and carbonate ions predominate. Inland salt lakes in mineral springs vary so much in composition and relative proportion of salt that generalizations are not possible.

Of the various ions studied, perhaps calcium has received the greatest attention. In nature, the changes in the ionic ratio of calcium occur in waters of reduced salinities. Relatively high proportions of calcium act to preserve protein structure and increase resistance of organisms to thermal stress (Schlieper and Kowalski, 1956). Ionic composition in fresh waters may vary considerably, depending on the type of soil that surrounds the lakes. Marked differences observed in the amounts of calcium, carbonate and bicarbonate ions have caused different degrees of hardness in these waters. Despite limited available information, certain generalities can be made about the influence of changes in the relative proportion of ions: (1) minor changes in ratios of ions probably are unimportant; (2) pronounced changes may reduce the capacity of organisms for reproduction, growth, stress compensation, and thermal stability; and (3) extreme changes may produce shock reactions and result in severe metabolic disturbances and increased mortality. For the most part, euryhalinic organisms apparently tend to be less affected by variations in the ionic ratios than do stenohalinic species.

Osmoregulation and Species Competition

Although many species are defined as marine, brackish, or fresh water on the basis of laboratory data, their distribution in an estuary may be different. This disparity may result from a number of reasons. Many individuals adapt to a wider range of salinities under laboratory conditions than they normally experience in their natural habitat. In other species, physiological limits coincide with ecological distribution. These different alternatives indicate the necessity of integrating laboratory and field studies in order to understand the ecology of a species. Factors other than osmoregulatory ability may be highly important in limiting the range of a species, and one that should not be overlooked is competition.

Although animals seldom actively compete for some specific feature of their environment, such as water, oxygen, or some chemical substance, they do actively compete for food and space, particularly in the case of sessile animals. Generally, if organisms live in a similar environment, they have evolved similar, major physiological adaptations. Within any one environmental complex, however, organisms may not be equally adapted, and then competition may be very important in defining the exact ecological boundaries of these respective groups. It is reasonable to assume that a species which is effectively and efficiently adapted to a specific environment is likely to have better competitive abilities than an inefficiently adapted species. An efficiently adapted species would be able to maintain itself better, reproduce a greater number of offspring, and ultimately would increase in numbers at the expense of the more poorly adapted species.

It has been suggested by Croghan (1961) that as organisms invaded brackish and estuarine waters, mechanisms of osmoregulation came into being. An organism in a dilute medium uses a considerable amount of energy for osmotic regulation, but if these organisms can increase the efficiency of osmoregulation, they require less energy. If less energy were required for osmoregulation, less food would be needed by the organism for this function so that more energy would be available for other phases of the life processes and, ultimately, the species could gain a competitive advantage. Croghan concluded that the major adaptations to a dilute medium all reduce the amount of energy required to maintain a steady state.

Physiological Races

Populations of any given species may be isolated from one another by various barriers. The environmental complex encountered by each population may differ significantly, especially if this species is widely distributed. With time, owing to the action of differential selective factors, the physiological capability of one population to respond to stresses of its particular environment may become distinct from that of a second population. This question of the existence of physiological races (or species) with respect to osmoregulative ability has received limited attention.

The isopod *Mesidotea entomon* lives in the Baltic Sea and in Scandinavian postglacial lakes that have been isolated from the Baltic. Population samples from the Baltic can acclimate to low salinities but not to the fresh water of the lake. Two principal differences in the ionic regulative ability of these two populations were noted: (1) The permeability of the fresh-water animals is considerably less than that of Baltic animals; and (2) in the fresh-water forms, the active transport mechanism has an increased affinity for the uptake of sodium from the medium over that of the Baltic animals (Croghan and Lockwood, 1968). In another example, the chief differences in the osmoregulatory responses of two populations of a species of polychaete (one from fresh water and the second from an estuary) were that the fresh-water population had a higher level of hyperosmotic regulation, a lower critical-salinity level, and a lower salt permeability of the body surface. These interpopulational differences, however, were small when compared to the osmotic responses of a less euryhalinic species (Oglesby, 1968).

EXCRETORY PRODUCTS

Aquatic Organisms

Along with the morpho-physiological adaptations for the maintenance of a water and salt balance compatible with organismic existence in a particular habitat, distinct biochemically adaptive alterations in the types of

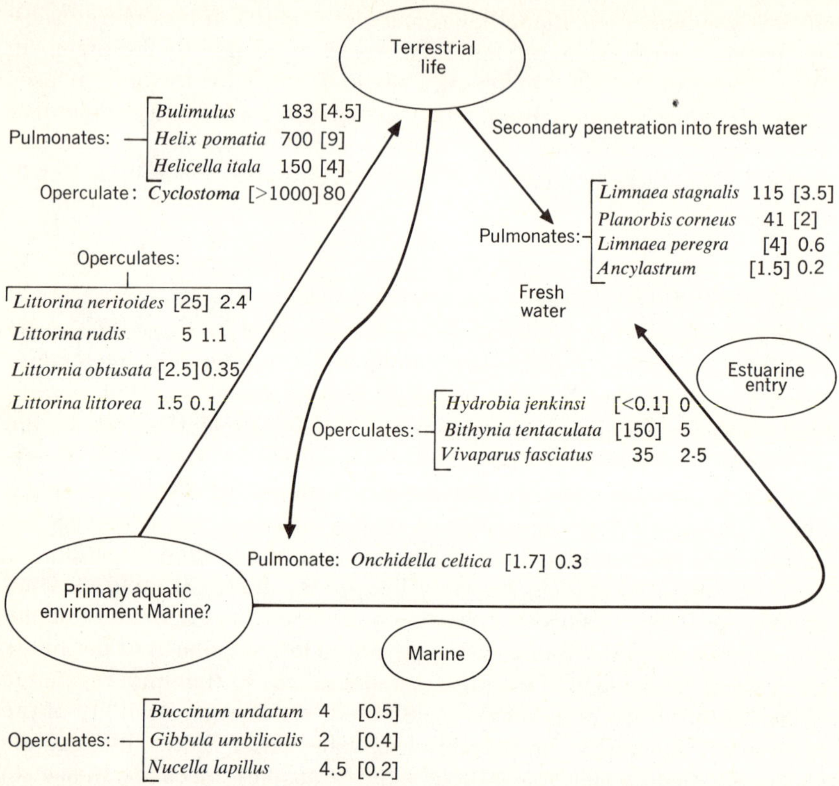

Figure 5-11. Uric acid retention in gastropod molluscs. Figures in brackets are calculated, the rest are experimental. (From Needham, 1938.)

excretory products have evolved. Three main types of nitrogen-containing compounds are excreted as metabolic end-products by animals: ammonia, urea, and uric acid. Each has certain chemical characteristics which undoubtedly are associated with an animal's habitat preference. Ammonia is relatively toxic to cells and must either be excreted rapidly or be converted to a less toxic substance before it is concentrated in the circulatory fluids. Since ammonia is soluble in water as well as readily diffusible, animals that excrete chiefly ammonia are present in aquatic environments and are called *ammonotelic*. Organisms that principally excrete urea are called *ureotelic*.

Urea has the main advantage of being much less toxic than ammonia. In general, few marine invertebrates excrete appreciable amounts of urea, but the amounts are significant in vertebrates. Some animals are *uricotelic* in that they convert nitrogenous wastes to uric acid, which is both less toxic than ammonia and relatively insoluble in water. This molecule can be stored in various parts of the body or may be excreted in crystalline form. Typically, animals such as land snails, insects, birds and terrestrial reptiles, which occupy habitats with limited water, excrete uric acid and thus are able to preserve body water.

The study on gastropods by Needham (1938) is a classic example of this correlation of habitat and type of excretory product (Fig. 5-11). The evolutionary trend for a shift from ammonotelism to uricotelism was observed when various snail species from the sea invaded fresh water or land. Since crustaceans as a group have not altered their nitrogenous excretory products in response to the availability of water, this lack of physiological plasticity might have contributed to their relatively minor success on land. Fresh-water fish are predominantly ammonotelic, but they excrete some urea. Although water retention is not one of the problems of fresh-water fish, it is a problem for marine fish. Many marine species excrete both ammonia and urea, but in different proportions; the starry flounder produced a greater percentage of ammonia and less urea, and the reverse was reported for the blue sea-perch. Together, these two compounds accounted for 75 and 98 percent of the total nitrogen excreted by the gills and the kidneys.

Another compound excreted is trimethylamine oxide (TMO). It is excreted in relatively small amounts (Wood, 1958), although earlier investigators reported that about one-third of the total nitrogen in the urine was TMO. Since only about 20 percent of the total nitrogen is excreted in the urine, and the remainder leaves the body via the gills, the low total amount of TMO excreted (Wood, 1958) is probably typical for most marine teleosts. The TMO in these fish appears to have been derived from their food rather than originating endogenously (Benoit and Norris, 1945).

Within a species, the type of nitrogenous waste may vary with the different habitats occupied by the various stages in the life cycle. This shift in excretory patterns with habitat is illustrated by the salamander *Triturus viridescens*. The aquatic larvae primarily excrete ammonia (about 75 percent of the total nitrogen), but also some urea (about 25 percent). After metamorphosis the animals live on land as the red eft stage, and then the pattern of nitrogen excretion is reversed: 87 percent is excreted as urea and the remainder as ammonia. When the mature animal returns to an aquatic habitat to mate, the amount of ammonia excreted increases again (Nash and Fankhauser, 1959). Similar shifts have been reported for other amphibians (Munro, 1953). Experimentally, a marked decrease in ammonia excretion by the aquatic amphibian *Xenopus* occurred when it was forcibly kept out of water, but upon its return to water, the rate of ammonia excretion returned to normal. Interestingly, much of the nitrogen wastes that accumulated during the animal's sojourn out of water had been retained in the form of urea (Balinsky *et al.,* 1961).

Nitrogen-containing compounds, some of which are normally regarded as waste products, may help to maintain internal osmoconcentration. Low molecular weight organic compounds, especially amino acids and urea, have been utilized for this purpose by many organisms. As a group, sharks illustrate well the importance of this physiological stratagem in meeting an environmental stress. The body fluids of elasmobranchs have a salt content not unlike that of many teleosts, but these fluids are slightly hyperosmotic to the surrounding medium, owing to a high concentration of urea (about one-third of the total osmoconcentration) and TMO. Osmotic water enters, as do salts, and these salts are removed against a concentration gradient by the kidneys and the rectal gland. The presence in fresh-water sharks of smaller rectal glands than those in marine sharks can be correlated with their regressive use in removing salts (Oguri, 1964). Urea, which is synthesized by most shark tissues, is strongly conserved by kidney reabsorption at a rate inversely related to water resorption. Excess urea is excreted largely via the gills.

Increased urea levels in sharks also greatly influence poikilosmotic tapeworms living in the shark's gut, for urea that diffuses into the body of the tapeworm from the gut fluid apparently is involved in maintaining salt balance. In the presence of urea-free sea water, the worm swells and then shrinks in size. Concentrations of urea similar to that present in elasmobranchs are lethal to trematodes and most tapeworms parasitizing other vertebrate classes, and must constitute a chemical barrier to these animals invading this habitat (Read *et al.,* 1959).

Terrestrial Animals

Conservation of water is the most important aspect of osmoregulation in terrestrial animals. Various stratagems have evolved to enable diverse groups of animals to invade land independently (Pearse, 1950). In general, organisms prevent or reduce the amount of water lost by behavioral means and/or by morpho-physiological adaptations.

Selection of microenvironmental niches where moisture and thermal stresses are reduced is common. For example, when the soil-surface temperature is high (56°C), small arthropods have been observed to burrow to a depth of 10 cm, where the temperature is 34°C. The peak time of locomotor activity may coincide with periods of less stressful environmental conditions, as demonstrated by salamanders. During the evening hours when the temperature is reduced and the relative himidity is minimal, terrestrial salamanders are observed in greatest number at the surface of the forest floor.

The major morpho-physiological adaptations of desert arthropods are as follows:

1. The integument of species living in xeric environments is structurally arranged to render it highly waterproof.
2. Uric acid is the principal waste product of nitrogen metabolism excreted by insects. In addition, the rectal wall and the Malpighian tubules function to regulate water balance. In times of a water shortage, dry fecal pellets may be produced, but if the animal is overloaded with water, a dilute urine may be produced.
3. The transpiration in desert species is less than in species from more moist environments.
4. The high lethal temperatures tend to be higher in desert arthropods than in other closely related species from mesic environments.
5. Free water may be obtained by oral and anal drinking from moist surfaces.
6. When food material is oxidized, metabolic water results, which is a major source of water to this animal.
7. Some arthropods, including the desert cockroach, are able to take up water from unsaturated air (Edney, 1967).

Although the problems of water conservation are resolved in different ways by various animal groups, the general methods are not unlike those of arthropods. For more details regarding other organisms, consult Chew (1961) and Potts and Parry (1964).

SUMMARY

The maintenance of a relatively constant internal chemical environment necessary for sustaining protoplasmic integrity is a basic problem confronting all organisms. It is not surprising that many mechanisms have evolved to solve this problem of internal chemical stability, since animals are found in habitats that differ widely in chemical composition. Aquatic animals are enveloped by an environment that contains water and ions, in contrast to terrestrial animals which must depend chiefly on food and drinking water for the input of both. The effectiveness of controlling the influx and outflux of water and ions is correlated with habitat. Those marine animals having limited regulatory ability (conformers) are restricted to oceanic waters, while excellent regulators live in estuaries or in dry terrestrial habitats. Both behavioral and physiological mechanisms are used to maintain a relative internal constancy.

Not only the concentration of water and ions is a major factor in regulation, but other environmental parameters such as temperature, food, and body size are also influential.

As a result of the intensive attention given to the ecological-physiological significance of water and ions, the groundwork has been laid for further studies of the cellular basis of transport and of the complex interaction between the environment and the organism governing its ecology.

References

Balinsky, J. B., M. M. Cragg, and E. Baldwin. 1961. The adaptation of amphibian waste nitrogen excretion to dehydration. *Comp. Biochem. Physiol.*, 3:236–244.

Beadle, L. C. 1934. Osmotic regulation in *Gunda ulvae*. *J. Exptl. Biol.*, 11:382–396.

Benoit, G. J., and E. R. Norris. 1945. Studies on trimethylamine oxide. II. Origin of trimethylamine oxide in young salmon. *J. Biol. Chem.*, 158:439–442.

Bliss, D. E., S. M. Wang, and E. A. Martinez. 1966. Water balance in the land crab, *Gecarcinus lateralis*, during the intermolt cycle. *Am. Zool.*, 6:197–212.

Brunson, R. B. 1949. The life history and ecology of two North American gastrotrichs. *Trans. Am. Micro. Soc.*, 68:1–20.

Chew, R. M. 1961. Water metabolism of desert vertebrates. *Biol. Rev.*, 36:1–31.

Conte, F. P. 1969. Salt secretion. *In* W. S. Hoar, and D. Randall (eds.), *Fish Physiology*, Vol. I, Ch. 3. Academic Press, New York.

Copeland, D. E. 1968. Fine structure of salt and water uptake in the land crab, *Gecarcinus lateralis*. *Am. Zool.*, 8:417–432.

Croghan, P. C. 1961. Competition and mechanisms of osmotic adaptation. In *Mechanisms in Biological Competition*, pp. 156–167. *Soc. Exptl. Biol. Symposium No. 15*.

Croghan, P. C., and A. P. M. Lockwood. 1968. Ionic regulation of the Baltic and fresh-water races of the isopod *Mesidotea* (*Saduria*) *entomon* (L.). *J. Exptl. Biol.*, 48:141–158.

Dehnel, P. A., and T. H. Carefoot. 1965. Ion regulation in two species of intertidal crabs. *Comp. Biochem. Physiol.*, 15:377–397.

Edney, E. B. 1967. Water balance in desert arthropods. *Science*, 156:1059–1066.

Fretter, V. 1955. Uptake of radioactive sodium (Na^{24}) by *Nereis diversicolor* Mueller and *Perinereis cultrifera* Grube. *J. Mar. Biol. Assoc. U.K.*, 34:151–160.

Fuller, H. J., and O. Tippo. 1954. *College Botany*. Holt, Rinehart and Winston, New York.

Gray, I. E. 1954. Comparative study of the gill area of marine fishes. *Biol. Bull.,* 107:219–225.

Gross, W. J. 1961. Osmotic tolerance and regulation in crabs from a hypersaline lagoon. *Biol. Bull.,* 121:290–301.

Gross, W. J. 1963. Acclimation to hypersaline water in a crab. *Comp. Biochem. Physiol.,* 9:181–188.

Gross, W. J., and P. V. Holland. 1960. Water and ionic regulation in a terrestrial hermit crab. *Physiol. Zool.,* 33:21–28.

Hall, F. G. 1922. The vital limit of exsiccation of certain animals. *Biol. Bull.,* 41:31–51.

Hickman, C. P. 1959. The osmoregulatory role of the thyroid gland in the starry flounder, *Platichthys stellatus. Can. J. Zool.,* 37:997–1060.

Hickman, C. P. 1968. Glomerular filtration and urine flow in the euryhaline southern flounder, *Parlichthys lethostigma,* in sea water. *Can. J. Zool.,* 46:427–437.

Hickman, C. P., and B. F. Trump. 1969. The kidney. *In* W. S. Hoar and D. Randall (eds.), *Fish Physiology,* Vol. I, Ch. 2. Academic Press, New York.

House, C. R. 1963. Osmotic regulation in the brackish water teleost, *Blennius pholis. J. Exptl. Biol.,* 40:87–104.

Houston, A. H. 1959. Osmoregulatory adaptation of the steelhead trout (*Salmo gairdnerii* Richardson) to sea water. *Can. J. Zool.,* 37:729–748.

Jones, L. I. 1941. Osmotic pressure relations of nine species of crabs of the Pacific coast of North America. *J. Cell. Comp. Physiol.,* 18:79–92.

Kalber, F. A., Jr., and J. D. Costlow, Jr. 1966. The ontogeny of osmoregulation and its neurosecretory control in the decapod crustacean, *Rhithropanopeus harrisii* (Gould). *Am. Zool.,* 6:221–229.

Kinne, O. 1958. Adaptation to salinity variations—some facts and problems. *In* C. L. Prosser (ed.), *Physiological Adaptation,* pp. 92–105. Am. Physiol. Soc., Washington, D.C.

Kinne, O. 1962. Irreversible nongenetic adaptation. *Comp. Biochem. Physiol.,* 5:265–282.

Krogh, A. 1939. *Osmotic Regulation in Aquatic Animals.* Cambridge University Press.

Lockwood, A. P. M. 1962. The osmoregulation of Crustacea. *Biol. Rev.,* 37:257–305.

Lockwood, A. P. M. 1964. *Animal Body Fluids and Their Regulation.* Harvard University Press, Cambridge, Mass.

McAlister, R. O., and F. M. Fisher. 1968. Responses of the False limpet, *Siphonaria pectinata* Linnaeus (Gastropoda, Pulmonata) to osmotic stress. *Biol. Bull.,* 134:96–117.

Munro, A. F. 1953. The ammonia and urea excretion of different species of Amphibia during their development and metamorphosis. *Biochem. J.,* 54:29–36.

Nash, G., and G. Fankhauser. 1959. Changes in the pattern of nitrogen excretion during the life cycle of the newt. *Science,* 130:714–716.

Needham, J. 1938. Contributions of chemical physiology to the problem of reversibility in evolution. *Biol. Rev.*, 13:225–251.

Oglesby, L. C. 1965. Steady-state parameters of water and chloride regulation in estuarine nereid polychaetes. *Comp. Biochem. Physiol.*, 14:621–640.

Oglesby, L. C. 1968. Responses of an estuarine population of the polychaete *Nereis limnicola* to osmotic stress. *Biol. Bull.*, 134:118–138.

Oguri, M. 1964. Rectal glands of marine and freshwater sharks: comparative histology. *Science*, 144:1151–1152.

Panikkar, N. K., and N. G. Sproston. 1941. Osmotic relations of some metazoan parasites. *Parasitol.*, 33:214–223.

Parry, G. 1960. The development of salinity tolerance in the salmon, *Salmo salar* L. and some related species. *J. Exptl. Biol.*, 37:425–434.

Parry, G. 1966. Osmotic adaptation in fishes. *Biol. Rev.*, 46:392–444.

Pearse, A. S. 1950. *The Emigrations of Animals from the Sea*. Holt, Rinehart and Winston, New York.

Potts, W. T. W., and G. Parry. 1964. *Osmotic and Ionic Regulation in Animals*. Pergamon, London and Oxford.

Prosser, C. L., J. Green, and T. J. Chow. 1955. Ionic and osmotic concentrations in blood and urine of *Pachygrapsus crassipes* acclimated to different salinities. *Biol. Bull. Mar. Biol. Lab., Woods Hole,* 109:99–107.

Read, C., L. T. Douglas, and J. E. Simmons, Jr. 1959. Urea and osmotic properties of tapeworms from elasmobranchs. *Exptl. Parasitol.*, 8:58–75.

Read, D. E., Jr. 1968. Salinity tolerance and osmoregulation of the pea crabs *Pinnotheres ostreum* Say and *Pinnotheres maculatus* Say. Unpub. M. A. Thesis, Duke University.

Reid, D. M. 1932. Salinity interchange between salt water in sand and overflowing freshwater at low tide. II. *J. Mar. Biol. Assoc. U.K.*, 17:299–306.

Schlieper, C. 1957. Comparative study of *Asterias rubens* and *Mytilus edulis* from the North Sea (30 ‰ S) and the western Baltic Sea (15 ‰ S). *Ann. Biol. t. 33,* 117–127.

Schlieper, C., and R. Kowalski. 1956. Quantitative Beobachtungen über physiologischen Ionenwirkungen im Brackwasser. *Kieler Meeres Forschungen,* 12:154–165.

Scholander, P. F., L. van Dam, J. W. Kanwisher, H. T. Hammel, and M. S. Gordon. 1957. Super cooling and osmoregulation in Arctic fish. *J. Cell. Comp. Physiol.*, 49:5–24.

Segal, E., and P. A. Dehnel. 1962. Osmotic behavior in an intertidal limpet, *Acmaea limatula*. *Biol. Bull.*, 122:417–430.

Smith, R. I. 1963. A comparison of salt loss rate in three species of brackish-water nereid polychaetes. *Biol. Bull.*, 125:332–343.

Smith, R. I. 1964. D_2O uptake rate in two brackish-water nereid polychaetes. *Biol. Bull.*, 126:142–149.

Vernberg, F. J. 1969. Acclimation of intertidal crabs. *Am Zool.*, 9:333–341.

Weil, E., and C. F. A. Pantin. 1931. The adaptation of *Gunda ulvae* to salinity. The water exchange. *J. Exptl. Biol.,* 8:73–81.

Wells, G. P., and I. C. Ledingham. 1940. Physiological effects of a hypotonic environment. I. The action of hypotonic salines on isolated rhythmic preparations from polychaete worms (*Arenicola marina, Nereis diversicolor, Perinereis cultrifera*). *J. Exptl. Biol.,* 17:337–363.

Williams, A. B. 1960. The influence of temperature on osmotic regulation in two species of estuarine shrimps (*Penaeus*). *Biol. Bull.,* 119:560–571.

Wood, J. D. 1958. Nitrogen excretion in some marine teleosts. *Can. J. Biochem. Physiol.,* 36:1237–1242.

6

Perception of the Environment

How is an animal able to locate a particular habitat, search out a specific type of food, detect enemies, and find a mate? Such information is relayed by the various sensory modalities, and since the perception of the external environment is so important to an animal, each of the modalities may greatly influence the behavior of the animal. The sensory modalities that are most important to an organism in the perception of its physical and biological environment, and the range of stimuli that will cause the receptors to respond, can be correlated with an animal's way of life. For example, light perception is less developed and less important to an animal living in the aphotic zone of the ocean than it is to a sea bird that lives in a well-lighted and unobstructed environment, where it uses vision in exploration of the environment, in location of prey, and in avoiding enemies.

VISION

Types of Receptors

Since animals depend on light for orientation, diurnal migrations, and synchrony of rhythmic activities, light reception probably is the most important sensory modality in the exploration of the environment. It is not surprising, therefore, that receptors for light are common to almost all animals; in those with better developed photoreceptors, light greatly influences behavior. Light receptors may be well-defined organs such as the vertebrate eye, the compound eyes of crustacea and insects, the simple eyes or ocelli of other arthropods and invertebrates, and the dermal light receptors. All these various receptors perceive only a small fraction of the total spectrum of radiation as light. The visible spectrum of all light receptors is limited to wavelengths between about 300 and 700 mμ. Because they are located in restricted regions of the body, the more complex optic receptors can perceive qualitative, temporal, and spatial properties of light, whereas photosensitivity is diffuse with dermal light receptors, which may extend over large parts of the body. It is thought that the photochemical systems responsible for all light reactions are similar, although actual sensitivity is several thousand times less in the dermal light receptors than in the more well-developed eyes.

Dermal Light Responses

Dermal light receptors are distributed widely throughout the animal kingdom and have been reported in every major phylum of Metazoa; they tend to be rare or absent, however, in groups having a hard or waterproof integument or eyes capable of image formation. They are most common to aquatic invertebrates, especially those having a simple epithelium or superficial nervous system. Specific dermal light receptors have not yet been identified. After an extensive review of preceding work, Steven (1963) concluded that not all dermal light reactions can be attributed to a single receptor system at the cellular level of organization; he suggested that the light response more probably results from accumulation of small amounts of photosensitive substances in the cytoplasm. Evidence shows that photosensitive substances may be localized in a particular part of certain types of cells.

The dermal light sense causes three major types of responses among animals: (1) orientation by bending the body; (2) the shadow reflex or withdrawal of exposed parts of the body in response to a sudden change in illumination; and (3) environment selection by free-moving organisms.

Bending Reactions

Bending reactions to light are called *phototropisms*. Orientation to light is particularly common among sessile animals such as hydroids and sea anemones; the column of these animals is bent toward the light by muscular contraction. Many sessile polychaetes and bryozoans accomplish orientation toward light by the action of their tube-secreting glands.

Shadow Reflexes

Shadow reflexes, which occur most commonly in sessile animals such as tubiculous worms, barnacles, and lamellibranchs, have obvious adaptive value to an animal, for they often function as a protective mechanism. Some animals can detect either a sudden increase or decrease in light intensity, but the response to decreased illumination generally is more vigorous. The complexity of shadow reflexes is illustrated by the responses of the spines of the sea urchin, *Diadema,* as reported in studies by Millott (1960). With a sudden change in illumination, the response is a sharp, swinging movement of the spines, but the reaction to a decrease in light intensity is much more easily elicited than reactions to an increase in light intensity. When an urchin is placed beneath a checkerboard of lamps and one lamp is turned off, the urchin's spines rapidly adjust to point at the lamp that has just been turned off. As far as is known, all these reactions are transmitted by the nervous system. Millott and Yoshida (1960) concluded that the shadow reflexes were not due to a primary effect of light on receptors, but were adaptive responses related to inhibition of the nervous system of the animal. Thus, in *Diadema,* the spontaneous movement of the spine is suppressed by light in direct proportion to its intensity. When this inhibition is removed, the animal responds.

Shadows may act as a stimulus to activity, however, as they do among parasites. Unfed leeches resting at the bottom of a pond respond to shadows by swimming upward (Herter, 1929), and the cercariae of some species of trematodes react similarly.

Environment Selection

Orienting mechanisms in free-living animals are classified into two main categories (Fraenkel and Gunn, 1940). If the direction of locomotion bears no relation to the direction of the stimulus, it is said to be a *kinesis*. If the orientation of locomotion is directed either toward or away from the source of stimulus, it is a *taxis*. Phototaxis is dependent upon a *localized photoreceptor;* a kinetic response is not. Among the invertebrates and the lower vertebrates kinetic responses to light are very important in the selection of the proper photic environment. Such a kinetic response is well illustrated in the study by Jones (1955) on the ammocoete larvae of the brook lamprey *Lampetra planeri.* The ammocoetes spend much time buried head downward in muddy streams or lakes. When they were dug up and exposed to light, the larvae swam with their heads pointed downward. When mud was put into an experimental tank, the downward directed movements became more active and the animals burrowed. Burrowing continued until the tips of the tails, where the light receptors are located, were covered. In an aquarium that has both lighted and darkened compartments, the larvae soon gather in the dimly lit section, where they remain unless disturbed.

In animals that have well-developed eyes as well as a dermal light sense, each type of photoreceptor may be concerned with a different type of behavior, as in the case of the fresh-water cladoceran *Daphnia* (Harris and Mason, 1956). As do many planktonic forms, this animal makes regular vertical migrations through the water column each day. Changing light intensity is the major determining factor in diurnal migrations, so that migratory planktonic animals aggregate in regions of optimal light intensity. In *Daphnia,* the migratory movement is independent of the direction of the light, and is entirely photokinetic. That is, at a certain intensity, light stimulates locomotion, but bears no relation to the direction of the light source. At high light intensities, however, the response is phototactic; that is, the direction of locomotion is governed by the light source. In *Daphnia,* the animal moves toward the light source when light intensities are reduced, and away from it when the light intensity is increased beyond a certain point. When *Daphnia* is blinded, the animal still responds photokinetically, but not phototactically. Thus the compound eye is necessary to *Daphnia* for directional responses, and dermal responses are necessary for the kinetic ones.

Response to light has also been reported in blind animals, particularly in cave-dwelling crustacea and fish. In the crayfish *Cambarus setosus,* it has been established that retinal elements and the lamina ganglionaries essentially are nonexistent (Fingerman *et al.,* 1964). Yet, when the head of this animal is illuminated, it exhibits a locomotor response (Larimer, 1966). Because this response is present even when the eye stalks are removed, either a dermal light sense or an anterior central nervous system receptor must be present. According to Steven (1963), most blind teleosts are able to detect light intensity and move away from it, but their responses are not distinct enough to be easily measured.

Other Light-Oriented Responses

Since animals commonly use light for primary orientation, some animals move horizontally, with the dorsal surface uppermost. This response is the dorsal light reaction, which is widespread particularly among invertebrates and fish. Reversing the direction of incident light can change the orientation of some animals such as the brine shrimp *Artemia,* which will turn over when the light is placed beneath it (Suffert, 1932). In animals with statocysts, the response is not regulated by photoreception alone. The shrimp *Palaemon,* which has statocysts that function in gravity orientation, normally keeps the dorsal surface uppermost regardless of the direction of incident light. If, however, the statocysts are removed, the animal no longer orients toward gravity, but instead orients itself solely toward the light. If the animal is held in a normal position and the light comes from the side, the shrimp makes pushing movements with its legs on the side opposite the light, turning the animal on its side.

Some animals can determine a direction by maintaining a constant angle of orientation between the body and the reference cue. This is termed the *light compass* reaction, and it forms the basis of sun-, moon-, and star-compass orientations reported in many homing and migratory species of animals.

Biotic Relationships

Predator-Prey Relationships

While taxic and kinetic responses are nearly universal in animals, only a few molluscs, some arthropods, and most vertebrates are able to perceive qualitative, temporal, and spatial qualities of light. In these animals with highly developed visual perception, vision largely controls their behavioral responses, as it does in predator-prey relationships, for example. Predatory animals have the best developed eyes in the animal kingdom, and prey animals have developed many characteristics that tend to hide them from the predator's view. Movement, primarily, evokes food-seeking behavior of many animals. If the object falls within certain size limits, the predators often will ignore color, shape, or pattern of the prey animal. For many predators, selection of prey is a function of the habitat where they hunt, or may be based on more specific discrimination by chemical or other means once the prey is captured. Many predators must approach their prey warily in order to avoid alarming them and causing them to scatter. The effectiveness of some predators is enhanced by their ability to take on the coloration of a nonpredatory animal or even of the prey itself (Brower *et al.,* 1960).

Prey animals, on the other hand, have numerous adaptations for concealing themselves from their predators. These include a general similarity to the colors of the environment; specific resemblance to objects that are commonly found in the environments, such as bark or leaves; countershading, which eliminates three-dimensional lighting effects; disruptive coloration, which serves to break up the outlines of the entire body or of a particular structure; and concealment of the shadow either through a low crouching position or through structures that hide the shadow from above. If the animal is caught, it may feign death, thereby removing the main stimulus-movement for prey-catching (Cott, 1957). The methods of concealment are illustrated in Figs. 6-1 to 6-4.

Some prey animals respond to the predator's presence by giving false visual-orienting clues. Squid, for example, eject a black cloud of ink that distracts the predator and gives the prey time to escape. Some salamanders and lizards can distract predators by autotomizing a wriggling tail. Ability to effect a sudden change in appearance offers another avenue of

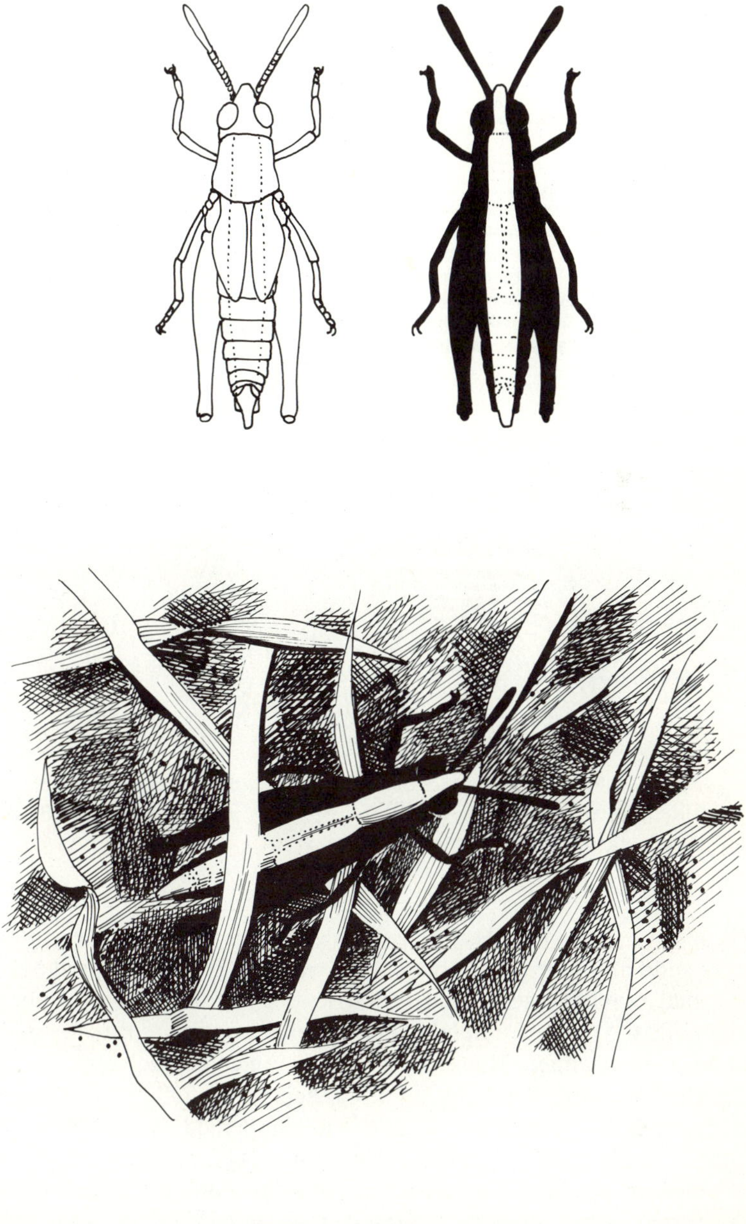

Figure 6-1. The relationship of the coincident pattern to its morphological background in *Chorthippus parallelus,* showing the disruptive effect against uniform and broken surroundings. (From Cott, 1957.)

Figure 6-2. The upper paired diagrams illustrate the principle of the use of disruptive patterns in concealment. The lower part of the figure illustrates the application of this principle. (From Cott, 1957.)

escape for prey animals. For example, animals such as the puffer fishes may suddenly increase in body size when startled by a predator. When alarmed, one of these fish can inflate its body until its form is almost globular (Fig. 6-5). Many amphibia and reptiles respond to threatening danger in this same manner. Mammals and birds are able to achieve a sudden and apparent increase in size by bristling their fur or fluffing their feathers. In some animals, the increase in size is accompanied by the sudden appearance of bright colors. Lizards of the genus *Anolis,* which exhibit this type of behavior pattern, have gular sacs that are normally folded up beneath the throat; when they are alarmed, their gular sacs fan out, exhibiting brilliant hues of red, yellow, and blue (Cott, 1957).

Other potential prey animals rely on chemical protection from predators, often coupled with warning coloration. In one group of tropical

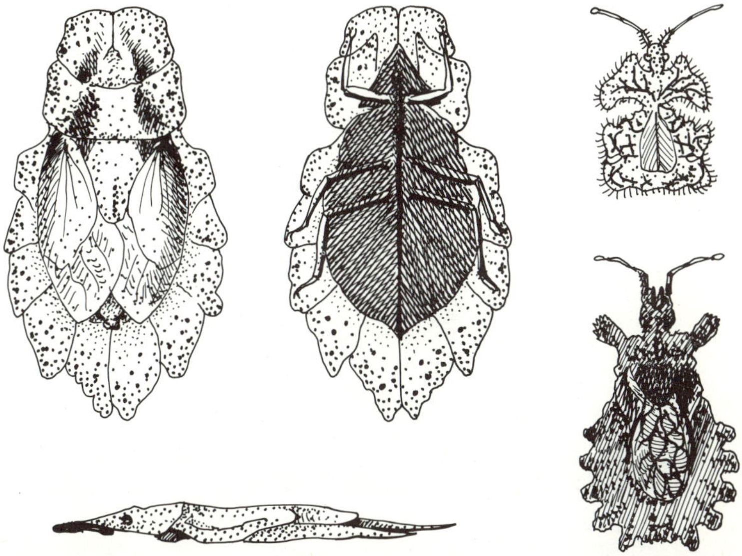

Figure 6-3. This figure illustrates the cryptic resemblances found in some of the bark-like heteopterous bugs: 1, 2 and 3, dorsal, ventral and lateral views of *Phloea subquadrata;* 4, *Craspedum phyllomorphum;* 5, *Dipodius lunatus.* (From Cott, 1957.)

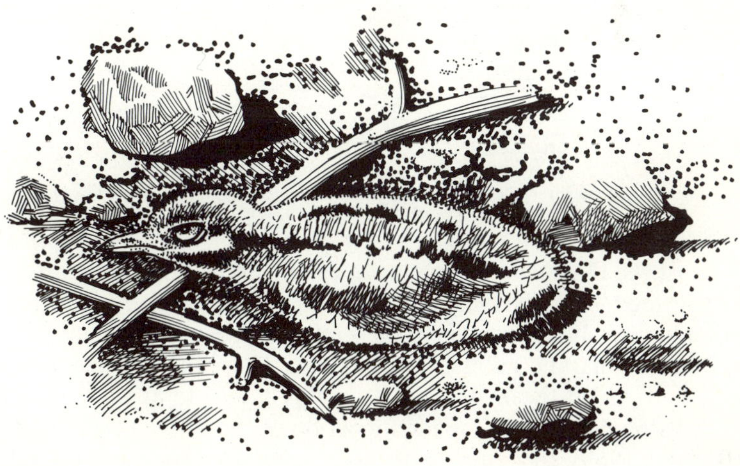

Figure 6-4. Concealment of a young stone-curlew in a squatting position. (From Cott, 1957.)

Figure 6-5. The puffer fish shown in an inflated and deflated posture.

insects, the Donainae, which includes monarch and queen butterflies, the larvae feed on milkweed plants that contain cardiac glycosides, compounds resembling digitalis in their effects. When a bird feeds on the conspicuously marked adults, it will vomit. Therefore, once a predator has attacked one of these animals, it quickly learns to avoid them, since the visual stimulus aids in the identification of these animals at another time (Brower, 1969). In a number of cases, several species of unpalatable organisms have very similar warning coloration patterns. Such convergent mimicry offers added protection, because once a predator has rejected one of this group, others marked in the same manner also will be rejected. This type of mimicry is called *Mullerian.* In another type of mimicry, termed *Batesian,* palatable prey species gain protection mimicing the visual patterns and behavior of animals that have chemical defenses.

Communication

Vision is crucial in the aggregation and dispersion of many animals. For example, schooling in many fish is largely dependent on vision, and schooling may be eliminated in blinded fish (Keenleyside, 1955). In a study of cichlid fishes, Baerends and Baerends-van Roon (1950) reported that mixed species of very young fish would school together, but as soon as they matured enough to have adult markings, the schools separated according to species; this indicates that vision is important in dispersing the fish. When reproductive markings are assumed, some older fish become very aggressive and claim as their own that part of the area formerly occupied by the school. Interterritorial fighting is common, and visual landmarks such as plants or stones help establish definite boundaries.

In many animals, visual signals facilitate coordination of reproductive activities, as illustrated by fireflies (Buck, 1937; Buck and Buck, 1968). During mating time, the male firefly *Photinus pyralis* flashes at regular intervals. The female, sitting on a blade of grass, flashes in reply to any flashing male that is nearby. It is not known how she can recognize his flashes, but the regularity of the flashing interval, about 5.8 sec, may be the recognition factor. The female flashes back at a time interval ranging from 1.4–2.4 sec, depending on the ambient temperature. The male will approach a light only if it flashes 2 sec after he does, regardless of size, color, or duration of the flash, and he will not respond to signals from other males. As the male approaches, the male and female exchange flashes until he reaches and mates with her.

Vision is often important in parent-young relationships. The stimulus that induces some young cichlid fish to follow their parents is the slow movement of the parent and, in some species, color. The parents apparently identify the young optically by the pattern of their markings (Baerends and Baerends-van Roon, 1950).

Visual signals, then, offer many advantages to animals. Since they are conspicuous and can be enhanced by color and movement, they are easy to localize. When the ecology permits, vision is ideal for close-range communication. However, it is of little use in long-distance communication, to many diurnal animals at night or to those that live in darkened habitats.

AUDITION

Types of Receptors

Many different structures have evolved which enable an animal to detect vibrations in the surrounding medium and the substrate. The perception of vibrations may enable an organism to distinguish between frequencies, to measure relative intensity, and to determine the direction of sound. It has been suggested that the term *phonoreception* be used to designate reception of sound, irrespective of the type of receptor that is stimulated or the medium through which the sound reaches the body, whether it be solid, liquid, or air (Frings, 1964; Frings and Frings, 1967).

A sound-receptor system can be one of two types. One type is sensitive to molecular displacement and consists of a membrane or hair that is suspended in such a manner that it responds to slight movements of the medium; if it is a membrane, sound waves must have free access to it from all directions. The other type of receptor is based on a pressure-sensitive system in which a heavy membrane is shielded on its inner surface from sound waves and the inner chamber is kept at some stable reference pressure that is usually equal to the atmosphere. Displacement receptors register directly the location of the stimulus; the pressure receptors register only indirectly.

Many invertebrates will respond to substrate vibrations. Frings (1964) has described a number of potential sound receptors in invertebrates, including innervated projections from the body wall, sensilla in the body wall itself, proprioceptors located deep in the muscles or external organs, and specialized sensillae. Only some arthropods, mainly insects, and vertebrates have a definite ability to detect airborne sound as well. In insects, the receptors do not seem to be responsive to changes in pressure, but instead detect molecular displacement. Insects perceive sound through special mechanoreceptors, such as the tympanal organs, which are often innervated by special groups of sense cells called *chordontal sensillae*. Johnston's organ in mosquitoes is an example of this type of receptor. It is a very specialized receptor, located at the base of the antennae in male mosquitoes, which vibrates only at the frequency of the wing-beat sounds of conspecific females.

The tympanal organ is especially characteristic of orthopterans. This organ consists of a membranous drum over a tracheal air sack; chordontal sensillae are so arranged that they are stimulated by movements of the drum or tracheal walls. Sound has free access to both sides of the drum, since the air sack connects with the tracheal system.

The ear, which functions as a pressure receptor, is the basic organ for sound perception in vertebrates. In terrestrial vertebrates and aquatic mammals, the ear responds to pressure changes in the surrounding medium. Fish that have a gas-filled swim bladder connected to the inner ear also respond in this manner. Fish may also utilize a displacement receptor, termed the *lateral line organ,* to perceive low-pitched vibrations, but the organ responds more directly to current flow of water on the body surface. For the most part, the lateral line organ is used as a "distant touch" receptor to detect and locate moving animals. The ability to detect mechanical disturbances is very important for the orientation of fish in space, particularly for those in murky waters.

Animals can respond to a very wide range of frequencies. Many insects have sensory hairs, which respond to frequencies between 100 and 800 cycles per second (cps), whereas insects with tympanal organs respond to frequencies from about 800 cps up into the supersonic range of about 100 kc per second. Some fish perceive frequencies from below 100 cps up to 2000 or 3000 cps. The range of frequencies for man is from about 20–20000 cps. Sound reception functions in the exploration of the environment, in finding food, avoiding danger, and most of all in social communication.

Environment Selection

As a general rule, animals use light to accomplish directional orientation in the environment. In environments with no light, however, many animals depend upon the emission of energy that interacts with the environment. Orienting information is thus obtained by the return of the signal. The generation of sound waves for orientation (echolocation) is a widespread phenomenon among animals that live in poorly lighted environments, such as fish, birds, bats, shrews, sea lions, porpoises, and whales. A good example of orientation by echolocation can be found in the study of Griffin (1953) on the oilbird *Steatornis caripensis,* which nests and spends the daylight hours in caves that are completely dark. These birds use sound orientation to find their way about by emitting bursts of clicking sounds separated by short intervals. In experimental studies if the ears of the birds were plugged, they could not avoid colliding with the walls in a dark room. When the plugs were removed, however, they were able to fly safely.

Biotic Relationships

Predator-Prey relationships

Although most predators rely on vision to locate potential prey, hearing can also play an important role. For example, owls, which hunt at night, rely on hearing to locate mice (Payne, 1962). Prey animals often depend on hearing to warn of approaching predators. Sharks are known to be able to detect and orient to sounds at least 600 ft away (Wisby and Nelson, 1964), and field observations suggest that sharks utilize sound to locate struggling fish (Limbaugh, 1963). In studies of porpoises in captivity, Kellogg (1961) demonstrated that porpoises can locate prey animals by echolocation. He found that these animals emitted high-frequency pulses every 15–20 sec. If a dead fish was dropped into the tank after one of these pulses, but before the next burst, then with the next series of pulses the porpoise seemingly became aware that the environment had changed and began emitting an almost continuous sound train until it had located and seized the fish. Bats and some whales also use echolocation to find prey.

Communication

The auditory sense is most important in animal communication, and a variety of structures may produce sound. Arthropods, for example, utilize frictional devices that involve almost every pair of movable parts of the body. They may also make sounds by muscular vibration of membranes, such as in the tymbal of cicadas, or by rapping parts of the body against the substrate. Some fish use the air bladder as a resonator, and others use it as a drum by either squirting out gas or passing gas from one section to another. Fish also can make stridulatory sounds with pharyngeal teeth. Terrestrial vertebrates usually produce sounds in association with the respiratory system. Membranes vibrated by air movements produce sounds amplified by resonating chambers. Each animal sound can be correlated with the auditory capacity of the animal; that is, the sounds produced contain frequencies that other members of the species can perceive. Sound production can be correlated with certain environmental, physiological, and behavioral states, and in addition, much of it is cyclical, both diel and seasonal. In some animals, sound production is limited to the reproductive season, but in others vocalization occurs during the nonreproductive season as well, in territorial defense, aggregating behavior, escape, and during migration.

Communication through sound is widespread among the insects. Cohabiting sympatric species that breed during the same season tend to have more complex sound patterns, with a correspondingly larger repertoire, than do species that are spatially isolated or breed at different seasons (Alexander, 1962). For example, among cricket species that are active at the same time and place, it has been demonstrated that the repertoire of the male cricket includes calling sounds to establish initial contact between the sexes, a courtship sound, a courtship interruption sound when a courting female moves away, an aggressive sound to repel other males, sounds that maintain contact between the pair after mating, and a

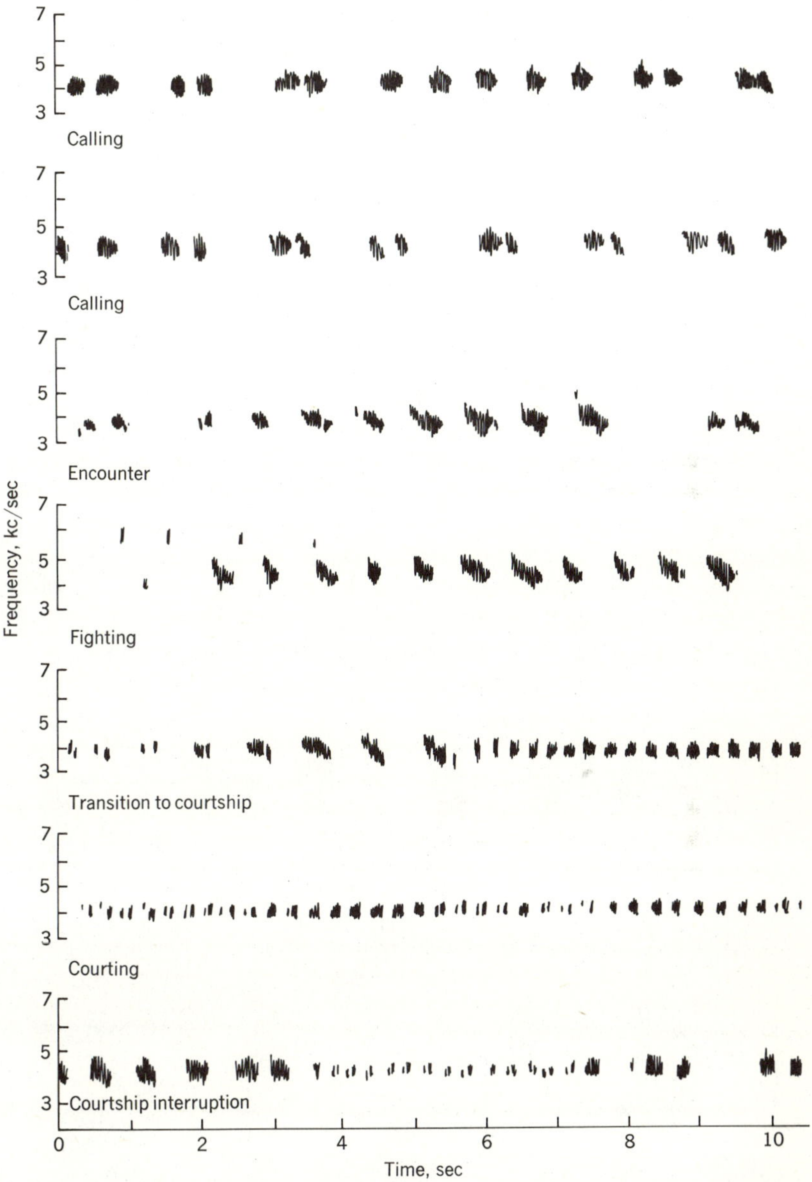

Figure 6-6. The vocal repertoire of the cricket, *Teleogryllus commodus* as analyzed on a sound spectograph. (After Alexander, 1962.)

recognition sound which is not fully understood in relation to function (Alexander, 1960, 1962). The vocal repertoire of one species of cricket, *Teleogryllus commodus,* is shown in Fig. 6-6 (Alexander, 1962).

It has been generally concluded that only terrestrial species of the arthropods use sound signals during their courtship interactions, but recent work by Salmon (1965, 1967) suggests that this may be a premature judgment. He observed that five species of the intertidal-zone fiddler crabs (genus *Uca*) use both visual and acoustical signals during the breeding season. During the day, the male fiddler crab waves his greatly enlarged chela at the female, but when the female approaches, the male ceases to wave and produces sounds either by rapping the major cheliped against the substrate or by vibrating its ambulatories, depending on the species. Sound alone was used in courtship when visual stimuli were ineffective at night or within the confines of the burrow during the day. As can be seen in Fig. 6-7, each species has a characteristic sound pattern.

Sound production in fish is not finely organized, and only major mood responses can be obtained (Winn, 1964) (Fig. 6-8). In the satinfin shiner *Notropis analostanus*, for example, during courtship and chasing the territorial males produced single knocks; during display fighting, a rapid series of knocks; and during courtship activities of approach, circling, and passing over the nest site, a purring sound. When the recorded knocks were played back, other males tended to respond aggressively, whereas purring sounds elicited courtship behavior (Stout, 1963).

Frogs and toads have distinctly specific calling behavior. At one

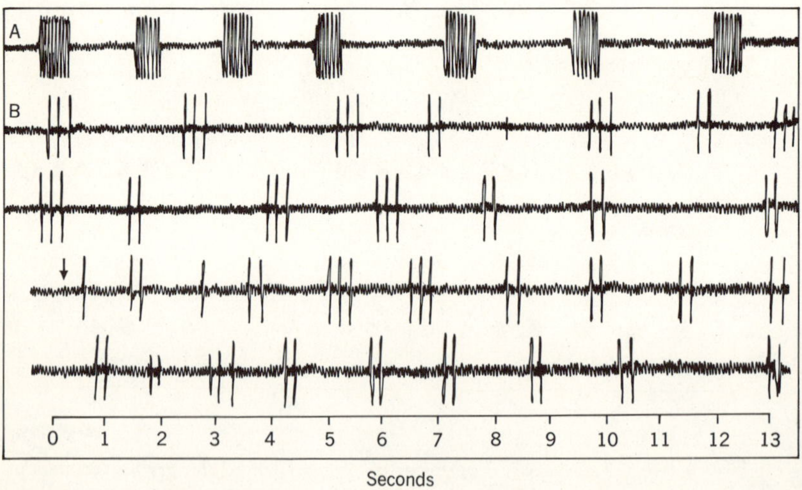

Figure 6-7. Oscillographs of sounds produced by two rapping species of fiddler crabs. *A* represents the sounds of *Uca pugilator* from Florida, *B* those of *Uca speciosa*. Sounds produced after the arrow are from the same male during tactile stimulation. (From Salmon, 1967.)

spawning site in Florida, Bogert (1960) reported that as many as 14 species of frogs could be heard, each one with a distinctive call.

The sounds produced by birds are as diverse and varied as any in the animal kingdom (Marler and Hamilton, 1966). Bird sounds may function

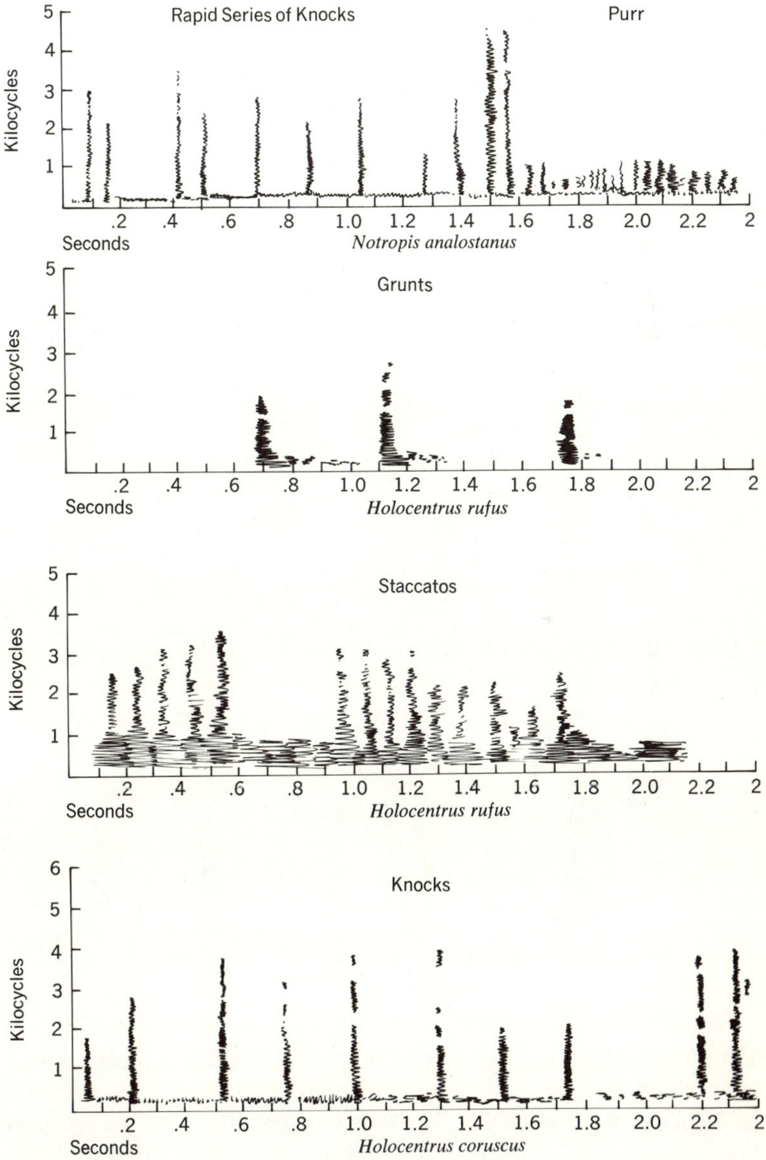

Figure 6-8A. Sonograms of various types of fish calls. (From Winn, 1964.)

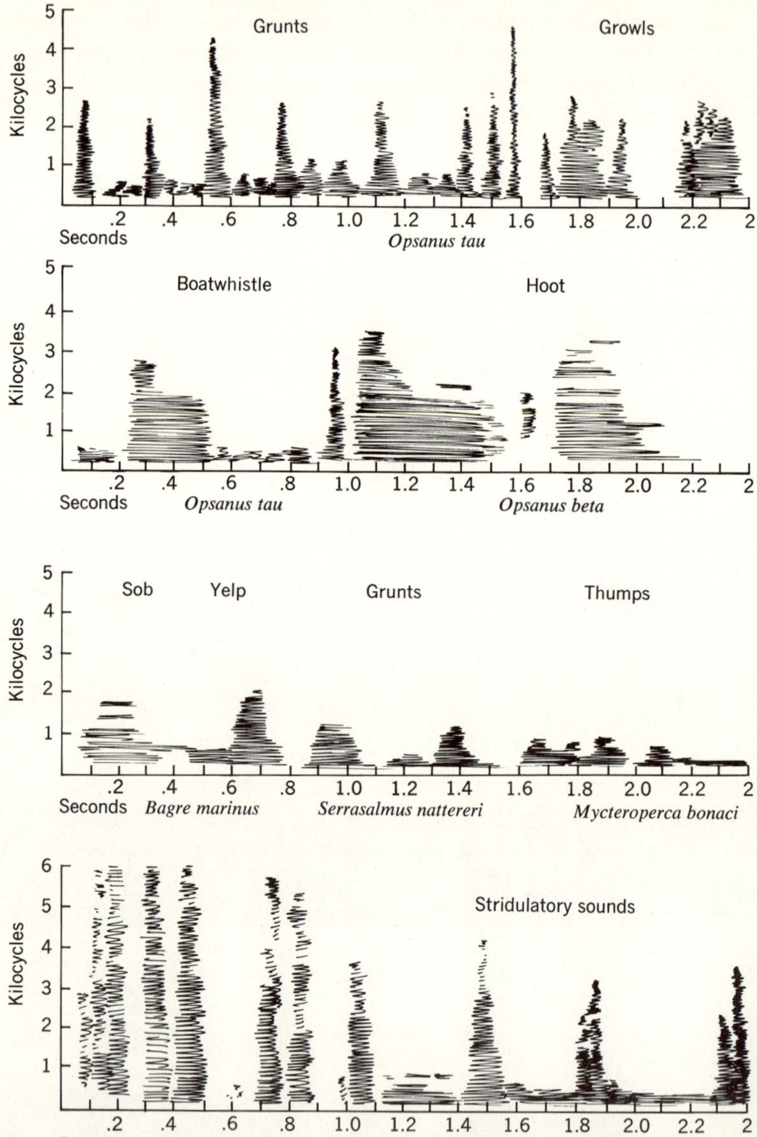

Figure 6-8B. Further sonograms of various types of fish calls. (From Winn, 1964.)

as confusion choruses, young-parental sound relations, alarm and all-clear signals, and recognition of individual call differences.

Because the auditory sense is very responsive to stimuli, it permits a rapid exchange of information, which is particularly important to highly mobile species. Sound is ineffective only when the background level is too high. Communication by sound has an added advantage in that exchange of information can occur simultaneously with other behavioral activities.

CHEMORECEPTION

Types of Receptors

Perception of the chemical characteristics of the environment is an attribute common to all living organisms. The chemical senses are utilized to select a habitat, to find food, to detect enemies, and to locate a mate.

There are three main types of chemoreceptors, which are classified on the basis of their relative sensitivity and on the location of their receptors on the body surface. The least sensitive of these is the "general chemical sense," which may extend over large parts of the body. It is mediated by free nerve endings and unspecialized receptors, which are only rarely identified, and is characteristic of both invertebrates and vertebrates. The response to stimulation of this sense is generally a simple avoidance reaction. The two other types of chemoreception, taste and smell, can evoke more varied responses. Taste, or the contact chemical sense, requires contact between relatively large quantities of water-soluble substances and specialized taste receptors. These receptors are intermediate in sensitivity between the general chemical sense and the sense of smell. The taste receptors of vertebrates generally occur in the mouth cavity and in papillae on the tongue, although they may be located over larger areas in aquatic animals. Catfish, for example, have taste receptors concentrated on their barbels. The sense of smell, olfaction, responds to low chemical concentrations and is the most sensitive of the chemoreceptors. The olfactory cells are found in the epithelium of the nasal cavity.

Environment Selection

The selection of an environment is often based on chemoreception, as illustrated by the home-finding of salmon in a stream, which is in part a response to the chemical characteristics of river streams. Young salmon hatch from eggs laid in fresh-water streams or rivers, but stay there only a short while before they swim downstream to the sea. After several years at sea the salmon are ready to spawn, and they start their migration unerringly back to the stream where they were hatched. How these fish can navigate through the sea is not known, but evidence indicates that a possible sun-compass type of orientation may be involved. Nonetheless, each generation of salmon returns to the same rivulet so consistently that different populations of salmon in relatively close streams have evolved along distinct lines. Once the salmon are in a stream system, home-finding may depend on recognition of an odor, for it has been shown that different streams have different odors, that the salmon do respond to these odors, and that they can learn to differentiate between odors (Hasler, 1967).

Aquatic animals are very sensitive to changes in the chemical characteristics of the water, such as the presence of pollutants, changes in oxygen tension, or carbon dioxide content, and their movements can be correlated with these changes. The adaptiveness of such a response can be illustrated by the experimental study on the burrowing shrimp *Callianassa* (Farley and Case, 1968), showing that these animals can detect external oxygen concentrations. In nature, this would have great adaptive value to the shrimp. During low tide, they can be found as much as a meter below the surface, where oxygen tensions are very low. With the incoming tide, oxygen tensions rise, and the shrimp are able to detect this increase so that they move toward the burrow openings and hyperventilate. Thus the anoxia during low tide is alleviated.

Dry gastropods from the intertidal zone respond positively to sea water that is splashed on the shell, or to immersion. The characteristic response is extension of the head, tentacles, and foot, followed by locomotion if exposure to the sea water continues (Arnold, 1957; Ohsawa and Tsukudo, 1955). This same response occurs in nature when the tide comes in and the spray progressively splashes over the molluscs. Sea water of varying salinities elicited graded responses (Fig. 6-9), with the

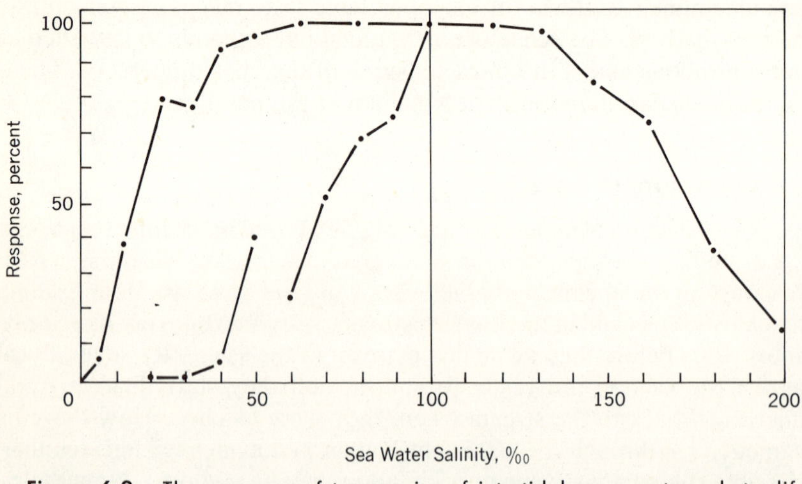

Figure 6-9. The responses of two species of intertidal zone gastropods to different salinities. The dots represent the high intertidal species *Patella vulgata,* the open circles *Nodilittorina granularis.* (From Ohsawa and Tsukuda, 1955.)

greater responses occurring at higher salinities. The fact that nonelectrolytes failed to elicit the response at any tonicity indicates that the response is not entirely due to osmotic pressure.

Biotic Relationships

Predator-Prey Relationships

Many animals rely on chemoreception in food searching, both in locating live prey or recently dead organisms. Some use chemoreception for long-range perception; for example, moray eels use smell in the location of distant food, but once the food is approached, other receptors are employed. It has been suggested that fish living in turbid waters, or bottom dwellers as well as nocturnal or crepuscular fishes, are more dependent upon chemical senses than visual clues in the location of distant food (Bardach *et al.,* 1959). In other animals, chemoreception is employed only after the prey has been located by other means. The pedunculate barnacle *Lepas anatifera,* a sessile animal that lives attached to pieces of wood, normally feeds on living plankton obtained from the surrounding sea water. The ability of these animals to detect amino acids was demonstrated by Crisp (1967), who suggested that barnacles use this sensitivity to determine whether the prey is living, by crushing the plankton with the sharp spines present on their cirri. The crushed plankton release stimulating substances, which set in motion the responses that cause the living prey to be eaten.

Many prey animals are sensitive to the smell of predators, as are the gastropods *Nassarius reticulata* or *N. mutabilis,* which respond dramatically in order to escape the presence of starfish (Weber, 1924). Many mammals that are hunted are so sensitive to the smell of their predators that the predator must approach them from downwind in order to capture them.

Pheromones

The term *pheromone* (derived from the Greek "pherein," to carry, and "horman," to excite) was proposed by Karlson and Butenandt (1959) for chemical signals exchanged between individuals of the same species. They defined pheromones as "substances which are secreted to the outside by an individual and received by a second individual of the same species, in which they release a specific reaction, for example a definite behavior or developmental process." In recent years, considerable work has been devoted to identifying these compounds and the behavioral and physiological reactions that they elicit.

Aggregation and Dispersal. Pheromones are sometimes involved in the initiation and maintenance of both temporary and persistent aggregations of animals. In insects, pheromones often govern "dancing" or "mating" swarms, and they are one of several factors involved in the persistent aggregations observed in social ants, bees, wasps, and termites. Wilson (1965) stated that the pheromone systems have reached their highest evolutionary development in these insects. Pheromones also influence the dispersal of animals. For example, when flour beetles (*Tribolium confusum*) reach high densities, they tend to disperse, and this dispersal is initiated by a pheromone containing quinones produced by the thoracic and abdominal glands (Naylor, 1959).

Alerting Pheromones. Another group of pheromones serves as an alerting signal of impending danger, and more of these pheromones have been identified than any others. Generally, the alarm substances have small, relatively simple molecules (Wilson, 1965). One of the pioneer studies of alarm substances was done by Frisch (1941), who showed that when a pike captures a minnow (*Phoxinus laevis*), the wounded minnow releases an alarm substance from the wound. Very small quantities of this substance are effective in inducing others in the school to flee for shelter. Although the minnows will react to alarm substances from closely related species, they do not respond so sharply as they do to the alarm substance from the epidermis of their own species (Pfeiffer, 1963). Similar responses have been observed in amphibian tadpoles (Kulzer, 1954).

Alerting pheromones have been found in many of the social insects, and in all those that have been studied, only the workers and the female reproductives (except for the queen honey bee), produce the alerting substances, never the males (Maschwitz, 1964). Wilson and Bossert (1963) studied in detail the alarm signal and response to it in ants of the genus *Pogonomyrmex*. A puff of alerting pheromone discharged by a worker occupies a sphere of air about 6 cm in diameter in less than 15 sec and then fades out in about 30 sec. Other ants are attracted to the source, and if sufficiently aroused, they will discharge more alerting pheromone, thus broadcasting it throughout the group. The concentration of the signal determines the response: at low concentrations an approach response is elicited, but at high intensities the response is one of aggression or retreat.

Many recent studies have attempted, through two approaches, to determine the feasibility of using pheromones for behavioral control of insects. One approach is the stimulation of behavior based on the ability of an insect to orient toward a pheromone source. In this method, enough pheromone must be released to compete successfully with pheromone released in nature by wild females. Consequently, the pheromone must be identified so that it can be synthesized in the laboratory. If the right pheromone level is reached, males will orient to a trap where they will either be destroyed or sterilized, thus preventing them from inseminating most of the wild females (Knipling and McGuire, 1966). Another approach is to saturate an area with such a high concentration of sex pheromone that it is impossible for the males to orient to the females in order to mate (Shorey *et al.*, 1967).

Trail Laying. Trail laying is another example of the use of chemical stimuli in intraspecific behavioral orientation. Often the substances involved are placed on the substrate where other members are likely to contact it. One of the most thoroughly investigated trail-laying behavior patterns is that of the fire ant *Solenopsis saenissima*. The trail is secreted by the Dufour's gland and released through the sting in a series of streaks. It is commonly made by ants that have discovered a profitable source of

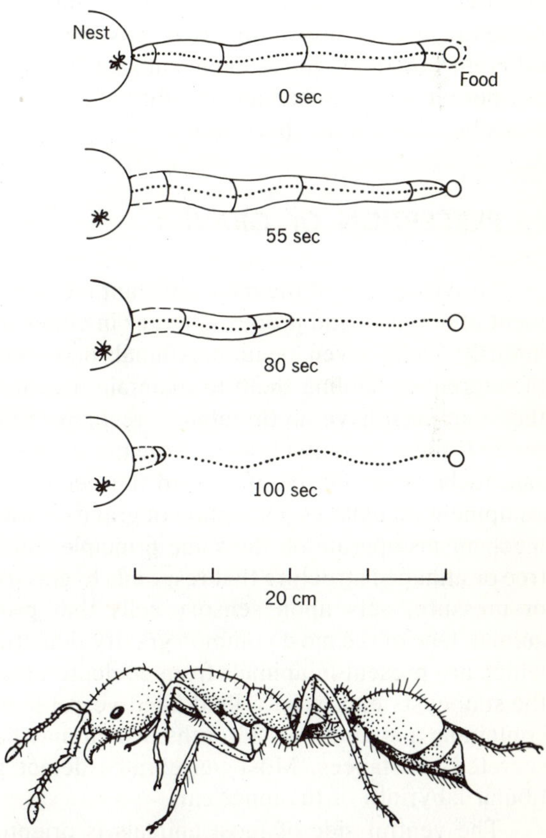

Figure 6-10. The pheromone trail laid down by the fire ant, *Selenopsis saevissima*. The trail, which is not continuous, is put down at intervals, and the effective shape of the trail changes as the secretion evaporates. (From Marler and Hamilton, 1967.)

food. The trail is used as a general guide, and the ants examine it with first one antenna, then the other. The life of the trail substance, which is volatile, is usually brief and within a few minutes it falls below the threshold for eliciting a response (Fig. 6-10). Thus a trail to a food supply that is exhausted soon ceases to be followed (Wilson, 1962; Bossert and Wilson, 1963). In many mammals such as dogs, glands present on the feet serve to make a trail.

Chemical signals offer several advantages over visual and auditory ones. They are particularly important to populations that occupy a stable position in space, for example (Marler, 1967). Their intensity can be raised above the background levels and their information continues to be conveyed in the absence of the animal. As a result, animals can communicate over great distances or, if vision is limited, the signals can be used in close-range communication. It is possible to have different signals for different functions. Since chemical signals are lasting, however, they do not permit rapid exchanges of information, nor can the signal characteristics be changed on short notice.

PERCEPTION OF GRAVITY

Gravity is one of the most constant mechanical forces in the environment of animals, and it does not vary in either the direction of pull or the intensity in any given location. Animals may use gravity as the basic plane of reference enabling them to maintain a definite attitude in space, and they may also have an orientation response to or from the gravitational force. If the orientation is away from the earth's center of gravity, they are said to be *geonegative;* if toward the center, *geopositive*. The sensory equipment that allows perception of gravity varies widely, although all the mechanisms operate on the same principle. Basically it consists of some free or attached structure that responds to gravity and, by a shearing force or pressure, acts upon sensory cells that provide information to the animal. One of the most common gravity-detection devices are statocysts, which are present in animals from coelenterates to mammals. Generally, the statocysts consist of a fluid-filled vessel lined with sensory hairs and containing a statolith or statoliths, which may be either grains of sand or secreted substances. Most vertebrates detect gravity through the vestibular labyrinth of the inner ear.

The ventral side of most animals is oriented toward the ground or other substrates, but often this is a very complex response that is controlled only in part by gravity, as in the dorsal light reactions discussed earlier in this chapter. Gravity is also one of the factors influencing the

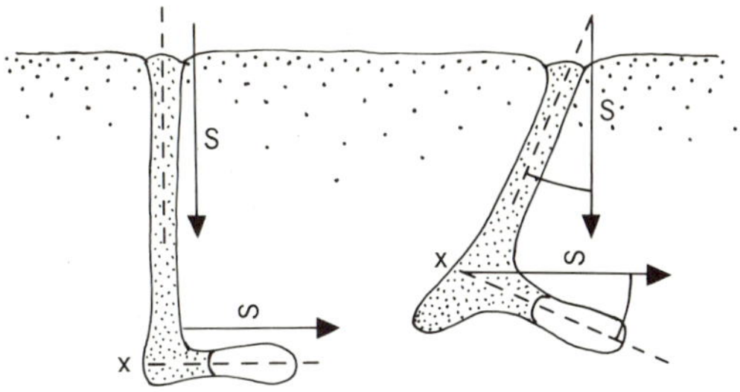

Figure 6-11. The shape of burrows made by two *Uca pugilator*. When the container was turned through 90° after the animal reached point x, then the crab dug at the same angle in relation to the direction of the force of gravity (S) as before. (After Schöne, 1961.)

vertical distribution of both marine and fresh-water animals. This distribution apparently is determined by the interaction between the pull of gravity and the support given by water. Many burrowing animals show positive geotaxis. The razor clam *Ensis ensis,* for example, will burrow downward as soon as it is placed on the surface of wet sand. Statocysts in the foot are responsible for initiating and directing this response (Fraenkel, 1927). The force of gravity also orients the direction of digging in many burrowing animals, as has been demonstrated (Dembowski, 1926) in the fiddler crab, *Uca pugilator* (Fig. 6-11).

Whether an animal is geopositive or geonegative may depend upon the stage of its life cycle and its physiological state. For example, the grasshopper *Poecilocerus hieroglyphicus* feeds on the green leaves of a plant that may reach a height of 10 ft. Because the leaves are from one to several feet above ground level, the animal must crawl upward against gravity in order to reach them. Thus, when it is hungry, the grasshopper is negatively geotactic. Once it has been fed, the grasshopper becomes positively geotactic and crawls down to the ground. Here, where it is protected against environmental fluctuation, the insect mates and deposits its eggs (Abushama, 1967).

PERCEPTION OF PRESSURE

In terrestrial organisms, pressure is not one of the major factors involved in ecological distribution. The maximum pressure at sea level is 14.7 psi or 760 mm Hg, and pressure decreases at heights above sea level. In an aquatic environment, however, pressure changes very rapidly as the depth of water increases; for example, pressure increases about 1 atm for every 10 meters, or 33 ft. Aquatic animals have become adapted

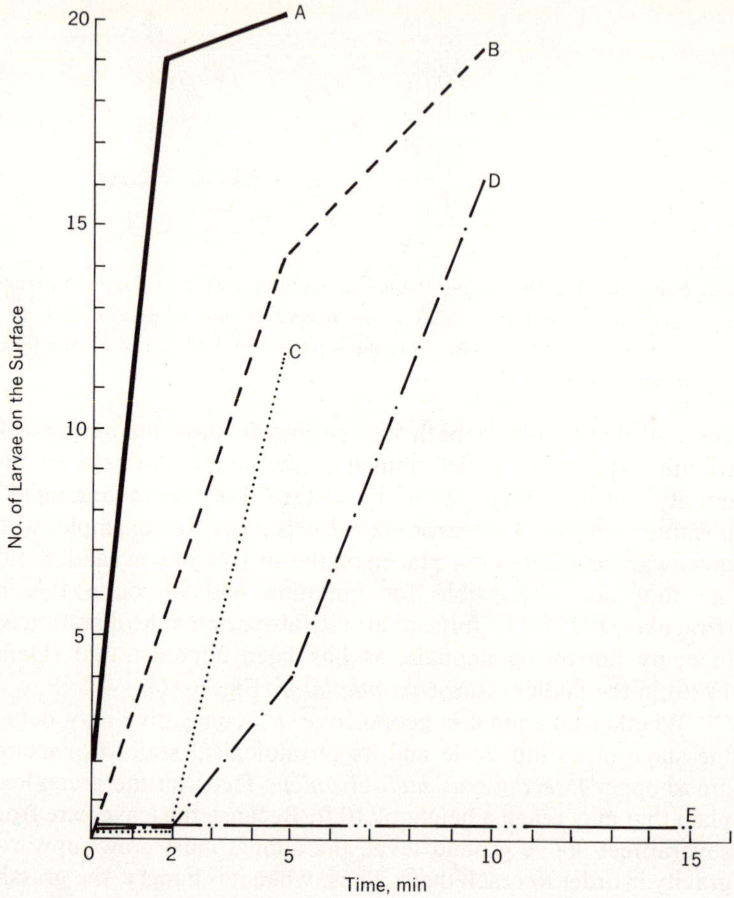

Figure 6-12. Response of *Mytilus edulis* to a pressure of 1.36 atmospheres above ambient. *A:* Early veliconcha experimentals; *C:* early veliconcha controls; *B:* eyed-veligers experimentals; *D:* eyed-veligers controls; *E:* pediveligers, experimentals and controls. (After Bayne, 1963.)

ecologically to live at certain depths, and life has been found to exist from the surface of the water down to 11,000 meters, where pressures are in excess of 1000 atm. Some organisms are restricted to a limited range of depth and others can move relatively freely from one level to another.

Many planktonic animals are sensitive to quite small changes in hydrostatic pressures, often to changes equivalent to less than a meter of water (Morgan *et al.*, 1964). Although the nature of the receptor organs in the invertebrates for perceiving pressure changes is not known, the reactions of the animals to such changes aid in environmental selection by regulating the depth and dispersal of the animals to the proper levels for settling and feeding. Generally, when the pressure is increased, the animals respond by swimming more strongly upward and/or toward the light. When pressure is reduced, the animals sink passively downward and/or away from light. The response of the animal to pressure commonly varies with the stage of the life cycle, as it does in the larva of the mussel *Mytilus edulis*, for example. The young larvae of *M. edulis* exhibit a positive response to increased pressure, which keeps the larvae in the surface layers of the water. This positive response to increased pressure decreases as the larvae ages and disappears in the pediveliger larvae that are approaching metamorphosis and settlement (Fig. 6-12). Thus, in the early larval stages, the positive response to increased pressure encourages dispersal, whereas the decrease to the positive pressure in the later stages allows the older larvae to come in contact with the substrate and settle (Bayne, 1963).

In fish, the swim bladder acts as a hydrostatic organ by compensating for density changes when the fish moves from one level to another. Thus the presence or absence of a swim bladder is one of the most important factors affecting the distribution of fish. Most fish without a swim bladder have difficulty moving in water when off the bottom and are bottom-dwelling species. The nature of the pressure receptors in fish is unknown, but it has been speculated that receptors are present in the wall of the swim bladder.

THERMAL RECEPTION

Temperature effects on animals have been discussed in detail in Chapter 3. Generally, terrestrial, fresh-water, and littoral animals are more sensitive to temperature changes than are marine animals, probably because these animals are subject to a greater range of environmental temperatures. Thermal information is also used to a greater extent in their behavior patterns.

Homeotherms have thermal receptors in the skin and a temperature-regulating center in the hypothalamus, which controls reflex responses to changes in blood temperature. The temperature-sensing devices in poikilotherms are not so well defined in all instances. Although fish are very sensitive to temperature changes, no distinct temperature receptors have been identified. Some of the most sensitive thermoreceptors are present in crotalid snakes. These sense organs, called *facial pits*, are on the head below and in front of the eyes. They function as directional distance receptors, which enable the snake to detect and strike at warm prey even when the eyes and nose are covered and the tongue cut out (Noble and Schmidt, 1937). In vertebrates, little is known of either the sensory structures or mechanisms involved in thermal sensitivity. In some insects, a hair sensillum has been identified as a thermal receptor. This consists of a simple hair with a sensory cell at the base and located on legs and antennae (Murray, 1962).

SUMMARY

The relative importance of each of the sensory modalities to an animal depends in large part upon the habitat of that particular species. Vision is common to nearly all animals and is very important in directional orientation in most, but animals that live in darkened habitats or that have limited vision may rely instead upon audition or a chemical sense; burrowing animals will orient through gravity perception, and animals living at great depths use pressure aids for orientation. Audition is the most important modality in animal communication; it is rapid and can be carried on simultaneously with other behavioral activities. It is not effective, however, if background noise levels are too great. On the other hand, vision can be used in close-range communication regardless of background noise, and chemical means allow communication over both long and short distances, with the additional advantage of having different signals for various functions. Animals that live in clear and open areas may use vision as the primary means for locating food, but in poorly lighted habitats, the audition, chemoreception, or thermal sense will be more useful. If food must be located over long distances, the animal is apt to rely primarily on chemoreception.

Thus, each modality can serve several functions, offering certain advantages and disadvantages for each one, and their use can be correlated directly with the needs of the animal living in its specific environment.

References

Abushama, F. T. 1967. Geotaxis in *Poecilocerus hieroglyphicus* (Acrididae: Pyrogomorphinae). *J. Zool., London,* 153:453–461.

Alexander, R. D. 1960. Sound communication in Orthoptera and Cicadidae. *In* W. E. Lanyon and W. N. Tavolga (eds.), *Animal Sounds and Communication,* pp. 38–92. A.I.B.S. Publ., Washington, D.C.

Alexander, R. D. 1962. Evolutionary change in cricket acoustical communication. *Evolution,* 16:443–467.

Arnold, D. C. 1957. The response of the limpet, *Patella vulgata* L., to waters of different salinities. *J. Mar. Biol. Assoc. U.K.,* 36:121–128.

Baerends, G. P., and J. M. Baerends-van Roon. 1950. An introduction to the study of the ethology of cichlid fishes. *Behaviour Suppl. 1.*

Bardach, J. E., H. E. Winn, and D. W. Menzel. 1959. The role of the senses in the feeding of the nocturnal reef predators *Gymnothorax moringa* and *G. vicinus. Copeia,* 2:133–139.

Bayne, B. L. 1963. Responses of *Mytilus edulis* larvae to increases in hydrostatic pressure. *Nature,* 198:406–407.

Bogert, C. M. 1960. The influence of sound on the behavior of amphibians and reptiles. *In* W. E. Lanyon and W. N. Tavolga (eds.), *Animal Sounds and Communication,* pp. 137–320. A.I.B.S. Publ. 7, Washington, D.C.

Bossert, W. H., and E. O. Wilson. 1963. The analysis of olfactory communication among animals. *J. Theoretical Biol.,* 5:443–469.

Brower, L. P. 1969. Ecological chemistry. *Sci. Am.,* 220:22–29.

Brower, L. P., J. van Z. Brower, and P. W. Westcott. 1960. Experimental studies of mimicry. 5. The reactions of toads (*Bufo terrestris*) to bumblebees (*Bombus americanorum*) and their robberfly mimics (*Mallophora bomboides*), with a discussion of aggressive mimicry. *Am. Naturalist, 94:343–355.*

Buck, J. B. 1937. Studies on the firefly. II. The signal system and color vision in *Photinus pyralis. Physiol. Zool.,* 10:412–419.

Buck, J., and E. Buck. 1968. Mechanism of rhythmic synchronous flashing of fireflies. *Science,* 159:1319–1327.

Cott, H. B. 1957. *Adaptive Coloration in Animals.* Methuen, London.

Crisp, D. J. 1967. Chemoreception in cirripedes. *Biol. Bull.,* 133:128–140.

Dembowski, J. B. 1926. Notes on the behavior of the fiddler crab. *Biol. Bull.,* 50:179–201.

Farley, R. D., and J. F. Case. 1968. Perception of external oxygen by the burrowing shrimp, *Callianassa californiensis* Dana and *C. affinis* Dana. *Biol. Bull.,* 134:261–265.

Fingerman, M., C. Oguro, M. Miyawaki, and R. G. McKinnell. 1964. The Neurosecretory system in the head of the blind cave crayfish *Cambarus setosus* Faxon. *Am. Midland Naturalist,* 71:415–421.

Fraenkel, G. 1927. Die Grabsbewegungen der Soleniden. *Z. vergl. Physiol.,* 6:167–220.

Fraenkel, G. S., and D. L. Gunn. 1940. *The Orientation of Animals.* Oxford University Press, London.

Frings, H. 1964. Sound production and reception. *In* W. N. Tavolga (ed.), *Marine Bio-Acoustics,* Vol. I, pp. 155–173. Pergamon, New York.

Frings, H., and M. Frings. 1967. Underwater sound fields and behavior of marine invertebrates. *In* W. N. Tavolga (ed.), *Marine Bio-Acoustics,* Vol. II, pp. 261–281. Pergamon, New York.

Frisch, K. von. 1941. Uber einen Schreckstoff der Fischhaut und seine biologische Bedeutung. *Z. vergl. Physiol.,* 46–145.

Griffin, D. R. 1953. Acoustic orientation in the oil bird, *Steatornis. Proc. Nat. Acad. Sci. U.S.*, 39:884–893.

Harris, J. E., and P. Mason. 1956. Vertical migration in eyeless *Daphnia. Proc. Roy. Soc. London, B,* 145:280–290.

Hasler, A. D. 1967. Underwater guideposts for migrating fishes. *In* R. M. Storm (ed.), *Animal Orientation and Navigation,* pp. 1–15. Oregon State University Press, Corvallis, Ore.

Herter, K. 1929. Vergleichende Bewegungsphysiologische Studien an Deutschen Egeln. *Z. vergl. Physiol.,* 9:145–177.

Jacobson, M., and M. Beroza. 1963. Chemical insect attractants. *Science,* 140:1366–1373.

Jones, F. R. H. 1955. Photo-kinesis in the ammocoete larva of the brook lamprey *J. Exptl. Biol.,* 32:492–503.

Karlson, P., and A. Butenandt. 1959. Pheromones (ectohormones) in insects. *Ann. Rev. Entomol.,* 4:39–58.

Keenleyside, M. H. A. 1955. Some aspects of the schooling behaviour of fish. *Behaviour,* 8:183–248.

Kellogg, W. N. 1961. *Porpoises and Sonar.* University of Chicago Press, Chicago.

Knipling, E. F., and J. V. McGuire, Jr. 1966. Population models to test theoretical effect of sex attractants used for insect control. *Agr. Inf. Bull.,* 308:1–20.

Kulzer, E. 1954. Untersuchungen über die Schreckreaktion der Erdkrötenkaulquappen (*Bufo bufo* L). *Z. vergl. Physiol.,* 36:443–463.

Larimer, J. L. 1966. A functional caudal photoreceptor in blind cavernicolous crayfish. *Nature,* 210:204–205.

Limbaugh, C. 1963. Field notes on sharks. *In* P. W. Gilbert (ed.), *Sharks and Survival,* pp. 63–94. Heath, Boston.

Marler, P. 1967. Animal communication signals. *Science,* 157:769–773.

Marler, P., and W. J. Hamilton, III. 1966. *Mechanisms of Animal Behavior.* Wiley, New York.

Maschwitz, U. W. 1964. Gefahrenalarmstoffe und Gefahrenalarmierung bei sozialen Hymenopteren. *Z. vergl. Physiol.,* 47:596–655.

Millott, N. 1960. The photosensitivity of sea urchins. *Sym. Comp. Biol.,* 1:279–293.

Millott, N., and M. Yoshida. 1960. The shadow reaction of *Diadema antillarum* Philippi. 2. Inhibition by light. *J. Exptl. Biol.,* 37:376–389.

Morgan, E., A. Nelson-Smith, and E. W. Knight-Jones. 1964. Responses of *Nymphon gracile* (Pycnogonida) to pressure cycles of tidal frequency. *J. Exptl. Biol.,* 41:825–836.

Murray, R. W. 1962. Temperature receptors. *In* O. Lowenstein (ed.), *Advances in Comparative Physiology and Biochemistry,* Vol. I, pp. 117–175. Academic Press, New York.

Naylor, A. F. 1959. An experimental analysis of dispersal in the flour beetle, *Tribolium confusum. Ecology,* 40:453–465.

Noble, G. K., and A. Schmidt. 1937. The structure and function of the facial and labial pits of snakes. *Proc. Am. Phil. Soc.,* 77:263–288.

Ohsawa, W., and H. Tsukudo. 1955. Extruding response of the periwinkle, *Nodilittorina granularis* (Gray). *J. Inst. Polytech., Osaka City University, Ser. D,* 6:71–96.

Payne, R. S. 1962. How the barn owl locates prey by hearing. *Living Bird,* 1:151–159.

Pfeiffer, W. 1963. The fright reaction in North American fish. *Can. J. Zool.,* 41:69–77.

Salmon, M. 1965. Waving display and sound production in the courtship behavior of *Uca pugilator,* with comparisons to *U. minax* and *U. pugnax. Zoologica,* 43:89–92.

Salmon, M. 1967. Coastal distribution, display and sound production by Florida fiddler crabs (genus *Uca*). *Anim. Behav.,* 15:449–459.

Schneider, D. 1962. Electrophysiological investigation on the olfactory specificity of sexual attracting substances in different species of moths. *J. Insect Physiol.,* 8:15–30.

Schone, H. 1961. Complex behavior. *In* T. H. Waterman (ed.), *The Physiology of Crustacea,* Vol. II, pp. 465–520. Academic Press, New York.

Shorey, H. H., L. K. Gaston, and C. A. Saario. 1967. Sex pheromones of noctuid moths. XIV. Feasibility of behavioral control by disrupting pheromone communication in cabbage loopers. *J. Econ. Entomol.,* 60:1541–1545.

Steven, D. M. 1963. The dermal light sense. *Biol. Rev.,* 38:204–240.

Stout, J. F. 1963. The significance of sound production during the reproductive behaviour of *Notropis analostanus* (family Cyprinidae). *Anim. Behav.,* 11:83–92.

Suffert, F. 1932. Phänomene visueller Anpassung, I, II, III. *Z. Morph. Ökol. Tiere,* 26:147–316.

Weber, H. 1924. Ein Umdreh- und Fluchtreflex bei *Nassa mutabilis. Zool. Anz.,* 60:261–269.

Wilson, E. O. 1962. Chemical communication among workers of the fire ant, *Solenopsis saevissima* (Fr. Smith). 1–3. *Anim. Behav.,* 10:134–164.

Wilson, E. O. 1965. Chemical communication in the social insects. *Science,* 149:1064–1071.

Wilson, E. O., and W. H. Bossert. 1963. Chemical communication among animals. *Recent Prog. Hormone Res.,* 19:673–716.

Winn, H. E. 1964. The biological significance of fish sounds. In *Marine Bio-Acoustics, Proc. Symp., Bimini, Bahamas, April 1963.* Pergamon, New York.

Wisby, W. J., and D. R. Nelson. 1964. Airplane observations of acoustic orientation in sharks. (*Abstr.*) *Am. Fish. Soc. Conf.,* Session on Fish Behavior and Sensory Physiology.

7

Energy Relationships

Our biosphere receives its basic input of energy from the sun, and the plants use this energy to produce organic material (food) photosynthetically. In turn, the food eaten by animals provides the source of energy needed to sustain the organism's metabolic machinery. Food provides energy for the maintenance of an organism at its existing level of performance and also for the synthetic activities associated with growth. Maintenance energy is required for those "treadmill" activities of organisms necessary merely to maintain a *status quo,* that is, for escaping from predators, for capturing food and digesting food, for glandular activity, and for maintenance of an osmotic gradient. Additional energy is required to grow and to synthesize the new and complex compounds from ingested substances that have been reduced to simpler "building blocks" during digestion.

Since energy is such a fundamental requirement, it is not surprising that morphophysiological mechanisms have evolved to ensure that organisms can reasonably expect to extract the necessary foodstuffs from their environment to sustain life. Not only are there well-defined correlations between habitat and the availability of food, but also between the type of food and the feeding mechanism, structure and shape of the alimentary canal, and nature of the digestive process, including types of enzymes.

The amount of time devoted to food gathering by various species is variable, but it is related to habitat and mode of feeding. Some animals, such as the filter-feeding invertebrates, feed almost continuously; at the other extreme, some animals do not feed at all during certain stages of the life cycle, as in the cercarial stage of trematodes. The feeding behavior of most animals, however, is somewhere between these two feeding patterns.

Processes of obtaining food and converting it to energy can be divided into a number of definable steps: perception of food in the environment; food gathering ingestive mechanisms; digestion and egestion of food; absorption; distribution to various metabolic sites; and utilization of energy in various processes, including respiration, reproduction, excretion or secretion. The perception of food was discussed in Chapter 6; the other processes will be included here.

Two aspects of food requirement that should be considered are: (1) the needs of the individual organism in terms of its particular ecological-physiological demands; and (2) food requirements of communities of organisms in terms of interspecific interaction and the flow of energy from one trophic level to another. The first aspect will be stressed in this chapter; the second is discussed in Chapter 11 (Ecosystems).

BIOENERGETICS

In recent years serious attempts have been made to estimate quantitatively the energy requirements of an organism in a demanding environment. The *energy budget* is one method of expressing the results of these energy balance studies (Fig. 7-1). Although no standardized expression of an energy budget has been formulated, the energy input into a living organism is generally computed and balanced against the total utilization and distribution of this energy, which may occur as follows: food (energy) ingested by an organism may be digested and assimilated by its body or egested as a waste. Assimilated energy is divided into various fractions:

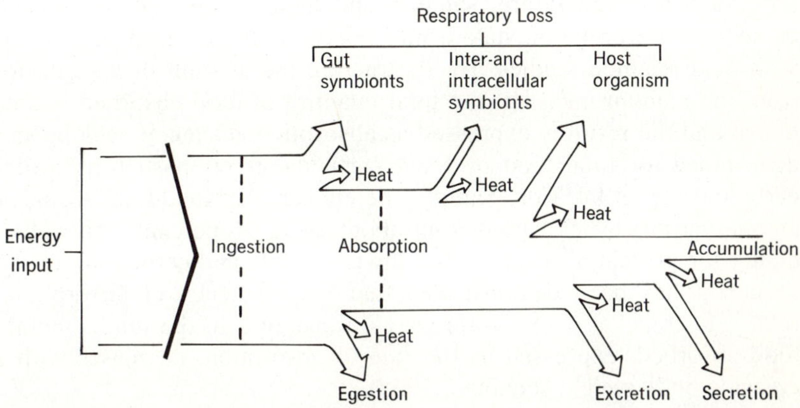

Figure 7-1. Model of energy flow through a single animal. The relative width of each arrow indicates the amount of energy loss. (After Darnell, 1968.)

some is used in energy-requiring functions necessary for the maintenance of an organism, and this fraction can be collectively measured and expressed under the general term *respiration.* Other energy is used and lost in such processes as secretion and excretion, or in such regular processes as molting or shedding of body structures. Additional energy is incorporated into reproductive products and is lost from the organism. The growth and increase in size of an organism represents storage of energy. Energy budgets may be calculated for an individual organism over a given period of time, for a population, or for an ecosystem. If the energy budget of a population or an ecosystem is to be estimated under natural conditions, obvious complicating factors must be considered, such as size distribution of a population or predation (Slobodkin, 1962; Phillipson, 1966).

The dynamics of energy flow in a species can be represented by measuring several physiological functions. The input of energy into the biotic system is determined by measuring food intake. Generally, food is introduced into the digestive tract by means of food-gathering mechanisms, but aquatic organisms can remove some dissolved organic compounds directly from their aqueous environment through their body covering (Stephens, 1968). At the other extreme, some endoparasites do not have a digestive tract and rely entirely on food taken in by their body-wall absorptive processes.

Feeding rates expressed as quantity of food per unit time have been determined for several animals through various techniques: analysis of stomach contents of field animals; experimental diets, including qualitative (food preference) studies and quantitative nutritional investigations; and experimental manipulation during feeding of environmental parameters, such as temperature, salinity, and light. To permit the broadest ecological interpretation, data from laboratory studies must be compared with field-oriented studies. By subtracting the amount of egested food from the amount ingested, the total quantity of food absorbed is determined and the result is expressed as absorption efficiency, which can be determined for total food or for a particular fraction such as protein, carbohydrate, or fat. For example, the efficiency of total food absorption among various fishes varied from about 86 to 96 percent; this value is higher for protein absorption, but lower for fats and carbohydrates. Although it has been demonstrated that the efficiency of absorption is relatively constant over a wide weight range of fish, the total amount of food absorbed, expressed as the rate of absorption, decreases with increasing body weight (Pandian, 1967).

By relating the amount of absorbed food to the amount converted into the growth of an animal, the efficiency of conversion and the conversion

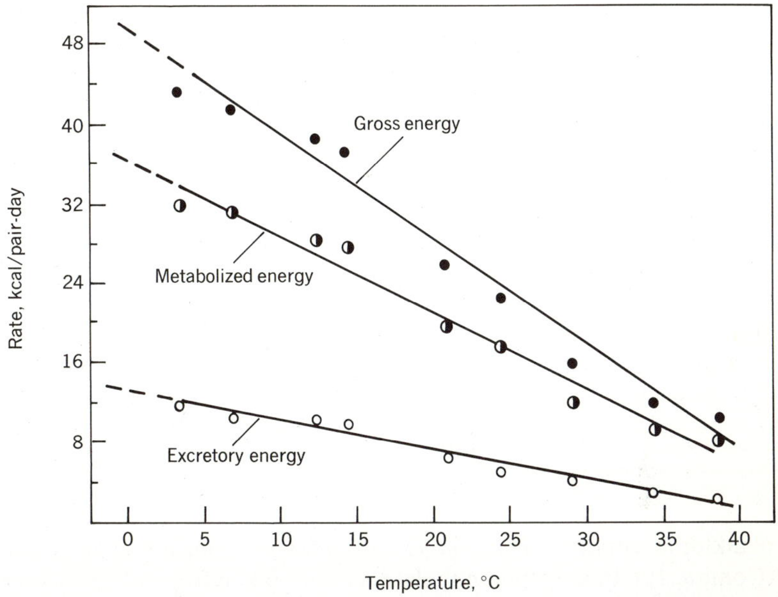

Figure 7-2. Energy budgets of Zebra finches in relation to air temperature. (From El-Wailly, 1966.)

rate can be determined. Comparison of these values for fish indicate the following trends: (1) the efficiency of food and nitrogen conversion decreases with increasing body weight; and (2) interspecific differences may be correlated with growth pattern, that is, fish that rapidly grow to a large size have a faster rate of food conversion than species which grow more slowly.

The energy budget of an animal can be modified by environmental parameters. For example, the gross energy intake, metabolized energy, and excretory energy of zebra finches are inverse linear functions of air temperature: as temperatures decrease, energy intake and outgo increase (Fig. 7-2). In these finches, which are present in dry areas throughout Australia, the food-utilization efficiency was maximum at 24.4°C. Another interesting point is the significantly increased efficiency at nearly all temperatures during nest-building, egg-laying, and incubation, when compared to the food-utilization efficiency of males and females that have been separated (El-Wailly, 1966).

Because complex factors influence energy flow through an organism, energy balance sheets have been prepared for few species, and not enough data are available for comparisons of species from diverse habitats. The results of a study on the annual energy budget for females of the

Table 7-1. The Energy Budget of *Diaptomus siciloides* by Generations and for a Year.

	Generation					
	I	*II*	*III*	*IV*	*V*	*Year*
Energy, cal						
Input	0.11	0.38	0.69	1.08	0.49	2.75
Respiration	0.12	0.29	0.31	0.58	0.19	1.46
Reproduction	0.03	0.025	0.042	0.077	0.008	0.18
Excretion and Egestion	—	0.065	0.333	0.42	0.29	1.11
Input, %						
Respiration	103	76	45.5	54	38.4	53
Reproduction	26	6.4	6.1	7.2	1.7	7
Excretion and Egestion	—	17.1	48	38.8	59.9	40

SOURCE: From Comita, 1964.

planktonic copepod *Diaptomus siciloides* are presented in Table 7-1 (Comita, 1964). Copepod populations were taken from a fresh-water lake in Minnesota, and the energy budget was determined for the ice-free period of the year; five generations were produced during this season. The amount of energy consumed throughout the year fluctuated considerably among copepods fed known quantities of algae, but feeding rates varied directly with temperature (Fig. 7-3). Seasonal fluctuations in respiration energy were also evident, but the energy involved in egg production fluctuated only slightly (Fig. 7-3). When energy input was high, egestive and excretory processes accounted for a larger percentage of the energy budget than when it was low. Furthermore, energy input differed between generations and was highest in the fourth generation, which was present during the more favorable summer conditions.

FEEDING PATTERNS

Some species of animals are very restricted in their diets; for example, the red tree mouse limits its diet to leaves of certain coniferous trees (Hamilton, 1962), and the koala feeds only on eucalyptus leaves. Food preferences in such animals are important to speciation in acting as a geographical barrier, for if the distribution of the food is altered, the distribution of the species feeding upon it is similarly altered—as illustrated by the distribution of the moth *Thera juniperata,* which lives on the juniper during its larval stage. At one time juniper was present throughout England, but now it is absent from the midland section and occurs only in

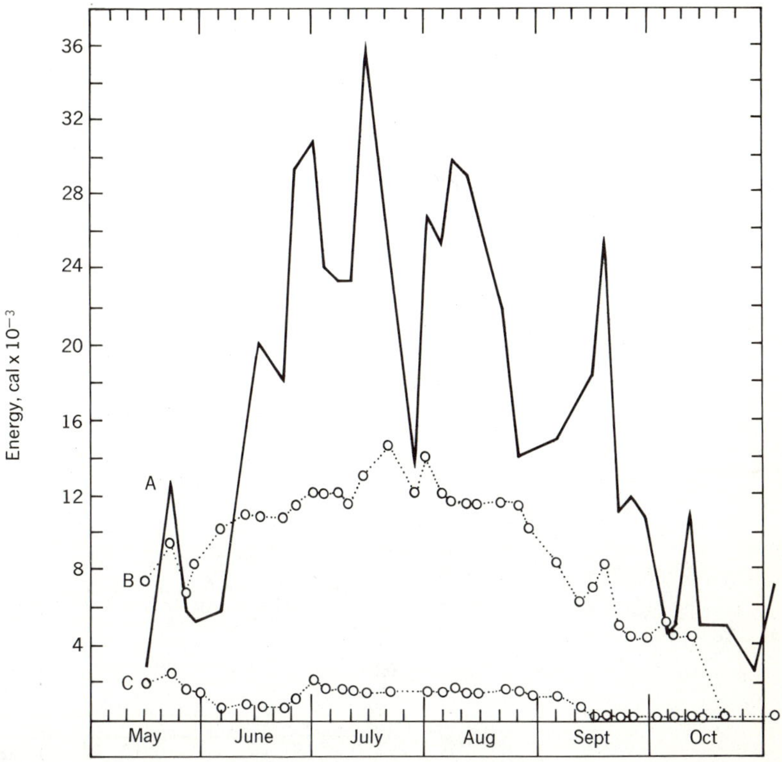

Figure 7-3. Seasonal fluctuations in energy budgets for a female *Diaptomus siciloides*. *A* represents energy input, *B* the energy of respiration, and C the energy of egg production. (From Comita, 1964.)

thc north and south. The moths disappeared from the midland section along with the junipers, and the distribution of moth populations followed that of the junipers. In time, the moth population in the northern sector became geographically isolated from that in the south, and now the two populations are considered to be distinct species (Huxley, 1942).

Most animals, however, have varied diets, even when food is superabundant, and although one food source may be limited or destroyed, the animals are able to utilize another. Nonetheless, animals do show definite food preferences that apparently are innate, as illustrated in laboratory studies of the water snake *Natrix sipedon.*A variety of foods was offered to adults captured in the wild and to their newborn offspring reared in captivity away from their parents. Both adults and newborn snakes would accept fish and frogs or tadpoles, but neither group would accept worms, mice, lizards, or miscellaneous small snakes (Dix, 1968).

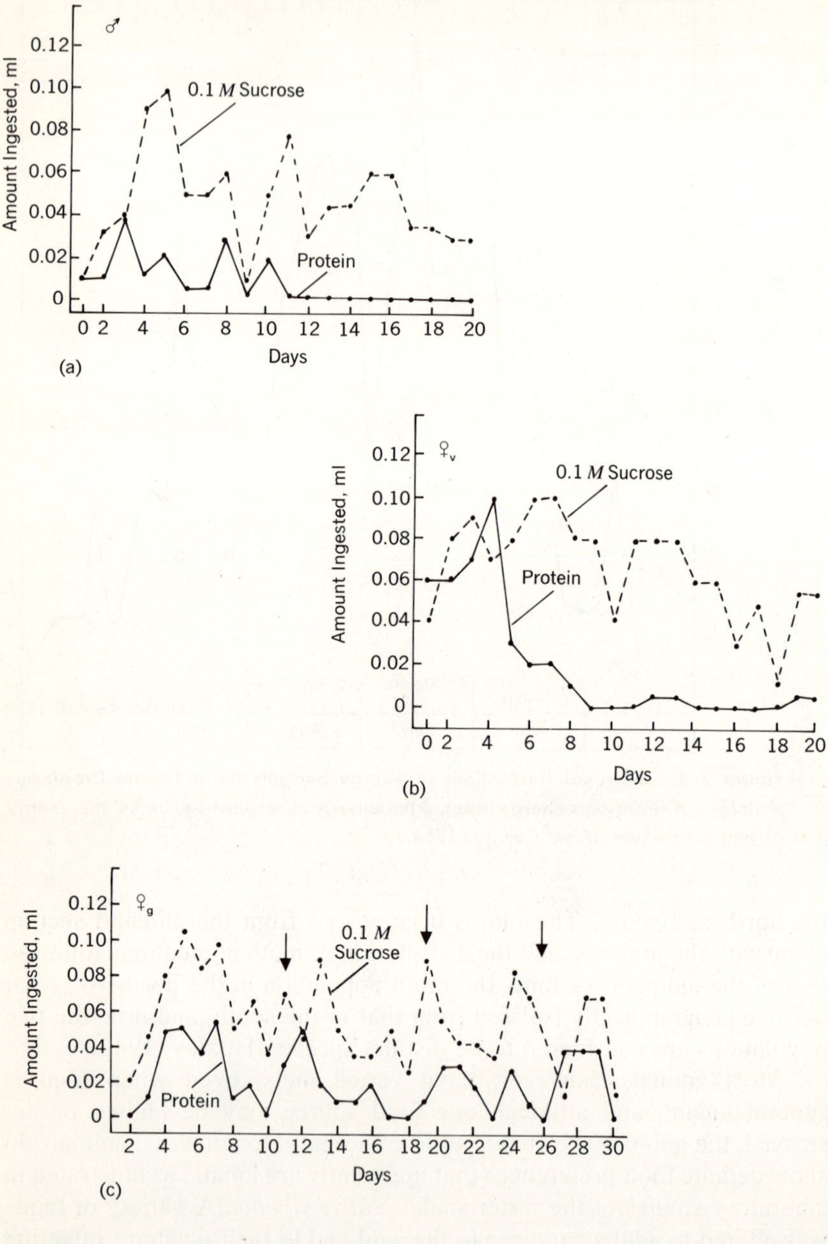

Figure 7-4. Daily intake of sucrose and brain-heart extract in blowflies in a two-choice situation. (*a*) male blowfly; (*b*) virgin female blowflies; (*c*) gravid female blowfly. (From Dethier, 1961.)

Food preferences and needs can change, furthermore, with different stages in the life cycle. For example, parasites that utilize several different hosts may have entirely different nutritional requirements during each stage of development. Immediately after emergence, blowflies, if given a choice, will ingest only a small amount of carbohydrate, but a large amount of protein. If they are denied protein temporarily, their preference for protein over carbohydrate in a choice situation is heightened. In males and nongravid females, little protein is taken after the first few days following emergence. Mated females, however, increase their protein intake after each period of oviposition (Fig. 7-4a, b, c) (Dethier, 1961).

Age and sex also influence the amount of food consumed, as shown by studies on the praying mantis *Hierodula crassa* (Holling, 1966). Food consumption of both male and female adults increased initially, but food consumption by males began to decline after the fourth day and finally stabilized at a very low level after the tenth day. In contrast, female food consumption increased up to the eleventh day and remained at that high level for at least another 23 days (Fig. 7-5). The higher food consumption of the female was presumably due to the demands of egg production.

The release of feeding behavior patterns depends on the internal state of the animal, so that fish which feed intensively one day will feed lightly

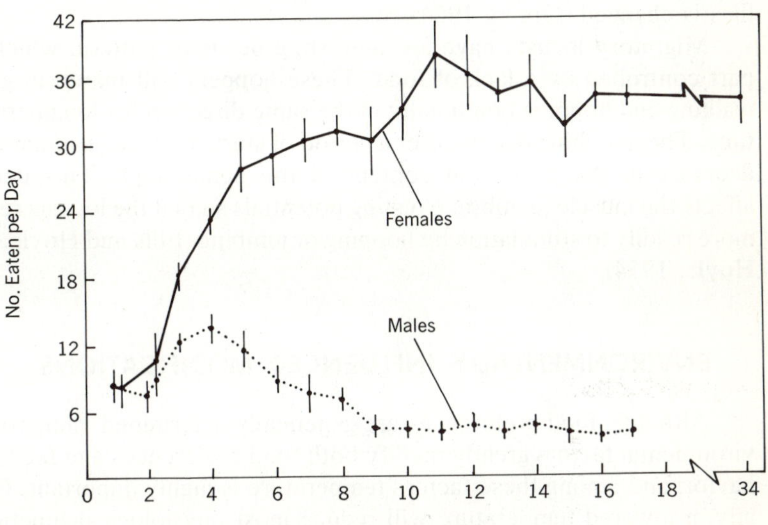

Figure 7-5. Effect of age on the number of prey eaten per day by male and female mantids, *Hierodula crassa*. (From Holling, 1966.)

the next (Pandian, 1967), and starved fish will eat more than individuals that have not been deprived of food (Rozin and Mayer, 1964). Different thresholds of hunger also may influence the feeding response, although the response is not necessarily the same in all animals. The praying mantis *Hierodula crassa,* for example, will not strike at prey until its hunger reaches a certain fixed level, and it will not pursue the prey until its hunger rises to the next threshold (Holling, 1966). The stickleback fish, on the other hand, may pursue a prey, but may not be hungry enough to capture it when it comes within reach. At another threshold, the stickleback might not be hungry enough to eat it, and the prey is discarded (Tugendhat, 1960).

What are the internal factors governing the initiation of feeding behavior? In vertebrates, evidence indicates that some parts of the brain respond to changes in the circulating blood, whereupon feeding behavior is motivated. Changes in concentration of glucose levels, decrease in body temperatures, increased availability of water, and differences in a hormone concentration, have all been suggested as eliciting stimuli (Marler and Hamilton, 1966). In invertebrates, much of the work has been done on blowflies. With increasing time of food deprivation, the activity of the flies increases, and as soon as a sugar solution is ingested, activity is immediately affected. It is believed that a hormone from the corpus cardiacum complex influences activity. As soon as receptors in the foregut detect the presence of food, a hormone that limits activity is released by the corpus cardiacum complex, and the activity pattern of the flies is changed (Green, 1964a,b).

Migratory locusts have an interesting behavior pattern, which is in part controlled by a lack of food. These hoppers will march in groups, walking and hopping continually in the same direction for long periods of time. The marching occurs after a period of starvation and is initiated by a decrease in the potassium content of the hemolymph. This decrease affects the muscle membrane resting potentials so that the locusts respond more readily to stimulation by hopping or jumping (Ellis and Hoyle, 1954; Hoyle, 1954).

ENVIRONMENTALLY INFLUENCED MODIFICATIONS

Although food preferences are genetically determined, numerous environmental factors greatly modify both food preferences and feeding behavior, and among these factors temperature is highly important. Generally, a lowered temperature will reduce most physiological functions in poikilotherms, including feeding activity; these reductions in rate functions are accompanied by reduced energy requirements. As a result, animals like suspension feeders and fish will feed only within a certain temperature range. Most poikilotherms need an optimum temperature for feeding, but it may not be the same for geographically separated populations of a single species. The pumping rate of the clam *Mytilus californianus,* from a northern latitude (48°27′ N), was noted to be the same at 6.5°C as that of a more southern population (38°31′ N) at 12°C or of a population from a latitude of 34° N at about 14°C (Rao, 1953) (Fig. 7-6). Thus, despite natural thermal differences experienced by widely separated populations, the rate of pumping water was about equal at their respective environmental temperatures. A similar adaptational response was reported for populations of this clam that occupied different regions of the tidal zone at one latitude. Animals from the cooler subtidal zone pumped water faster than clams from the warmer region of the intertidal zone, when the pumping rate was measured at a common temperature (Segal *et al.,* 1953).

Mammals generally consume food at a rate that varies inversely with the ambient temperature, as illustrated in studies (Sealander, 1952) on field mice, *Peromyscus* (Fig. 7-7). When these animals were acclimated to one temperature and then suddenly exposed to either a higher or lower temperature, they underwent a period of adjustment in rate of food consumption: a lowered air temperature initially increased the rate of food intake, followed by a steady level of consumption after about ten days; with an increase in temperature, the rate of food consumption decreased at first, but was stabilized after an average of 15 days. Temperature also influenced the type of food selected by rats. Cold-acclimated rats chose diets composed of about 25 percent protein, 25 percent fats, and 30 percent carbohydrates, but at high temperatures, glucose consumption increased to about 70 percent of the total intake and lipid uptake decreased correspondingly (Dugal *et al.,* 1945).

Many parameters of the environmental complex interact with varying degrees of significance at different periods in the lifetime of an organism, as observed in the developmental stages (cysticercoids) of the cestode *Hymenolepis diminuta,* parasitizing the beetle *Tribolium confusum.* If infected beetles are exposed to 39.5°C after the parasites have undergone 3–5 days of development, few morphologically normal cysticercoids result, and most are not infective to rats. If, however, the beetles are maintained on a diet of pure sugar rather than whole wheat flour, the harmful effects of high-temperature stress on this tapeworm are markedly reduced (Voge, 1959).

The digestive system also may show an adaptive response to temperature; for example, the proteolytic activity of the digestive secretion of the snail *Helix pomatia* partly depends on the previous thermal history of

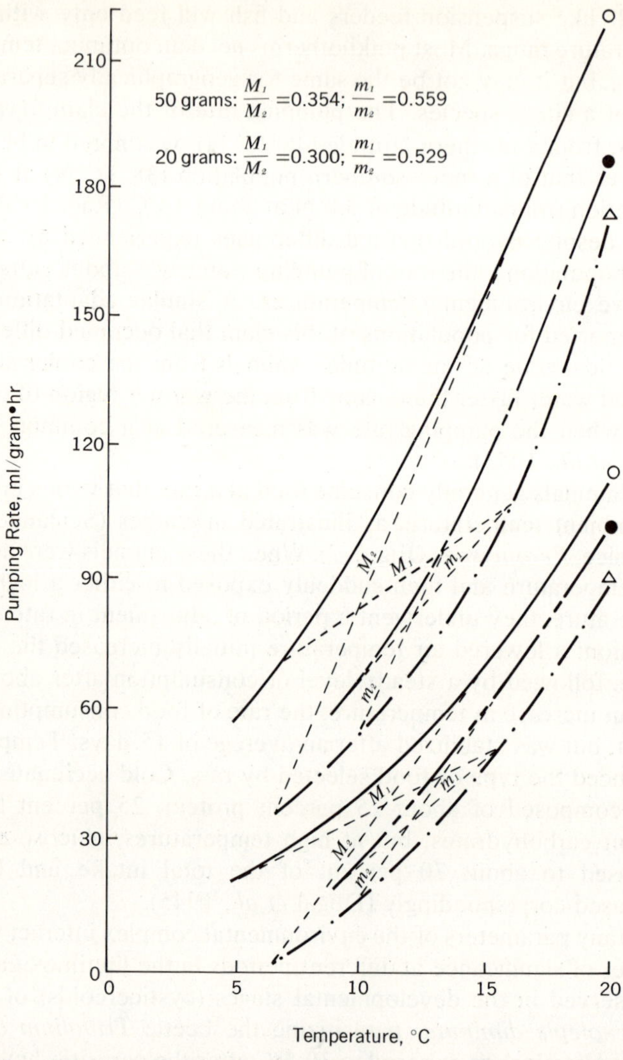

Figure 7-6. Rate of water propulsion in latitudinally separated populations of *Mytilus californianus.* △: Los Angeles; ● Fort Ross; ○: Friday Harbor. Variation in rate in either direction not greater than ½N, where n is ml/gram/hr. (From Rao, 1953.)

the animal. Snails maintained at 7°C for eight days had more proteolytic activity than did animals maintained at 25°C for an equal period of time, when measured at a common temperature. An intermediate value was noted for animals maintained at an intermediate temperature of 16.5°C

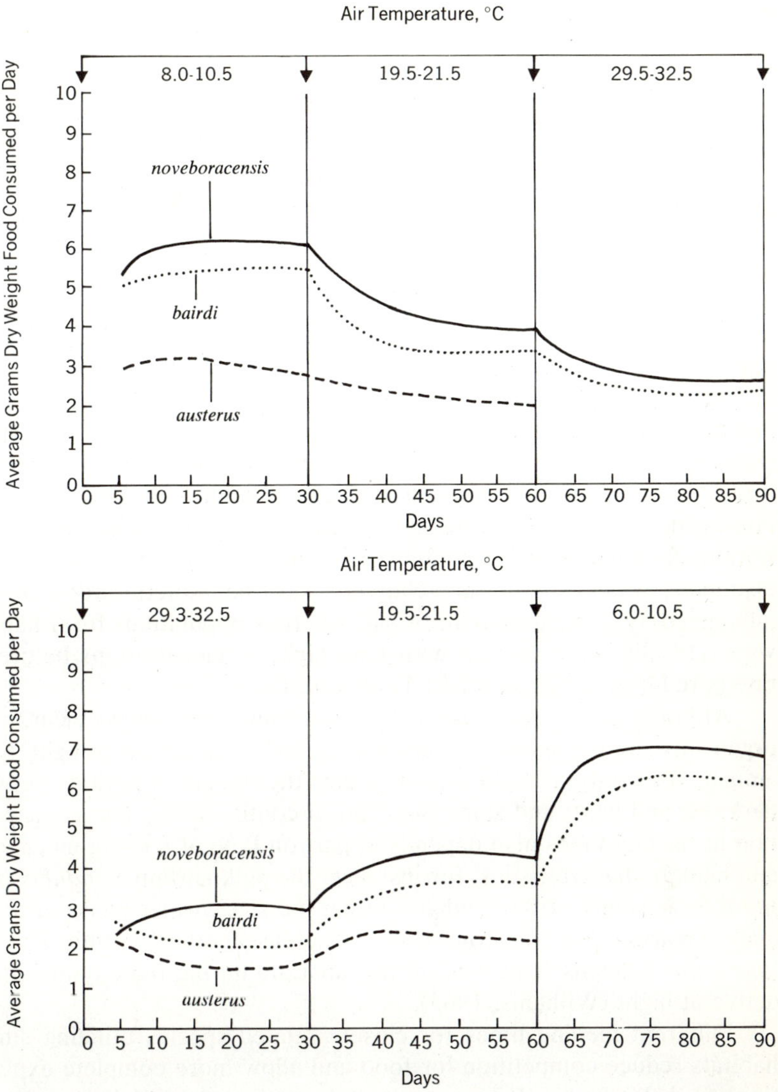

Figure 7-7. The relationship between temperature and food consumption in *Peromyscus*. In the upper chart temperatures were periodically raised, and in the lower chart, temperatures were periodically lowered. (From Sealander, 1952.)

(Mews, 1957). This compensatory response is similar to the metabolic response of some animals described in Chapter 4. It is noteworthy that lipase activity, which occurs at different seasons of the year, was not dependent on temperature, whereas no protease activity was detected in hibernating winter animals.

Changes in salinity and pH effect feeding behavior of marine animals. The commercial oyster *Crassostrea virginica,* for example, stops pumping water effectively when salinity drops to 13 ‰, but normal rates occur over the range of 25–39 ‰ (Hopkins, 1936). Food intake by the euryplastic fish *Cyprinodon macularius* is maximal at 35 ‰ and decreases directly with a decrease in salinity (Kinne, 1960). Furthermore, at pH 4.25 the pumping rate of oysters, *Crassostrea virginica,* is greatly reduced; a sudden drop in pH from 7.75 to 6.75–7.0 resulted in a temporary increase in the pumping rate followed by a rate lower than the original value (Loosanoff and Tommers, 1947).

In general, the rate of water transport in suspension feeders is widely independent of concentration and quality of the suspended particles (Jørgensen, 1966). At very high concentrations, however, the reduced rates reported for some animals appear to result from clogging of the filtering devices. Silt depresses the feeding rate in oysters, but the concentration of silt necessary to inhibit the feeding varies with populations; for example, oysters living in relatively silt-free waters are adversely influenced by a small amount of silt, whereas populations from habitats with typically more silt can withstand higher concentrations before the rate is reduced (Loosanoff and Tommers, 1948).

Although the cause-and-effect relationship is not always clear, many organisms show a periodicity of feeding that correlates with light. Some animals feed only at night; others prefer the transition periods between darkness and light; and some feed almost continuously. The great variation in feeding times also depends in part on type of feeding mechanism and habitat characteristics; for instance, the pink shrimp *Pandalus montagui* feed mainly in the daylight during the slack-water phase of a tidal cycle (Warren and Sheldon, 1967), whereas another shrimp, *Penaeus duorarum,* remains burrowed in the substrate during the daytime and is active at night (Williams, 1965).

Different feeding times for closely related species utilizing similar habitats reduce competition for food and allow more complete exploitation of a habitat type. For example, five species of fish living together in one lake have different feeding peaks during a 24-hr period; one species is diurnal, one is nocturnal and diurnal, and the remaining three are predominantly diurnal. In one species, the pumpkinseed *Lepomis gibbosus,* the type of food consumed shifts with the time of day, and this shift corresponds in general with the known activity cycle of the prey species (Keast and Welsh, 1968). Hence, there are at least two significant consequences of differences in feeding periodicity: (1) reduced interspecific interaction, and (2) increased utilization or harvesting in one habitat of a potential energy source by a number of predator species, which may occur rhyth-

Table 7-2a. Selective Feeding by *Rhincalanus nasutus* Offered a Mixture of Food Organisms.

Mixture	*Food Organisms* *Artemia salina* nauplius	*Coscinodiscus wailesii*	*Ditylum brightwellii*	*Thalassiosira fluviatilis*	*Cyclotella nana*
Approx initial concn	0.13/ml	0.95/ml	120/ml	1080/ml	7090/ml
Approx vol, μ^3/organism	2.9×10^7	1.6×10^7	5.0×10^4	4.3×10^3	313
Approx carbon, $\mu\mu$g C/organism	5×10^5	1.5×10^5	1.9×10^3	3.0×10^2	41
Contrib. to total C available in mixture	6%	14%	22%	30%	28%

Table 7-2b. Food Distribution.

R. nasutus Stage	*Total* μ*g C Ingested/ 18 hr*	μ*g C Ingested/ 18 hr-*μ*g C of Copepod*	*Contribution of Each Food Organism to Total C Ingested by Each Stage (%)*				
Nauplius II and III	0.2	0.6	—	0	100	0	0
Nauplius IV	1.9	1.7	—	0	46	54	0
Nauplius V and VI	1.3	0.7	—	62	37	0	0
Copepodite I	2.2	1.2	0	14	49	37	0
Copepodite II	4.4	1.2	0	13	51	32	4
Copepodite III	5.7	0.4	0	0	28	40	31
Copepodite IV	8.2	0.2	0	95	3	2	0
Copepodite V	21.8	0.4	4	54	23	19	0
Adult male	4.7	—	25	75	0	0	0
Adult female	16.5	0.1	37	45	18	0	0

SOURCE: Mullin and Brooks, 1967.

mically. Similar ecological relationships can be noted within a species where the feeding requirements may change with different stages of the life history. For example, the food of laboratory-reared planktonic copepods changed during development from the naupliar to the adult stage (Table 7-2).

Seasonal differences in rate of food intake have been reported, but their cause and effect are not always fully understood. Variation in the intensity of some physical factor such as temperature or light might act directly on the animal and/or the source of food might consequently be influenced.

FOOD-GATHERING MECHANISMS

Various distinctive mechanisms that have evolved enable animals to capture and ingest food material. About 40 years ago, C. M. Yonge (1928) proposed a classification of feeding mechanisms of invertebrates—although it could include vertebrates as well—and with some minor alterations, it is still widely accepted (Table 7-3). The principal developments since then have been in descriptive studies of previously undescribed animals and experimental studies on quantitative and qualitative aspects

Table 7-3. Classification of Feeding Mechanisms.

1. Mechanisms for dealing with small particles
   - A. Pseudopodial
   - B. Tentacular
   - C. Mucoid
   - D. Muscular
   - E. Setous
   - F. Ciliary
2. Mechanisms for dealing with large particles or masses
   - A. For swallowing inactive food (e.g., bottom deposits)
   - B. For scraping and boring
   - C. For seizing prey
     - (1) For seizing and swallowing only
     - (2) For seizing and masticating before swallowing
     - (3) For seizing followed by external digestion before swallowing
3. Mechanisms for taking in fluid or soft tissues
   - A. For piercing and sucking
   - B. For sucking only
   - C. For absorption through surface of body

SOURCE: Yonge, 1928

of feeding. The types of feeding included in this classification are based on ecologic rather than systematic considerations and represent convergent, or at least parallel, lines of evolution. The methods of food gathering are unlimited, and different animals have independently evolved feeding mechanisms most appropriate to their particular ecological circumstances. The diversity of feeding problems and subsequent solutions are almost overwhelming. Therefore no attempt will be made to cite all the various types of mechanisms, but a few will be given to illustrate significant ecological correlations. The student is referred to the following references for further fact finding: Yonge, 1928, 1954; Jørgensen, 1966.

Small-Particle Feeders

Small particulate substances that can serve as food for organisms occur in bottom deposits or are suspended in water. Typically, mechanisms for removal of these microscopic particles are present only in aquatic or semiaquatic organisms that are either sessile or have limited locomotor ability and which have some type of appendage used to pick up particulate matter and transfer it to the mouth. Generally, these animals are quantitative feeders and do not discriminate beyond particle size so that some filtered material will have no energetic value for the organisms. Little or no mastication is necessary in this type of feeding, and salivary glands are absent. Organisms that feed on large masses of sand, mud, detritus, or earth, however, are included in another category.

The aquatic environment tends to be a highly dilute medium of organic material; therefore, organisms dependent on this source of food must be able either to filter a great deal of water or to concentrate molecules against a concentration gradient, thereby extracting organic molecules directly from a dilute medium without also taking in vast quantities of water. Various estimates of organic composition indicate great geographic and seasonal variation; these estimates have been summarized for sea water by Jørgensen (1966) and for fresh water by Hutchinson (1957). Reliable comparative values for many bodies of water are lacking because of the inadequacies of present-day sampling techniques. It can be assumed, however, that sufficient organic material generally is available to support the various animals depending on this source of food.

Probably more is known about suspension feeding in bivalve molluscs than in any other group. This method of feeding apparently arose secondarily within bivalves as a modification of the water transport and cleaning mechanism of the gill and mantle. Extremely elaborate mechanisms that have been described are based on the generalized scheme presented in Fig. 7-8 (see Jennings, 1965). Various morphological modifications of

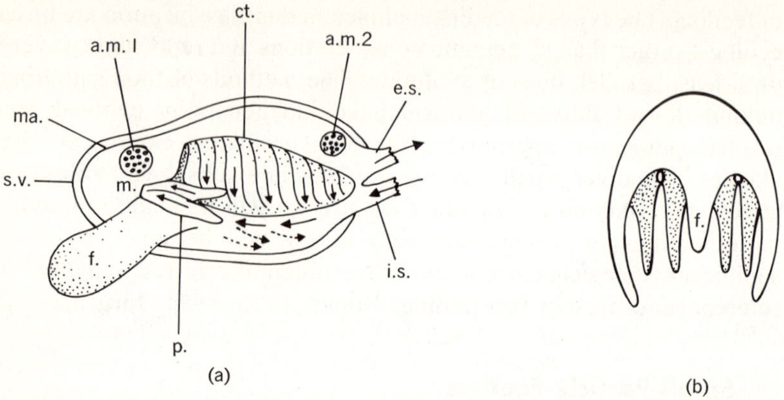

Figure 7-8. Generalized view of a lamillibranch mollusc. (*a*) Side view with left shell value and ctenidium removed. a.m.l.: anterior adductor muscle; a.m.2: posterior adductor muscle; ct.: ctenidium; e.s.: exhalant siphon; p.: palp; s.v.: right shell value. Water currents and paths followed by food particles are indicated by arrows; path of rejected particles are indicated by broken arrows. (*b*) Vertical section to show the position and arrangement of the ctenidia relative to the foot (f.) and shell valves. (From Jennings, 1965.)

the three principal organs of feeding (the mantle, gills, and mouth palps) allow different species to exploit various habitats. The mantle, which covers all internal organs of the animal, forms the outer edges of the mantle cavity. The edge of the mantle is modified to form an opening and exit (siphons) through which the ambient waters pass. Part of the mantle, particularly in the region of the incoming water, has a sensory function in that it "tests" the water for noxious substances so that water of undesirable quality is rejected. Epifaunal species tend to have unspecialized siphons, whereas deeper burrowing forms have long siphons that function to bring food and oxygen-enriched surface water to the animal and to remove "used" water. Because the opening of the incurrent siphon generally is large in active filter-feeding species and that of the excurrent siphon is relatively small, filtered water can be jetted away from the water source of the in-current siphon.

Tentacles formed from the mantle edge and located at the opening of the in-current siphon also differ in species from different habitats. In regions of abundant coarse material, the organisms have long tentacles that serve to strain out these undesirable particles. In contrast, species from relatively clean habitats having little suspended material typically have poorly developed tentacles (Ansell, 1961). The structure of the gills and palps also varies, and species living in turbid regions tend to have more strongly developed mechanisms for sorting suspended particles than do species inhabiting water with little suspended matter (Ansell, 1961).

Complex ciliary mechanisms for food sorting are characteristic of molluscs, which begin feeding when water passing over the gills is drawn through the interfilamentary slits by lateral cilia action. Both mucus and straining action of the laterofrontal cilia help to trap particulate matter in the gill, where the particles are thrust on the frontal ciliary tracts by the beating laterofrontal cilia and subsequently are carried to various sorting tracts, such as a groove along the gill axis leading to the oral groove and the mouth. In the presence of heavy and/or large particles, ventrally beating tracts may be stimulated to carry these undesirable particles to superficial ciliary tracts where they are rejected in the mantle cavity as pseudofeces. From the gills the food may pass to equally complex ciliary tracts on the palps, whose mucus formation and muscular action are important in the sorting process. The ciliary tracts on the palps may be of three general types: rejection currents, acceptance currents, and resorting currents. Although ciliary pathways vary greatly, in all molluscs the small particles are sorted out preferentially and carried to the labial palps and the mouth along protected pathways, whereas large particles are transported along superficial tracts and rejected (Jørgensen, 1966).

Ciliary-feeding mechanisms are common among lower chordates as well. In Ascidians, water enters through a siphon (mouth) into a ciliated branchial sac (pharynx) and passes through numerous openings (ostia) into an atrial sac; then it returns to the ambient environment through an atrial siphon. The branchial sac functions for feeding as well as for respiration. The endostyle, located along the ventral side of the branchial sac, produces a continuous layer of mucus on each side of the pharynx, which is carried dorsally by the cilia to a structure called the *dorsal lamina* where it is formed into strings of mucus that are passed to the gut. The mucus sheet is no longer formed, however, when these animals are disturbed, but the ostia filter out larger particles which become trapped in the small amount of mucus still produced. Spontaneous rhythmic squirting may be as important in creating feeding currents as ciliary currents (Hoyle, 1953). Other Tunicates have similar feeding mechanisms.

Semiaquatic organisms also have mechanisms for handling small particles; fiddler crabs, for example, capture food material in spoon-tipped "hairs" on the mouth parts. Interspecific differences are correlated with both habitat selection and food availability. The fiddler crab *Uca pugilator,* which lives in the intertidal zone in sandy areas, feeds on beaches where food is not abundant and the substratum is heterogeneous. Material from the beach is picked up and passed to the buccal cavity, where it is sorted by a flotation action, that is, a flushing of respiratory water from the gill cavities back and forth over the ingested material. The spoon-tipped hairs retain much of the small organic matter and discard the larger inorganic particles. On the other hand, another species of fiddler crab, *U. minax,* which lives in tidal marshes where abundant food material is present on the surface of a silty substratum, has fewer, less specialized flat-tipped hairs on the mouth parts than do species living in sandier areas (Miller, 1961).

Large-Particle or Mass Feeders

Different types of feeding mechanisms have evolved to utilize large particles or masses of food ranging from inactive food, such as bottom deposits, to active food, represented by highly mobile prey species. These mechanisms are employed not only by animals of many phyla, but also by species occupying the major ecologic domains.

Inactive Food

Inactive food generally is utilized by animals feeding on mud, sand, or bottom deposits, where the organic material usually consists of biological remains, such as body parts or feces, microorganisms, and some burrowing species. These animals tend to be sluggish and/or burrowing forms that may swallow everything indiscriminately by means of extroversible gullets, pushing tentacles, or similar "shoveling devices." Because of the loose nature of the ingested material, these omnivorous organisms have poorly developed organs of mastication.

Some echinoderms feed almost exclusively on bottom deposits, which they may glue together with mucus and shovel into the mouth by means of buccal tentacles; others use tentacles directly. The stomach contents of these species have the same character as the bottom deposits, thus indicating little selection of food. The tube feet of other echinoderms transport bottom material along the body surface to the mouth. Various tubiculous and burrowing worms also rely on bottom deposits or soil for much of their organic material, and frequently these species leave telltale worm casts after swallowing large amounts of inert material. Since the stomach contents of some gastropods and crustaceans are similar in composition to that of the bottom material, it appears that these animals also feed on bottom deposits.

Many estuarine vertebrates and invertebrates have been observed to feed on detritus sometime in their life history. Although detritus appears to be an important source of food, the observation that it is a universally abundant resourse and that many organisms ingest detritus does not afford proof of its nutritive value (Darnell, 1967).

Other animals have evolved mechanisms that enable them to bore into wood surfaces or hard surfaces of animals in order to feed on the flesh or to scrape off encrusting animals or plants. Organisms exhibiting these mechanisms tend to be slow-moving and possess powerful buccal apparatus such as radulae, jaws, teeth, modified shell edges, or mandibles. Frequently they also have organs for holding food masses, which may be tube feet, a muscular foot, or other appendage. Typically, they cannot swallow food until it is greatly reduced in size and therefore their feeding rate is slow but continuous. These animals generally are selective feeders that restrict themselves to certain species or types of food, such as the wood fibers ingested by *Limnoria* and *Teredo,* which are wood borers (Ray, 1959). Many insects, echinoderms, and molluscs have jaws or radulae that enable them to eat encrusting animal or plant species (Yonge, 1928). The radula of these animals is a movable, elongated structure containing numerous teeth, and it is associated with the mouth region of gastropods. When active, the radula is extended and worked back and forth on the surface of the food, much like the rasping action of a file, and particles ripped off are swallowed. The worn-out anterior part of the radula is gradually replaced at its base posteriorly.

The ability to penetrate the calcareous exoskeleton of another organism is a widespread biological phenomenon, and is common among fungi, algae, bacteria, and at least seven animal phyla. The methods of penetration, which may be both chemical and/or physical, are not fully understood, but the chemical action of such compounds as CO_2, mineral acids, organic acids, enzymes, or sequestering compounds can cause dissolution of the shell. Many animals can physically penetrate other organisms because of anatomical modifications such as the specialized regions of the edge of shells of molluscs, bristles in various worm phyla, mouth parts of some insects and echinoderms, and the radulae in gastropods. Radulae are particularly complex, since they function both chemically and mechanically. Experimental data indicate that the radula removes parts of the shell by abrasion after the shell has been softened by secretions from the accessory boring organ; the pH of the secretion from the accessory boring organ of the boring snail *Urosalpinx,* for example, has been reported at 3.8. Such secretions may serve to weaken the shell during boring (Carriker *et al.,* 1967).

Active Food

Predatory animals employ many different methods for capturing prey species: a hawk uses talons to grasp its prey; an ant lion digs a pit in the sand to trap prey; snakes may inject a paralyzing substance to immobilize prey; and a spider traps his prey in a web. The behavior pattern of a particular predator is fixed and predictable, as it is for the praying mantis *Hierodula crassa,* which feeds on the common housefly, *Musca domesticus.* Hanging upside down from a suitable substrate, the mantid will wait in ambush for its prey and will not move except to stalk or strike at a fly, which it locates with well-developed compound eyes. The predator does not move until the fly approaches closely, and then it begins stalking and pursuing the prey in a slow, deliberate movement characterized by swaying of the body. When the fly is within reach of the grasping forelegs, the mantid strikes (Holling, 1966). The feeding processes that follow capture of the prey may vary, but several characteristics are common to all predators: (1) often they have salivary glands that are used for preliminary digestion; (2) the food is usually introduced quickly into the site of digestion; and (3) feeding tends to be irregular and intermittent.

Fluid or Soft-Tissue Feeders

Most fluid or soft-tissue feeders are parasites on plants or animals, and feeding may be accomplished by piercing and sucking, sucking only, or by absorption through the surface of the body. The life modes of these animals range from free-living to ectoparasite and endoparasite. Typically, these animals are sluggish, with the obvious exception of insects, and selective feeders. Certain of the host-parasite relationships are highly specific.

Many different devices are utilized for piercing prey among them chitinous teeth or armored jaws that pierce or lacerate the host tissue. Such devices are commonly found in some endoparasitic nematodes and leeches. Styletto-sharp mouth parts that penetrate the body covering are characteristic of other animals such as insects, particularly mosquitoes and the blood-sucking bats. Many species may release an anticoagulant at the time of feeding, which facilitates the feeding process by preventing the fluid from clotting. The amount of time spent with a prey by animals that feed by piercing and sucking may vary; some predators are more or less permanently attached to their prey, as are endoparasites, but others, such as leeches, feed only occasionally.

Another aspect of this type of feeding has added ecolocial significance, for in addition to obtaining food from a prey, the predator establishes new biologic interaction by transmitting microorganisms to the prey species. Various degrees of complexity have evolved in the interaction of this predator-prey system in which the predator serves as a prey and *vector* (transmitter) for a variety of endoparasitic organisms. One relatively simple method of transmission is for the endoparasitic organism to pass through the digestive tract of the predator species along with the predator's feces. The endoparasite may then penetrate the prey's body through any cuts or wounds. The causative agent for typhus fever,

Rickettsia prowazeki, is transmitted by the human body louse to humans in this manner. At the other extreme are some complex life cycles, such as malaria, during which the endoparasite undergoes marked multiplication in the vector either by sexual or asexual reproduction.

Animals that feed exclusively by sucking rely on readily accessible fluids, such as external plant secretions or the fluid medium available to endoparasites, so that they are not required to penetrate any type of barrier. A vast majority of nematodes, butterflies, and moths feed by sucking in food. Great variation in the length of the proboscis in Lepidopterans is associated with adaptations for different food preferences.

Many parasites living in fluid habitats rich in nutritive substances feed by absorption through the body surface. Although they possess neither a complete digestive tract nor specialized structures for obtaining food, there is evidence that in cestodes the absorption of soluble sugars involves active transport mechanisms (Phifer, 1960).

MODIFICATION OF THE DIGESTIVE SYSTEM

Morphological Adaptations

Following ingestion of material from the environment, a number of processes must take place before the energy source can be utilized by an organism. Not all the ingested material is useful, and this part is discarded as fecal material, or is regurgitated. Before being absorbed, the ingested material (which is usually in a complex chemical form) must be broken down by digestive enzymes into smaller chemical units. In unicellular animals the site of the digestive process obviously occurs within a single cell, where there are varying degrees of cellular differentiation. Among the metazoans, the tendency has been for specialization of cells that allow the organism to extract energy efficiently from its environment. The vast diversification of animal and plant life during the long course of evolutionary history has been accompanied by a progressive increase in availability of food. As a result, not only is there a direct correlation between the morphological adaptation of food-gathering devices and ecological requirements of an organism, but similar correlations are also present in the digestive tract (Yonge, 1937).

One of the dominant trends in the digestive process has been a shift from intracellular to extracellular digestion. Where intracellular digestion has persisted, a large ingestive surface is necessary to supply the organism's body needs as the animal increases in size. A reduction in the relative size of the alimentary tract is usually accompanied by greater extracellular digestion.

There has also been a resultant differentiation and specialization of the digestive tract, and structural modifications can be correlated with the mode of existence. For example, grain-eating birds and fish-eating birds have large crops, whereas insectivorous birds do not. The intestines of herbivores tend to be larger than those of carnivores, and this tendency can be observed both on an interspecific basis and during ontogenetic development when changes in food habits occur; for example, the intestine of the herbivorous tadpole stage of a common frog is coiled and much longer than that of the carnivorous adult.

The elaboration of the digestive tract has obvious value to the animal concerned. In some instances it has also created new habitats for other species. This development of the digestive tract is ecologically important, offering habitats for many different types of symbionts.

Enzymatic Adaptations

The character and relative strength of an animal's digestive enzymes can be correlated with its food preferences. In general, omnivorous animals have enzymes that digest the three main types of food (fats, carbohydrates, and proteins) with equal ease. Carnivores have powerful proteolytic enzymes, and herbivores have relatively more powerful carbohydrases. Differences in the kinds and levels of digestive enzymes may reflect both genetic differences between species and all the nongenetic changes associated with feeding within an individual. For example, snails of the genus *Achatina* show an increase in amylase and saccharase content in the midgut gland when maintained on a high-starch diet, but the level of proteolytic activity does not increase when the animals are placed on a meat diet (Weel, 1959). Young, immature snails of this genus are generally herbivorous, but become omnivorous and have strong carnivorous preferences when they become sexually mature. With this change in natural diets, significant changes occur in the enzymes. The mature individuals then produce less amylase and more protease than do the immature snails (Smith and Weel, 1960). Immediately following a blood meal, the protease activity of the midgut of the mosquito *Aedes aegyptii* drops, probably owing to depletion of the enzymes. A rapid increase follows until a maximum is reached at 18 hr after feeding (Shambaugh, 1954). Different functional groups of bees from one hive appear to have different composition of enzymes; for instance, nurse bees have more proteinases than forager bees, whereas wax bees have no proteinases (Inglesent, 1940).

SUMMARY

Energy is the coinage of all biotic activity. Numerous morphological and physiological mechanisms have evolved that enable animals to extract the necessary energy (food) from very diverse habitats. The various types of feeding mechanisms are classified on the basis of ecological rather than taxonomic affinities. There is a distinct correlation between the method of feeding, the type of food, the structure of the digestive tract, the type of enzymes, and the habitat of animals.

The energy budget of an animal subjected to different environmental conditions gives an indication of the animal's adaptive ability. Energy not required for maintenance and survival can be used for synthetic activities associated with growth.

An intensification of the comparative study of bioenergetics involving species from diverse habitats is necessary before a comprehensive understanding of energy requirements of the organismic-environmental interaction is possible.

References

Ansell, A. D. 1961. The functional morphology of the British species of *Veneracea* (Eulamellibranchia). *J. Mar. Biol. Assoc. U.K.,* 41:489–515.

Carriker, M. R., D. Van Zandt, and G. Charlton. 1967. Gastropod *Urosalpinx:* pH of accessory boring organ while boring. *Science,* 158:920–922.

Comita, G. W. 1964. The energy budget of *Diaptomus siciloides* Lilljeborg. *Verh. Internat. Verein. Limnol.,* 15:646–653.

Darnell, R. M. 1967. Organic detritus in relation to the estuarine ecosystem. *In* G. H. Lauff (ed.), *Estuaries,* pp. 376–382. A.A.A.S. Publ. No. 83, Washington, D.C.

Darnell, R. M. 1968. Animal nutrition in relation to secondary production. *Am. Zool.,* 8:83–93.

Dethier, V. G. 1961. Behavioral aspects of protein ingestion by the blowfly *Phormia regina* Meigen. *Biol. Bull.,* 121:456–470.

Dix, M. W. 1968. Snake food preference: innate intraspecific geographic variation. *Science,* 159:1478–1479.

Dugal, L. P., C. P. Leblond, and M. Thérien. 1945. Resistance to extreme temperatures in connection with different diets. *Can. J. Res., Sect. E, Med. Sci.,* 23:244–258.

Ellis, P. E., and G. Hoyle. 1954. A physiological interpretation of the marching of hoppers of the African migratory locust (*Locusta migratoria migratoroides* R. and F.). *J. Exptl. Biol.,* 31:271–279.

El-Wailly, A. J. 1966. Energy requirements for egg-laying and incubation in the zebra finch, *Taeniopygia castanotis*. *Condor,* 68:582–594.

Green, G. W. 1964a. The control of spontaneous locomotor activity in *Phormia regina* Meigen. I. Locomotor activity patterns of intact flies. *J. Insect Physiol.,* 10:711–726.

Green, G. W. 1964b. The control of spontaneous locomotor activity in *Phormia regina* Meigen. II. Experiments to determine the mechanism involved. *J. Insect Physiol.,* 10:727–752.

Hamilton, W. J., III. 1962. Reproductive adaptations of the red tree mouse. *J. Mammal.,* 43:486–504.

Holling, C. S. 1966. Response of invertebrate predators to prey density. *Memoirs Entomol. Soc. Canada,* 48:3–86.

Hopkins, A. E. 1936. Adaptation of the feeding mechanism of the oyster (*O. gigas*) to changes in salinity. *Bull. U.S. Bur. Fish.,* 48:345–363.

Hoyle, G. 1953. Spontaneous squirting of an ascidian, *Phallusia mammillata* Cuvier. *J. Mar. Biol. Assoc. U.K.,* 31:541–562.

Hoyle, G. 1954. Changes in the blood potassium concentration of the African migratory locust (*Locusta migratoria migratoroides* R. and F.) during food deprivation, and the effect on neuromuscular activity. *J. Exptl. Biol.,* 31:260–270.

Hutchinson, G. E. 1957. *A Treatise on Limnology,* Vol. I. Wiley, New York.

Huxley, J. 1942. *Evolution the Modern Synthesis*. G. Allen, London.

Inglesent, H. 1940. Zymotic function of the pharyngeal, thoracic and post-cerebral glands of *Apis mellifica*. *Biochem. J.,* 34:1415–1418.

Jennings, J. B. 1965. *Feeding, Digestion and Assimilation in Animals*. Pergamon, New York.

Jørgensen, C. B. 1966. *Biology of Suspension Feeding*. Pergamon, New York.

Keast, A., and L. Welsh. 1968. Daily feeding periodicities, food uptake rates, and dietary changes with hour of day in some lake fishes. *J. Fish. Res. Bd. Canada,* 25:1133–1144.

Kinne, O. 1960. Growth, food intake and food conversion in a euryplastic fish exposed to different temperatures and salinities. *Physiol. Zool.,* 33:288–317.

Loosanoff, V. L., and F. D. Tommers. 1947. Effect of low pH upon rate of water pumping of oysters, *Ostrea virginica*. *Anat. Rec.,* 99:112–113.

Loosanoff, V. L., and F. D. Tommers. 1948. Effect of suspended silt and other substances on rate of feeding of oysters. *Science,* 107:69–70.

Marler, P., and W. J. Hamilton, III. 1966. *Mechanisms of Animal Behavior.* Wiley, New York.

Mews, H. H. 1957. Temperature adaptation and resistance in the digestive enzymes. *Z. vergl. Physiol.,* 40:345–355.

Miller, D. C. 1961. The feeding mechanism of fiddler crabs, with ecological considerations of feeding adaptations. *Zoologica,* 46:89–100.

Mullin, M. M., and E. R. Brooks. 1967. Laboratory culture, growth rate, and feeding behavior of a planktonic marine copepod. *Limnol. Oceanog.,* 12:657–666.

Pandian, T. J. 1967. Intake, digestion, absorption and conversion of food in the fishes *Megalops cyprinoides* and *Ophiocephalus striatus. Mar. Biol.,* 1:16–32.

Pearson, O. P. 1948. Metabolism and bioenergetics. *Sci. Monthly,* 66:131–134.

Phifer, K. O. 1960. Permeation and membrane transport in animal parasites: on the mechanism of glucose uptake by *Hymenolepis diminuta. J. Parasitol.,* 46:145–153.

Phillipson, J. 1966. *Ecological Energetics. Institution of Biology's Studies in Biology No. 1.* St. Martin's, New York.

Rao, K. P. 1953. Rate of water propulsion in *Mytilus californianus* as a function of latitude. *Biol. Bull.,* 104:171–181.

Ray, D. L. (ed.) 1959. *Marine Boring and Fouling Organisms.* University of Washington Press, Seattle.

Rozin, P., and J. Mayer. 1964. Some factors influencing short-term food intake of the goldfish. *Am. J. Physiol.,* 206:1430–1436.

Sealander, J. A., Jr. 1952. Food consumption in *Peromyscus* in relation to air temperature and previous thermal experience. *J. Mammal.,* 33:206–218.

Segal, E., K. P. Rao, and T. W. James. 1953. Rate of activity as a function of intertidal height within populations of some littoral molluscs. *Nature,* 172:1108–1109.

Shambaugh, G. F. 1954. Protease stimulation by foods in adult *Aedes aegypti* Linn. *Ohio J. Sci.,* 54:151–160.

Slobodkin, L. B. 1962. Energy in animal ecology. *In* J. B. Cragg (ed.), *Advances in Ecological Research,* Vol. I, pp. 69–101. Academic Press, New York.

Smith, A. C., and P. B. van Weel. 1960. On the protease and amylase production in the midgut gland of young and mature African snails. *Experientia,* 16:60.

Stephens, G. C. 1968. Dissolved organic matter as a potential source of nutrition for marine organisms. *Am. Zool.,* 8:95–106.

Tugendhat, B. 1960. The normal feeding behavior of the three-spined stickleback (*Gasterosteus aculeatus* L.). *Behaviour,* 15:284–318.

Voge, M. 1959. Sensitivity of developing *Hymenolepis diminuta* larvae to high temperature stress. *J. Parasitol.,* 45:175–181.

Warren, P. J., and R. W. Sheldon. 1967. Feeding and migration patterns of the pink shrimp, *Pandalus montagui* in the estuary of the River Crouch, Essex, England. *J. Fish. Res. Bd. Canada,* 24:569–580.

Weel, P. B. van. 1959. The effect of special diets on the digestion processes (enzyme production and resorption) in the African Giant Snail, *Achatina fulica* Bowdich. *Z. vergl. Physiol.,* 42:433–448.

Williams, A. 1965. Marine decapod crustaceans of the Carolinas. *Fish. Bull. U.S. Fish and Wildlife Service,* 65:1–298.

Yonge, C. M. 1928. Feeding mechanisms in the invertebrates. *Biol. Rev.,* 3:21–76.

Yonge, C. M. 1937. Evolution and adaptation in the digestive system of the Metazoa. *Biol. Rev.,* 12:87–115.

Yonge, C. M. 1954. Feeding mechanism in the Invertebrata. *Tabulae Biologicae 21,* Pt. 3, No. 22.

8

Population Continuity

> Probably the most fundamental evolutionary requirements of any organism are to take full advantage of opportunities for reproduction and to possess and maintain mechanisms for providing enough genetic variability to meet the demands imposed by changes in the environment.
>
> —T. M. Sonneborn

The continuity of a species is dependent upon the ability to reproduce itself, and to this end both asexual and sexual modes of reproduction have evolved in the animal kingdom. Considerable genetic variability and a large genetic pool useful for meeting the stresses of a changing environment are characteristic of sexual reproduction. Despite the absence of genetic recombination in asexual reproduction, two major mechanisms enable asexually reproducing animals to achieve a degree of variability and adapt to a changing environment: (1) mutation, which operates successfully in haploids that are able to produce large populations rapidly, and (2) multiple loci in the genotypes for traits that are mutually exclusive and rapidly transformable. Multiple loci may be present also in sexual organisms having a well-developed asexual phase and in the asexual phase of development in higher organisms. Even with these means of achieving variability, animals that reproduce only asexually usually cannot withstand a stressful environment as well as those which reproduce sexually. On the evolutionary time scale, some of the most successful asexually reproducing animals are those that have adapted to very slowly changing environments, such as organisms that live in large bodies of water, or some of the symbionts (Sonneborn, 1957).

Exclusively asexual reproduction is not widespread among animals. Among the lower forms, sexual reproduction usually alternates with several generations of asexual reproduction, and in many of the Protozoa, the long-term survival of a population may depend upon the sexual processes, as it does for the protozoan *Paramecium aurelia*. If a population of *P. aurelia* reproduces only asexually for a prolonged period of time, death (either genetic, somatic, or both) eventually will result. Sexual reproduction in these protozoans may occur by: (1) conjugation, which is the union of one individual of one mating type with another individual of the opposite mating type (normally a different clone of the same variety); (2) selfing, which is conjugation with an individual of the same strain; and (3) autogomy, or the formation of a fertilization nucleus by the union of male and female gametic nuclei in one individual. Cross conjugation is primarily important, since it may be essential to long-term survival, but because this form of reproduction depends on contact between animals of a different clone, it cannot always be guaranteed. On the other hand, selfing and autogomy, which are dependable processes of fertilization, can ensure only short-term survival (Sonneborn, 1957).

The complex reproductive process of sexually reproducing animals can be subdivided into three phases, all of which must function to ensure continuity: (1) gonadal development; (2) spawning and fertilization; and (3) development and growth.

REPRODUCTIVE CYCLES

Most higher animals reproduce on a predictable annual basis during a restricted period, which can occur at any time during the year and which may be limited to a short time or may extend over several months. In general, the shortest reproductive periods are characteristic of animals from the polar regions, and the longest of animals from the tropics, with those of temperate-zone animals intermediate. Although most animals do have predictable breeding seasons, there are some striking exceptions, such as some birds and amphibians in arid regions, which breed only after a rainfall, and some marine organisms that spawn almost continuously throughout the year.

Reproductive cycles generally consist of a vegetative or resting stage, which is a period of little gonadal activity, and a reproductive stage, which is initiated by the beginning of gametogenesis followed by the maturation and release of gametes.

As an aid in assessing the influence of environment on reproduction, various techniques are used to determine the stage of reproductive cycle in a population of animals. One common method is to determine the ratio of gonad size to body weight; this ratio is termed the gonad index (GI). As

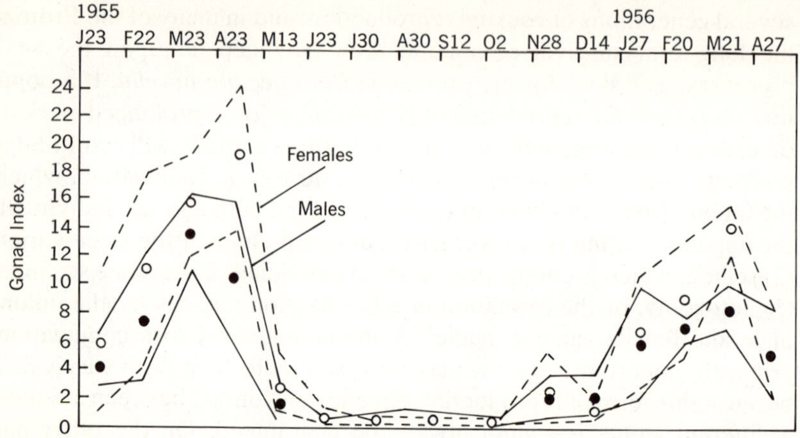

Figure 8-1. Gonad index for *Pisaster ochraceus* showing seasonal variability for the population. The points are the averages for the males (●) and females (○) in a sample where the bands are the 95% confidence limits. No essential difference exists between the relative gonad index of males and females. (From Farmanfarmaian *et al.*, 1958.)

the gonads increase in size, the gonad index increases. The relationship between the GI and various phases of the reproductive cycle is represented in Fig. 8-1. Quantitative and qualitative data obtained from examination of pieces of gonadal tissue are useful in classifying the stage of gametogenesis (Fig. 8-2). The percentage of animals in each of these stages at different times of the year will give an indication of the duration of the reproductive cycle. Observations on the breeding and spawning

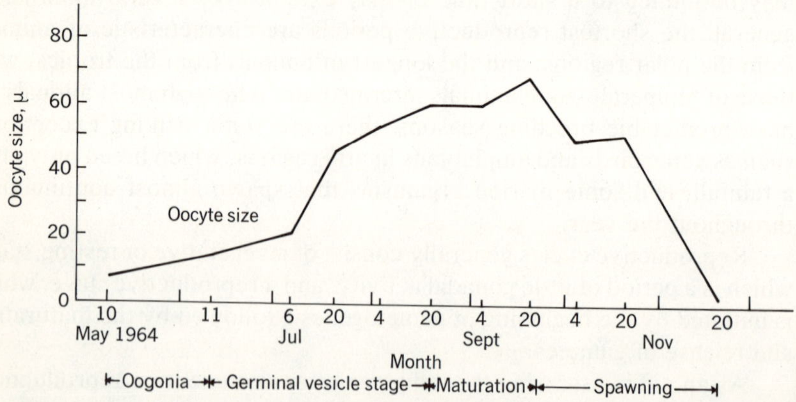

Figure 8-2. Changes in average oocyte size of scallops collected at intervals during the reproductive period. Reproductive conditions of the population is shown on the bottom of the graph. (From Sastry, 1966.)

periods, and/or the occurrence of larvae, also provide information about stages of the reproductive cycle.

FACTORS INFLUENCING THE REPRODUCTIVE CYCLE

In most animals, both endogenous and exogenous factors are involved in the timing for breeding. Marler and Hamilton (1966) have postulated that breeding is generally triggered by external stimuli in animals with erratic and unpredictable breeding seasons, and that endogenous factors have only a permissive function in the timing mechanism of these animals. Endogenous factors are, however, more important to animals with annual breeding seasons.

Several environmental factors influence the initiation of reproductive cycles and may affect certain phases of the reproductive cycle—among them, food, ions and water, temperature, oxygen, photoperiod, and lunar cycles.

Food

It has long been recognized that when the maximum food supply is present in the environment, many species of animals reproduce. The direct cause-and-effect relationship, however, has not yet been established, since experimental evidence is lacking to show that the seasonal appearance of any one particular type of food acts as a breeding stimulus.

A close connection between food supply and egg production is apparent in many animals, however. For example, in the planktonic copepod *Calanus,* spawning and breeding times correlate rather closely with spring and summer phytoplankton blooms. During laboratory experiments with *Calanus,* a ripe female that is starved will stop laying eggs, but when food is provided again, she resumes egg laying. The relationship between number of eggs laid and quantity of available food is shown graphically in Fig. 8-3 (Marshall and Orr, 1964). Reproductive periodicity has been linked to food supply among other marine animals, such as two species of brittlestars, *Ophiura ljungmani* and *Ophiomusium lymani,* that occur at depths between 1100 and 3800 meters in the western North Atlantic, where the physical environment is quite uniform and temperature, salinity, and bottom type are free of seasonal variation. Yet, studies on these brittlestars have shown a seasonal peak in gonad development, and presumably in reproduction, with a greater production of young during

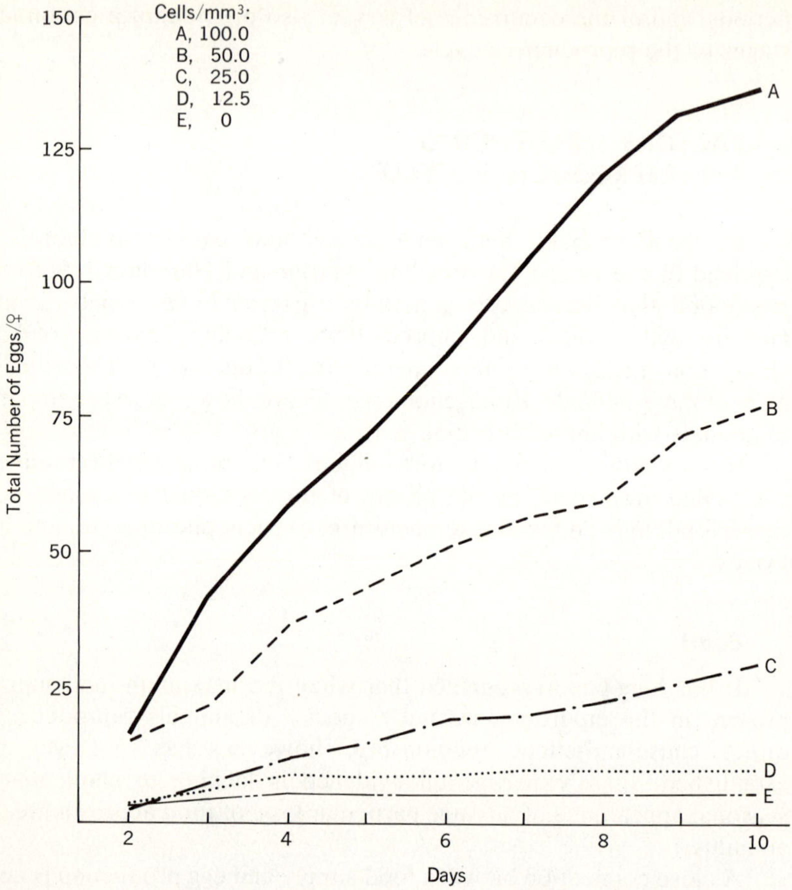

Figure 8-3. Egg-laying by *Calanus* fed on different concentrations of *Skeletonema*. (From Marshall and Orr, 1964.)

the summer months. Therefore it appears that the reproductive cycle of these organisms may be linked to a periodic increase in food in the environment (Schoener, 1968).

For some animals, gonad development and activity also may depend on adequate food supply. In controlled experiments on the boreo-arctic barnacle *Balanus balanoides*, an adequate food supply was shown to be necessary for ovarian development, since an interruption of the food supply resulted in a regression of ovarian tissue. In nature these barnacles will store food when it is most abundant and subsequently utilize it for gonad development. Once ovarian development is initiated, however, future development is dependent on current food supplies (Barnes and Barnes, 1967, 1968). Dependency on food for normal gonad activity is not limited to invertebrates, however, for if the normal food intake of domestic birds is reduced by a fourth, then gonad activity almost ceases (Breneman, 1955).

The effects of food deprivation may interact with other factors in the environment to influence the reproductive cycle. When the scallop *Aequipecten irradians* is maintained at 20°C in the absence of food, oocytes do not develop, but if these scallops are fed and kept at 20°C, oocyte development occurs. When the temperature is lowered to 15°C, however, no oocyte development takes place even when food is present (Sastry, 1968).

Water and Ions

Rainfall or its effects have been shown to influence the reproductive cycle of numerous animals, particularly those living in dry environments. In general, these species have no definite breeding season, but breed in response to either the rainfall or the environmental effects that follow rain. Among these animals is a bird, *Quelea quela,* in Tanganyika, which uses green grass in nest weaving. The reproductive cycle of first-year birds is influenced by the appearance of the green grass that grows after a rain. The reproductive cycle of older birds, however, responds to the rainfall itself or to the high humidity accompanying the rainfall (Marshall and Disney, 1957). The frog *Scaphiopus h. holbrooki* has maintained a xeric breeding pattern even though its distributional range now extends to non-desertlike areas. From the center of distribution near northern Mexico, this amphibian has spread northward as far as Massachusetts. Initiation of the reproductive cycle in these animals occurs only after a heavy rainfall of 1 in. or more falling within 24–48 hr. In the more northern areas of its distribution, however, rainfall interacts with temperature and is ineffective until temperatures reach a certain level (Gosner and Black, 1955).

The influence of salinity on reproduction is most dramatically demonstrated in the migration of many adult fish and crustaceans to water of suitable salinity before they spawn. That reduced salinity decreases the reproductive potential of various marine animals is well documented in species inhabiting the Baltic Sea, where there is a marked salinity gradient. Many such animals can and do live as adults in low-salinity waters, but they fail to reproduce there and their repopulation depends on the invasion of larvae or adult migratory populations from higher-salinity waters (Remane and Schlieper, 1958). In contrast, certain Indian marine organisms will not reproduce unless inshore waters are diluted by winter rains (Panikkar and Aiyar, 1939).

Field observations have indicated that lower salinities cause a loss in reproductive potential in marine organisms, but experimental verification often is lacking; it has been demonstrated, however, that the birth rate of the marine fish *Cyprinodon macularius* is higher in sea water than in fresh water. On the other hand, reproductive performance of many brackish-water animals is reduced in high-salinity waters (Kinne, 1964).

Temperature

Numerous field studies have indicated the importance of temperature to the initiation and completion of the reproductive cycle. Temperature in some cases may act as a trigger setting off reproduction when a certain thermal level is reached. In other animals, gametes will develop at lower temperatures, but the development rate is retarded if temperatures rise slowly. Orton (1920) suggested that the breeding season of each species occurs at a characteristic and constant temperature that prevails throughout the entire distributional range of the animal, and that this breeding season is longer and begins earlier in the year in the warmer parts of the geographical range. For example, the oyster *Ostrea edulis* begins to breed when the water temperature reaches 15–16°C and continues breeding until the temperature drops below this level. At Naples this temperature is reached earlier in the year and persists longer than in England, so that the oyster has a longer breeding season at Naples.

Not all species obey Orton's dictum, since the temperature necessary for breeding may vary in some species with the typical habitat-temperature range. For example, in many species the more southerly distributed population breeds at higher temperatures than does the northern population. Furthermore, population differences may occur at a given latitude, as in southern Spain where one population of oysters breeds in March or April when the temperature is between 13 and 14°C, whereas other neighboring populations require higher temperature (Korringa, 1957). This difference in reproductive periodicity might be an important factor in preventing genetic exchange between the two populations.

It has not been fully established whether differential responses to temperature reflect endogenous periodicity differences between populations or whether these responses are environmentally induced. Sea urchins, *Arbacia punctulata*, from Massachusetts typically have two reproductive peaks during the year, whereas those from North Carolina have only one (Fig. 8-4). When Massachusetts animals in the nonreproductive phase are transplanted to North Carolina and are maintained under identical environmental conditions with North Carolina *Arbacia*, identical reproductive patterns are observed. These population

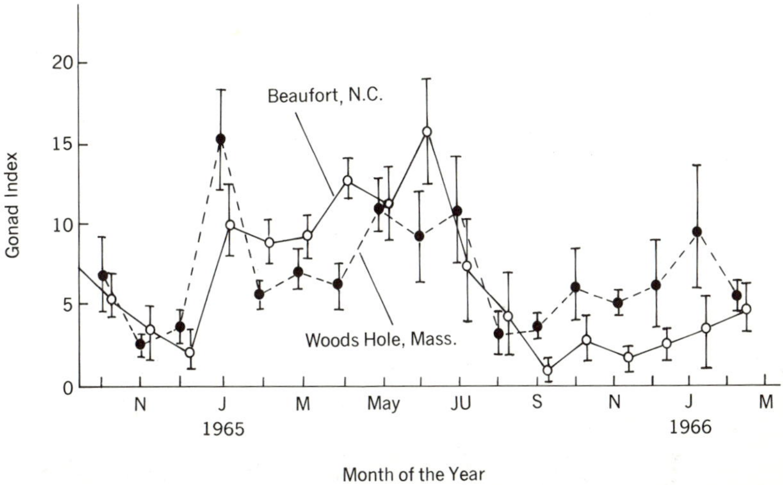

Figure 8-4. The monthly mean gonad index values of latitudinally separated population of sea urchins (*Arbacia punctulata*) are similar when maintained under comparable environmental conditions. Vertical lines represent the 95% confidence intervals. (After Sastry, 1966b.)

differences, then, are environmentally controlled and do not reflect genetic differentiation. When scallops, *Aequipecten irradians,* were subjected to similar environmental manipulation (Sastry, 1966b), however, the reproductive periodicity of scallops from Massachusetts still did not coincide with that of the North Carolina population (Fig. 8-5). These results suggest that since the breeding periodicities cannot be modified by environmental factors, there are genetic differences between these two populations.

In contrast to many other physiological functions, reproductive processes occur over a narrow thermal range. Although scallops require a temperature of 20°C in order to spawn, activity and survival can occur at lower temperatures (Sastry, 1966a). Similarly, the boring gastropod *Urosalpinx cinerea* is capable of locomotor activity at 10°C and will drill oysters at 15°C, but oviposition will not take place until temperatures reach 20°C (Stauber, 1950). This difference between organismic and reproductive thermal limits has ecological significance, for an organism living in a habitat with a narrow temperature range might have to migrate to another habitat with different thermal characteristics in order to reproduce. Extension of the geographic range of a species in terms of establishing a breeding population is also dependent on the reproductive ability at new thermal levels.

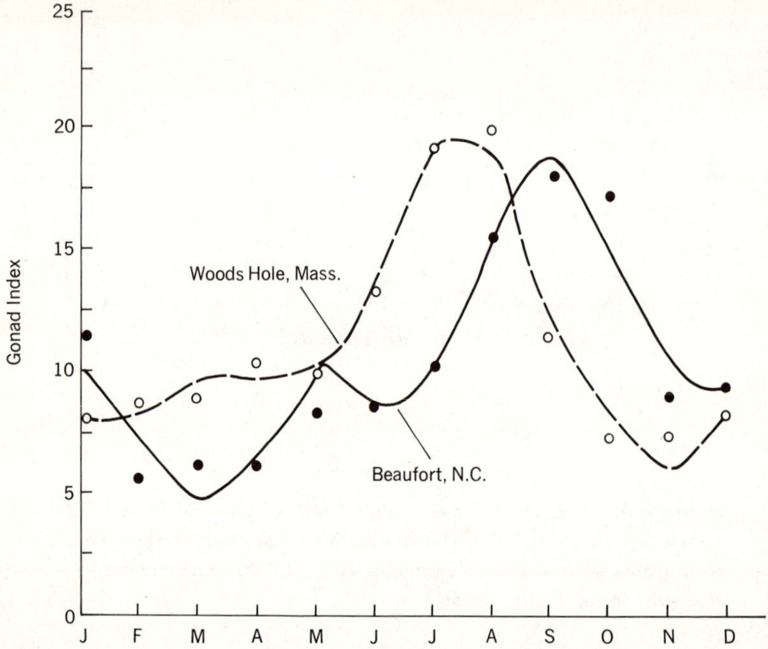

Figure 8-5. Differences exist in the monthly mean gonad index values of latitudinally separated populations of scallops (*Aequipecten irradians*) which suggest genetic diversity. (From Sastry, 1966b.)

Because spawning is one of the more readily observed phases in the reproductive cycle, this process is better understood than are less overt phases of reproduction. Although many environmental factors can trigger spawning, either singly or through interaction, perhaps the most important single factor is temperature. Some species, such as the bivalve *Mytilus edulis,* spawn when temperatures rise to a certain level, but others, such as the bay scallop *Aequipecten irradians* from North Carolina, spawn only when a temperature rise is followed by a decrease (Sastry, 1963). The bay scallop normally spawns during summer and fall months, but it can be experimentally induced to spawn throughout most of the year when the temperature regime is manipulated. Because gametes are liberated at lower temperatures during the cooler months of the year rather than in the warmer months, it appears that seasonal differences may affect the triggering response to temperature (Sastry, 1963).

Geographically separated populations of the bay scallop also exhibit differences in the time of spawning; the Massachusetts population spawns when the water temperature reaches 14–16°C, but North Carolina and

Florida populations require higher temperatures. The Massachusetts population starts spawning when the temperature is increasing, but more southerly populations spawn only as the temperature decreases following a temperature rise (Belding, 1910; Sastry, 1963, 1966a). Many aquatic animals do not release their gametes when the temperature is too low to allow early development of the young, but the causal effect is not known (Thorson, 1950). Reduced temperatures often cause resorption of gonads in the adults.

In some areas, particularly marine waters, faunal assemblages with high- or low-temperature affinities overlap, and frequently these groups spawn at different times of the year. An excellent example are the benthic animals in the Adriatic Sea. Of 97 species of these animals, 64 were spring spawners; 49, summer; 26, autumn; 18, winter; and some spawned in two seasons (Vatovo, 1949). Differences in spawning times and reproductive cycling allow a more complete exploitation of the resources in the area and therefore may profoundly influence the trophic dynamics of an ecosystem (see Chapter 11).

Oxygen

In comparison to other environmental parameters, relatively little is known about the influence of oxygen on the reproductive processes. Studies on the polychaete *Capitella capitata* indicate that low oxygen tensions suppress the reproductive cycle. Worms suspended in water having little or no dissolved oxygen did not feed or reproduce, although they survived for several days; in water with an oxygen content of 2.9 ppm, the animals fed but did not reproduce. Only when the oxygen content rose to 3.5 ppm did reproduction occur (Reish and Barnard, 1960).

The level of oxygen concentration greatly influences other phases of reproduction. For example, the sperm of *Arbacia,* which are highly sensitive to oxygen lack, are irreversibly damaged and incapable of fertilizing the eggs after 3–4 hr without oxygen. Eggs of *Arbacia* are slightly more resistant to oxygen deprivation and will remain normal for about 8 hr in the absence of oxygen, but oxygen lack tends to stimulate parthenogenetic development (Harvey, 1956). In a number of species of sea urchins, reduced oxygen levels retard the rate of cleavage, and anaerobic conditions will block cleavage. In fish, anaerobic conditions differentially affect the viability of eggs of various species. The eggs of *Fundulus,* an extremely hardy species of fish that tolerates a greatly fluctuating environment, are highly resistant to oxygen lack; in contrast, the eggs of many other species of fish show a large number of abnormalities following exposure to low oxygen tensions.

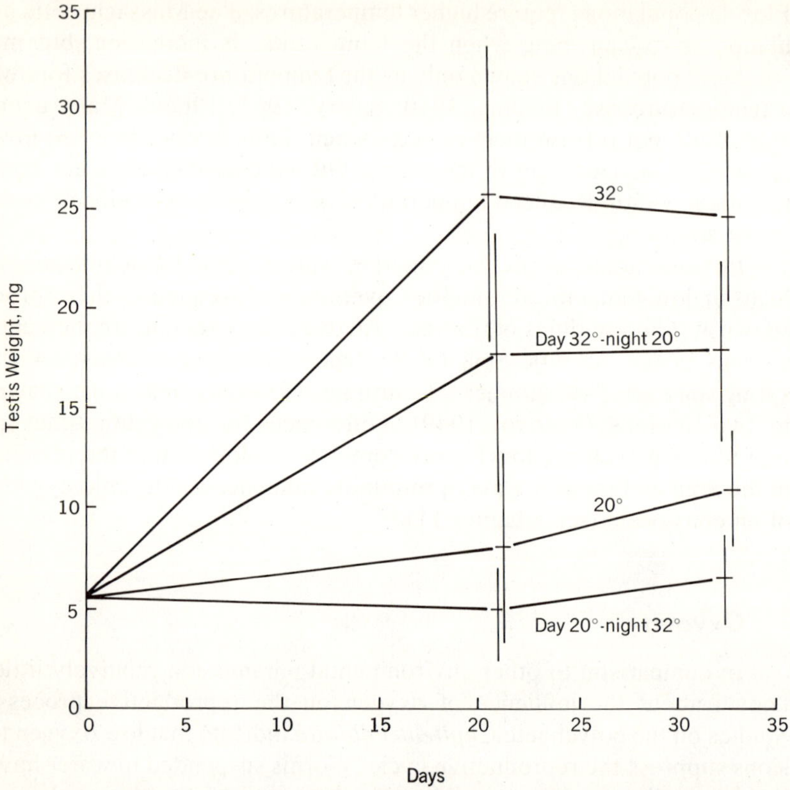

Figure 8-6. Effect of various thermal conditions on testis size in *Anolis*. The animals were exposed to a photoperiod of 14 hr light—10 hr dark. The experimental temperature regimes are indicated on the figure. Each vertical line indicates the 95% confidence limits. (From Licht, 1966.)

Photoperiod

The reproductive cycle of most temperate-zone animals typically follows the day length of the season, and the peak of the breeding season generally occurs in the spring. During the remainder of the year, breeding behavior is suppressed and the gonads regress. In general, endocrine mechanisms that initiate gonad growth are activated by an increasing day length among spring breeders and by a decreasing day length in animals with fall-breeding cycles.

Many birds have a refractory period during the breeding season when changes in day length cannot initiate a breeding condition. Such a period prevents breeding at an unfavorable time, such as late summer and fall, and is particularly important to transequatorial migratory birds that breed in the spring. Owing to reversal of the seasons, these birds would be subject to two long day-length cycles, so that without the refractory periods they could be triggered into breeding in their winter quarters (Marshall, 1960, 1961). The refractory period is not always entirely successful, however, as demonstrated by experimental manipulation of photoperiod exposure in bobolinks, a transequatorial migrating species. These birds can overcome a postnuptial photorefractory period, but regrowth of the testes proceeds at a much slower rate during the long days of the southern hemisphere than in species restricted to the temperate zone (Engels, 1961).

In general, animals living in northern latitudes have fewer broods of young per year than do species at more southerly latitudes. The number of young per brood, however, appears to increase with latitude. An increase in daylight during the nesting season has apparent survival value in that there is a longer period for feeding the developing young.

Even under constant conditions, several species of birds will undergo cyclic changes during the reproductive process, and although a precise annual rhythm is not evident, a clear capacity to cycle seems to operate independently of seasonal changes in photoperiod (Marler and Hamilton, 1966).

Photoperiod alone is ineffective in initiating the reproductive cycle of some animals if the temperature has not reached a certain level. For example, the photoperiodic responses of the testes are very temperature-sensitive in the lizard *Anolis carolinensis,* and if body temperatures do not approach the preferred level (32°C) for at least part of the daily light period, long day lengths are ineffective in stimulating testicular recrudescence. Further evidence (Licht, 1966) of the complex interaction between temperature and light is that continuously low temperatures retard testicular growth less than high night temperatures or a 20°C day temperature (Fig. 8-6).

Lunar Cycle

The reproductive cycle of numerous marine organisms is timed to coincide precisely with certain phases of the moon, and the behavior of these animals demonstrates the interaction of lunar, annual, and daily cycles. Lunar periodicity is well illustrated in the reproductive behavior of the palolo worm, particularly the Atlantic palolo worm *Leodice fucata* and the Pacific species, *Eunice viridis* (Clark and Hess, 1942). These worms live in the crevices and burrows of corals and rocks, and they feed at night by extending a part of their body from the hole. During the spawning season, millions of these worms leave their burrows and swim freely on the surface of the water. Subsequently, their bodies break into two parts, and the posterior segments, which are filled with eggs or sperm, are left behind in great masses on the surface of the water; the other half returns to the burrows in the rock or coral, where it apparently regenerates the reproductive parts. The bulk of the Pacific palolo worms swarm on a single day, seven to nine days after a full moon and sometime between November 7 and December 22. The Atlantic palolo worms swarm during July, about eight days after a full moon. Environmental factors can alter this pattern to some extent, for if the wind is stronger than 8 mph, swarming is inhibited in populations from unprotected regions; worms that are sheltered from the wind action may spawn on schedule, and the two groups will be out of synchrony. Nevertheless, swarming almost always occurs within a 30-day period, and most often at dawn.

Reproductive periodicity in the grunion *Leuresthes tenuis* is even more precisely timed than in the palolo worm. The grunion, which is a small edible fish common in waters off southern California, has a semilunar cycle extending from March through August, and the heaviest spawning occurs in late spring and summer. During a spawning run, the fish swim ashore just after the tide begins to fall, and quickly dig pits 2–3 in. deep in the sand, where they deposit eggs and sperm; the fish return to the ocean with the next wave. The success of this activity is determined by extremely precise timing, for the spawning runs must coincide not only with the highest waters of each tide, but also with the highest tides each month. The eggs take about two weeks to develop in the wet sand, so that their emergence is concurrent with the next high tide that occurs within 12 to 14 days. Hence, if the eggs are not laid at the proper time, they cannot complete development before being washed out to sea.

REPRODUCTIVE TIMING

Timing is of utmost importance to species that aggregate and promiscuously shed their gametes, and various strategies have evolved to ensure gametal union. Often, as in the palolo worms, the mating period is limited to one or two days a year and to a particular time of day. The mating swarms of many fish are synchronized to a particular time of day. In contrast, the males of some pelagic invertebrates release gametes throughout the breeding season, thereby eliminating the need for precise synchronization. Certain chemical substances in the testes and sperm will stimulate female spawning, and similar excitants are present in the ovaries and eggs. The radial nerves of starfish have a neurochemical substance that stimulates the maturation of gametes and also initiates spawning, and this substance has been found at a constant level throughout the year. At first, its significance in triggering spawning was not clear, but further investigation revealed that an inhibiting substance, termed *shedhibin,* is present until the time that the starfish matures, but subsequently is apparently reduced to the point where it no longer prevents the action of the neurochemical shedding substance from the radial nerve (Chaet, 1966).

Interplay between the sexes is often observed in species that swarm. Early spawning males excite the females to release eggs, and in turn the release of the eggs stimulates refractory males to shed in a chain reaction. The result of this reciprocal excitation is a mass synchrony in the release of gametes, which thereby ensures a greater degree of fertilization than would be otherwise possible.

Communal spawning occurs also among vertebrates. For example, a large spawning aggregation is formed by the glassy darter *Etheostoma vitreum,* since small groups of this species seem to attract numbers of both sexes. Eggs are deposited in the path of a strong current on a solid surface such as rocks or spillways. Ordinarily, these eggs would be exposed to many potential predators, but because the communal spawning takes place in early spring before temperatures have warmed up, in water that is silt-laden and where the current is swift, it is both physically and visually difficult for predators to feed on the eggs (Winn and Picciolo, 1960).

The males of some fish establish territories where gravid females come to spawn, but the fish perform a well-defined ritual before spawning. For example, the male toadfish *Opsanus tau* finds a suitable nesting site such as a tin can or large empty shell, and stands guard. When the male is ready to spawn, he emits a characteristic boat-whistle call that acts as a stimulus to attract females that are ready to lay eggs. As soon as the female lays a clutch of eggs, she leaves the nest, but the male stays in residence until the young attain the free-swimming stage (Gray and Winn, 1961). In most fish, however, there is little extended mutual interaction between the sexes, as is common among birds and mammals.

A more protracted reproductive relationship is common between males and females of animals having internal fertilization, and is most characteristic of terrestrial animals. Often pairs are formed that extend through a breeding cycle, a whole season, or even years. These species tend to have fewer young and to be more involved with the postnatal young than are those species having external fertilization.

BIOCHEMICAL CHANGES

The metabolic requirements of an animal shift markedly with the transition from one phase of the reproductive cycle to another. Although biochemical processes are known to be influenced by physical and chemical factors, the dynamic interaction of the external environment and the biochemistry of reproduction has scarcely been studied. In many invertebrates, two tissues in particular are affected: the gonads and the digestive gland. Typically, during the vegetative phase of the reproductive cycle, the digestive gland index is high in relation to the gonad index, but

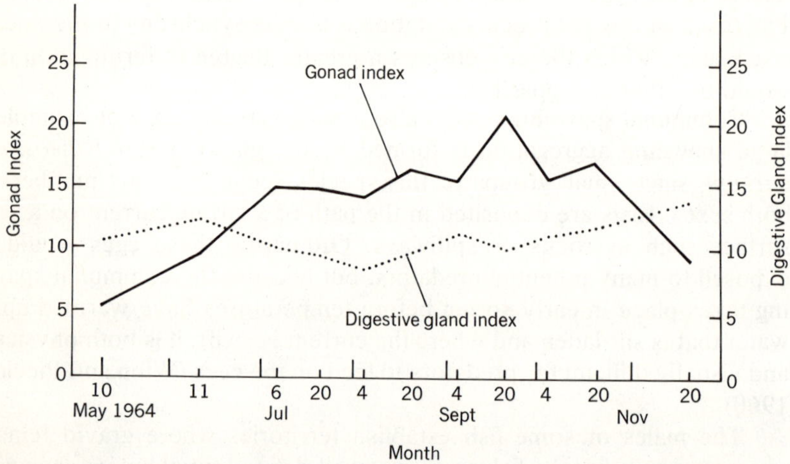

Figure 8-7. The inverse relationship between the gonad index and the digestive gland index of scallops during the reproductive period. (From Sastry, 1966.)

during the reproductive phase, the relationship is reversed. Such changes in the gonad and digestive-gland index of scallops are shown in Fig. 8-7 (Sastry, 1966). In addition to the relative indices, the chemical composition of gonad tissue can also change with the phase of the reproductive cycle (Table 8-1). For example, the gravid gonads of limpets have a high-protein and lipid content, whereas the nongravid gonads are high in carbohydrates, but low in protein and lipids (Giese and Hart, 1967). Similar dramatic changes in the chemical composition of tissues have also been reported in some vertebrates.

In both vertebrates and invertebrates, modifications of the reproductive tract result from hormonal production. For a discussion of hormone action, consult a text in endocrinology.

GROWTH AND DEVELOPMENT

The early stages of embryonic development represent a rapidly changing system that is very sensitive to environmental influences which may, during this period of differentiation, irreversibly influence the structural and physiological capabilities of the organism. In a more static system (that is, an adult stage) many responses represent only temporary deviations from a relatively fixed standard of normality. Studies of developing fish illustrate the influence of a physical factor on an anatomical feature of a changing system. For example, the number of vertebrae of many species of fish varies with the temperature at which the organism is reared, being lowest at about 6°C. Gastrulation is typically the most thermally sensitive period, but once the vertebrae have differentiated, thermal experiences will not effect a change in their number. Thus a phenotypic expression can become fixed and unvaried (Taning, 1952).

The physiological capacity of an adult may depend upon environmental influences during development. For example, the salinity lethal limits of the amphipod *Gammarus* depend upon the salinity at which the animal is reared, so that animals grown at a lower-salinity level can withstand fresh water longer than those reared at high salinity (Kinne, 1953). Similar adaptive responses to temperature have been reported for various species. Environmental factors, then, can have a profound effect on developing young. The following environmental parameters have been shown to influence growth and development.

Temperature

An animal's rate of development will increase as temperatures rise, up to a certain level. As might be expected, the range of temperatures tolerable to a developing larva varies with the species and can be correlated with the thermal characteristics of the habitat. Not all stages in the development of an organism are equally sensitive to temperature; in many marine invertebrates the cleavage stages are more sensitive to temperature fluctuations than are the larval stages.

For the most part, populations from higher latitudes develop at lower temperatures and have a lower thermal lethal range than do populations from warmer habitats, even if two different populations of the same species are being compared. The series of studies on the North American frog *Rana pipiens* (Moore, 1946, 1947, 1949a,b, and 1950) showed pronounced physiological differences between the latitudinally separated populations in the geographic range, which extends from northern Canada to Panama and includes a variety of climatic conditions. Although the

Table 8-1. Chemical Constituents in Spent and Gravid Gonads of the Purple Sea Urchin (in % dry wt).

Sex and Condition	*Gonad Index*	*NPN*	*Protein*	*Lipid*	*Glycogen*	*RNA*	*DNA*	*TN*	*Water*	*Reducing Sugar*
♂ spent	2.3	2.0	23.01	—	—	3.25	0.0014	5.7	—	0.007
	—	2.0	27.4	—	—	3.46	0.0010	6.4	—	
	3.5	2.2	23.7	—	—	2.59	0.0018	6.0	—	
Av	2.9	2.1	24.7	—	—	3.10	0.0014	6.0	—	
Gravid	17.8	1.3	35.0	—	—	5.0	0.0007	6.9	—	0.037
	18.2	2.1	34.4	—	—	4.8	0.00065	7.6	—	
	18.9	1.0	39.4	—	—	3.8	0.00059	7.3	—	
Av	18.3	1.5	36.3	—	—	4.5	0.00065	7.3	—	
♂ spent	1.52	4.1	25.9	—	—	2.23	4.7	8.3	—	0.0036
	2.84	2.1	36.0	—	—	2.4	4.1	7.9	—	
	2.43	2.2	36.0	—	—	2.1	4.4	8.0	—	
Av	2.26	2.8	32.6	—	—	2.3	4.4	8.1	—	
Gravid	21.5	0.66	43.0	—	—	1.3	9.8	7.6	—	0.034
	19.2	1.20	39.8	—	—	0.9	8.0	7.6	—	
	1.77	0.59	42.5	—	—	0.9	9.3	7.4	—	
Av	19.5	0.82	41.8	—	—	1.0	9.0	7.5	—	
♀ spent	1.18	3.75	23.2	9.5	0.41	—	—	7.5	63.2	
	1.28	3.07	20.1	18.7	3.35	—	—	6.3	78.8	
	1.49	2.84	28.2	15.5	2.77	—	—	7.4	—	

Av	1.32	3.22	23.8	14.6	2.18	—	—	7.1	71.0
Gravid	21.8	1.83	33.5	18.1	7.98	—	—	7.2	—
	17.0	1.92	26.8	19.9	2.35	—	—	6.2	64.9
	17.7	2.41	42.1	21.3	4.54	—	—	9.1	70.0
Av	18.8	2.05	34.1	19.8	4.96	—	—	7.5	67.4
♂ spent	1.43	3.03	38.2	17.4	2.63	—	—	9.2	—
	3.0	3.14	33.4	14.2	6.41	—	—	8.5	76.2
	1.42	3.34	32.4	20.5	1.48	—	—	8.5	74.9
Av	1.95	3.17	34.7	17.4	3.51	—	—	8.7	75.5
Gravid	21.3	1.63	37.5	13.0	4.24	—	—	7.9	—
	21.0	1.91	24.4	11.0	9.86	—	—	5.8	65.1
	21.6	4.47	35.2	12.6	4.43	—	—	10.1	66.3
Av	21.3	2.67	32.4	12.2	6.51	—	—	7.9	65.7

SOURCE: Giese *et al.*, 1959.

animals from the more northerly populations tolerate a lower temperature during embryonic development than do the more southerly populations, the reverse is true for high-temperature tolerance. Thus the temperature range in which northern populations will develop normally can be as low as 5°C and as high as 27.5°C, whereas in more southerly populations the range is from a low of 9–12°C to a high of 32 to 33°C. The northern frogs also develop faster, have larger eggs, and breed earlier than frogs from more southerly areas. When crosses are made between individuals from latitudinally separated populations having markedly different temperature regimes, the hybrids are defective, but the reproductive incompatability between populations apparently is caused by differences in temperature regimes rather than by wide geographical separation. Crosses of populations from Vermont and Colorado, for example, result in normal hybrids, but crosses between Vermont and Florida populations give abnormal hybrids. This same trend can also be observed in altitudinally separated populations of frogs and toads from the same latitude; animals from the colder high-altitude areas respond physiologically to temperature as do the high-latitude, low-temperature populations. Because genetic differences have evolved in some species, it is not possible to crossbreed different populations. For example, when frogs, *Rana pipiens,* from Vermont were crossed with the same species from Mexican lowlands, the embryos were abnormal; in contrast, crosses between populations of this species from Vermont and the highlands of Mexico produced viable embryos (Ruibal, 1955). See Fig. 8-8.

The developmental stages of some species of animals have a very labile thermal response, and latitude- and habitat-temperature regimes do not alter their response. The copepod *Calanus finmarchius* is present in both Arctic and temperate-zone waters and will develop at temperatures ranging from 0–20°C (Marshall and Orr, 1957).

Temperature can influence not only the rate of development of an animal, but also its growth pattern and form. Naturalists long ago observed that animals from colder environments tend to be larger than those from warmer climates. Based on these observations, two generalizations or rules have been postulated:

1. *Bergmann's rule* states that smaller-sized individuals of a species generally occur in the warmer part of a species range, whereas the larger-sized individuals occur in the cooler regions.
2. *Allen's rule* states that the protruding parts of the body are relatively shorter in animals from the cooler parts of their range than in those from warmer regions.

Although it has been demonstrated that these rules apply to a majority of birds and mammals, their adaptive significance has been in question.

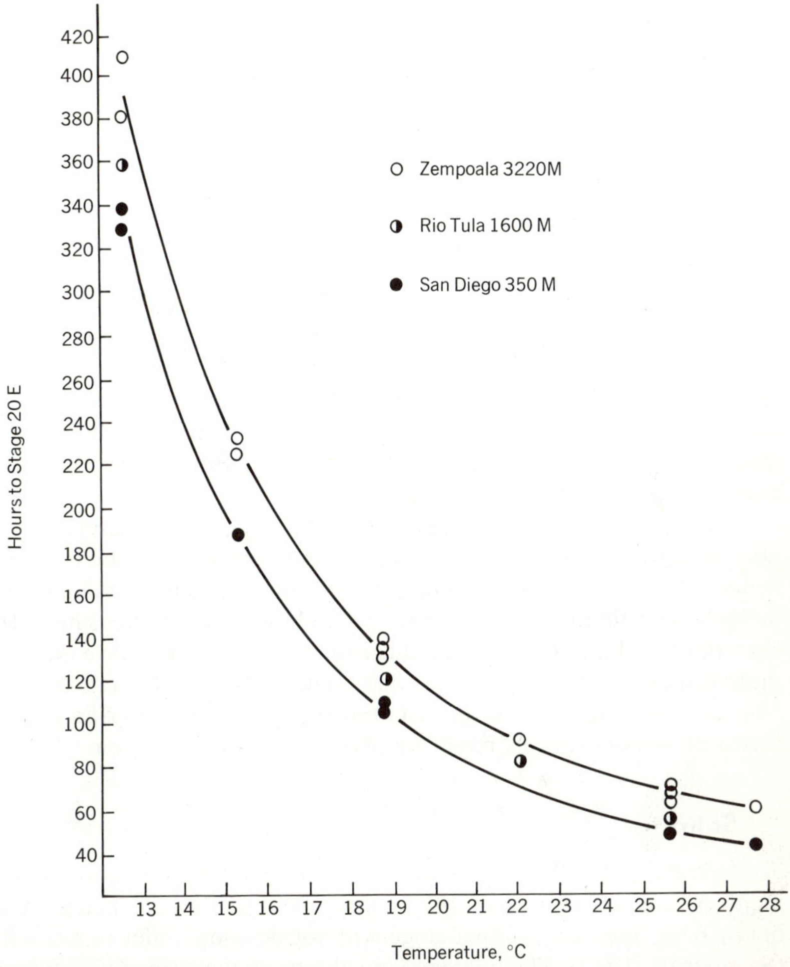

Figure 8-8. Development rates to stage 20E of frog embryos (*Rana pipiens*) from three localities at different altitudes in Mexico. (From Ruibal, 1955.)

Usually, the physiological explanation of Bergmann's rule in terms of heat conservation is that as body size is enlarged, the body bulk increases by the cube, whereas the body surface increases by the square so that the relative amount of body surface decreases. Since the metabolic rate tends to be proportional to body surface, a smaller body with its relatively large surface would lose more heat, an advantage in a hot climate but a physiological liability in a cold environment. Opponents of this interpretation point out that in warm-blooded animals, such devices as fur, feathers, and circulatory mechanisms are more efficient in preventing heat loss than are shifts in surface/volume ratios (Scholander, 1955). Since burrowing mammals are protected from the cold, they seldom obey Bergmann's rule. Body size is more dependent on the availability of food. Exceptions such as these have led Mayr (1965) to speculate that some selective factor in the environment, other than temperature, overrides the size trend in all cases of exception to Bergmann's rule.

The application to poikilothermic animals of these ecogeographical rules has been categorized as follows: (1) The rules are generally applicable. (2) An inverse relationship exists; for example, tropical poikilotherms are larger than northern animals. (3) Since poikilotherms produce no significant amount of metabolic heat, the rules are meaningless. Evidence, however, indicates that these rules are applicable to some poikilotherms. When 12 species of poikilotherms were grown in the laboratory at different temperatures and 4 additional species were studied in the field, about 75 percent of the species conformed to Bergmann's rule and 70 percent to Allen's rule (Ray, 1960).

Temperature may influence other physical factors, which in turn may affect the growth response of animals. For example, since the viscosity of water is lowered with increasing temperature, an organism will have more difficulty remaining near the surface when H_2O temperature is high. It has been observed that many organisms from warmer waters have more elaborate spines or other appendages that facilitate floating than do organisms from colder waters. Field and laboratory studies of winter and summer forms of a given species have corroborated this observation.

Salinity

Salinity may differentially influence various stages of larval development. Although the jellyfish *Aurelia aurita* will produce normal germ cells below 6‰, the scyphistoma stage will not develop under this condition (Segerstråle, 1957). The influence of salinity on the early development of a marine animal was studied on two populations of the polychaete worm *Nereis diversicolor* (Smith, 1964), and the resulting data (Fig. 8-9) demonstrate certain trends: (1) The population from the low-salinity waters of Finland are better adapted for the low-salinity environment in terms of reproduction and development than is the Swedish population; (2) the most critical developmental stage in response to salinity extremes is cleavage; and (3) the salinity range for cleavage is 9–27‰ for Swedish populations and 5–14‰ for Finnish animals.

Rate of development can also be affected by salinity. At temperatures above 8°C, the rate is progressively retarded in the amphipod *Gammarus duebeni,* both in 2‰ and 30‰, in relation to values obtained in 10‰

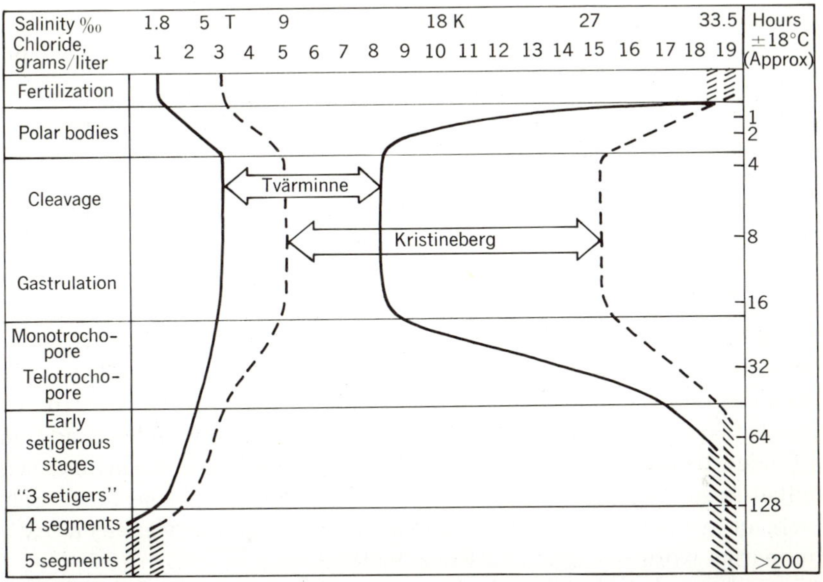

Figure 8-9. Salinity limits and developmental time for larvae of *Nereis diversicolor,* a polychaete, from two geographically separated areas—Tvarminne, Finland, and Kristineberg, Sweden. *T* and *K* represent approximate surface salinity of these areas. The stages and times are approximations at about 18°C. (From Smith, 1964.)

(Kinne, 1953). In fish populations from different saline conditions, the rate of development may be increased or decreased, depending on origin of the population (Heuts, 1956).

Various reports indicate that salinity also influences the growth of organisms, but not always in a predictable way: (1) The largest specimens of the clam *Rangia cuneata* were found at low salinities; (2) the smallest specimens of the clam *Mytilus* were observed at low salinities; and (3) the smallest specimens of the snail *Neritina virginea* were found at either salinity extreme. Furthermore, salinity apparently influences the coloration and thickness of molluscan shells, but not in a consistent manner in all species.

Dissolved Gases

Fertilized eggs of many fish are more sensitive to oxygen lack or low oxygen levels only during certain periods of their development. For example, the time of greatest sensitivity to hypoxia in developing eggs of the chum salmon is between 100 and 200 Celsius-degree days. (A Celsius-degree day is an expression of time and temperature, and is calculated by multiplying the number of days by temperature.) The chief compensatory mechanisms of the eggs to reduced oxygen tensions are a slowing down of the rate of development and a reduced metabolic rate (Alderice *et al.*, 1958).

Hatching of eggs is also influenced by the concentration of other dissolved gases. Hydrogen decreases hatching times in salmon, and excess oxygen inhibits hatching in *Fundulus*. When salmon eggs are exposed to low dissolved-oxygen levels just prior to hatching, they hatch prematurely.

The amount of dissolved oxygen can affect the growth pattern of organisms, and some polychaete worms, for example, attain a larger size when maintained at low oxygen tension than those kept at higher oxygen levels (Fox and Taylor, 1955). In many animals grown under conditions of reduced concentration of oxygen, the respiratory surface may increase, either by an enlargement in size of the respiratory organs, as in salamanders, or by the development of more numerous respiratory appendages, as in marine worms (Kinne and Rosenthal, 1967).

The development of fish eggs often is influenced by an interaction between oxygen tension and temperature, as illustrated by the mortality rate of the eggs of the fish *Cyprinodon macularius*. At temperatures above 20°C, the mortality is higher in 70 percent air-saturated water than in fully saturated water. At 27.5°C, there is 100 percent mortality in hypoxial water, but in fully saturated water the 100 percent mortality level is not reached until temperatures rise to 36°C. Furthermore, the rate of development is influenced, for in fully saturated water at 27°C, egg development takes 5.5 days, whereas in 70 percent air-saturated water, development takes 6.6 days (Kinne and Kinne, 1962).

Pollutants

Pollutants have been shown to significantly affect reproductive and growth phenomena. Two recent studies, one of the effects of a common industrial waste (sulfuric acid) and the second of the accidental spillage of oil at sea, will illustrate the harm that pollutants can cause.

In a concentration of 1:24,000 and 1:32,000, sulfuric acid has only limited effects of fertilization and development of the herring *Clupea harengus*. Under more concentrated conditions, however, harmful effects are noted: (1) The percentages of successful fertilization and of egg survival are reduced; (2) embryonic growth rate is retarded; (3) duration of incubation is shortened; (4) the percentage of successful hatching is decreased; and (5) more structural abnormalities are observed in freshly hatched larvae. Although a concentration of 1:32,000 had little effect on development, young larvae failed to perform prey-catching maneuvers at this concentration. Death resulted within a few days at a concentration of 1:8000 (Kinne and Rosenthal, 1967).

The accidental introduction of oil into the aquatic environment has long been a problem, but recent accidents have received world attention after being dramatically televised to a world audience. One prominent case was the release of crude oil from the oil carrier *Torrey Canyon* when the ship was stranded off the Cornish coast of Great Britain. To remove the spilled oil from the beach area where marine life was being harmed, detergents were introduced. Experimental studies of one prominent intertidal-zone species, the barnacle *Eliminus modestus,* demonstrated that detergents are more toxic than oil; it was observed that the young developmental stages were particularly more sensitive than adults (Corner *et al.*, 1968).

Pollutants, which have now become a basic biological problem throughout the environment, are discussed in more detail in Chapter 12.

SUMMARY

Both sexually and asexually reproducing animals have evolved mechanisms to permit genetic variability for meeting the demands of the environment, although sexually producing animals have the added advantage of genetic recombination.

The environment greatly influences all aspects of the reproductive process. The initiation of the reproductive cycle depends on the interaction between endogenous factors and such exogenous factors as food, ions and water, temperature, oxygen, photoperiod, and lunar cycles. In animals that have external fertilization, timing is very important, and in many of these animals, environmental factors interact with behavioral and physiological mechanisms to permit synchronization of release of gametes.

Early stages in the growth and development of an animal are particularly sensitive to environmental influence, and in some animals the adult response to a particular environmental factor may reflect the environment in which the animal was reared. Because of the unusual sensitivity of larval stages to exogenous factors, chemical and physical pollutants may harm them at levels that would not affect the adult stages.

References

Alderice, D. F., W. P. Wickett, and J. R. Brett. 1958. Some effects of temporary exposure to low dissolved oxygen levels on Pacific salmon eggs. *J. Fish. Res. Bd. Canada,* 15:229–250.

Barnes, H., and M. Barnes. 1967. The effect of starvation and feeding on the time and production of egg masses in the Boreo-arctic cirripede *Balanus balanoides* (1.). *J. Exptl. Mar. Biol. Ecol.,* 1:1–6.

Barnes, H., and M. Barnes. 1968. Egg numbers, metabolic efficiency of egg production and fecundity; local and regional variations in a number of common cirripedes. *J. Exptl. Mar. Biol. Ecol.,* 2:135–153.

Belding, D. L. 1910. A report upon scallop fishery of Massachusetts, including the habits, life history of *Pecten irradians,* its rate of growth and other factors of economic value. *Spec. Rept. Comm. Fish and Game, Mass.,* 1–150.

Breneman, W. R. 1955. Reproduction in birds: the female. *Mem. Soc. Endocrinol.,* 4:94–113.

Chaet, A. B. 1966. Neurochemical control of gamete release in starfish. *Biol. Bull.,* 130:43–58.

Clark, L. B., and W. N. Hess. 1942. Swarming of the Atlantic palolo worm, *Leodice fucata* (Ehlers). *Pap. Tortugas Lab.,* 33: 21–70. Carnegie Inst. Washington, Publ. 524.

Corner, E. D. S., A. J. Southward, and E. C. Southward. 1968. Toxicity of oil-spill removers ("detergents") to marine life: an assessment using the intertidal barnacle *Elminius modestus. J. Mar. Biol. Assoc. U.K.,* 48:29–47.

Engels, W. L. 1961. Photoperiodism and the annual testicular cycle of the bobolink (*Dolichonyx oryzivorus*) a transequatorial migrant, as compared with two temperate zone migrants. *Biol. Bull.,* 120:140–147.

Farmanfarmaian, A., A. C. Giese, R. A. Boolootian, and J. Bennett. 1958. Annual reproductive cycles in four species of West Coast starfishes. *J. Exptl. Zool.,* 138:355–367.

Fox, H. M., and A. E. R. Taylor. 1955. The tolerance of oxygen by aquatic invertebrates. *Proc. Roy. Soc. London, B,* 143:214–225.

Giese, A. C., and M. A. Hart. 1967. Seasonal changes in component indices and chemical composition in *Katharina tunicata. J. Exptl. Mar. Biol. Ecol.,* 1:34–36.

Giese, A. C., L. Greenfield, H. Huang, A. Farmanfarmaian, and R. Lasker. 1959. Organic productivity in the reproductive cycle of the purple sea urchin. *Biol. Bull. Mar. Biol. Lab., Woods Hole,* 116:49-58.

Gosner, K. L., and I. H. Black. 1955. The effects of temperature and moisture on the reproductive cycle of *Scaphiopus h. holbrooki. Am. Midland Naturalist,* 54:192–203.

Gray, G., and H. E. Winn. 1961. Reproductive ecology and sound production of the toadfish, *Opsanus tau. Ecology,* 42:274–282.

Harvey, E. B. 1956. *The American Arbacia and Other Sea Urchins.* Princeton University Press, Princeton, N.J.

Heuts, M. J. 1956. Temperature adaptation in *Gasterosteus aculeatus* L. *Publ. staz. Zool. Napoli,* 28:44–61.

Kinne, O. 1953. Zur Biologie und Physiologie von *Gammarus duebeni* Lillj., I. *Z. wiss. Zool.,* 157:427–491.

Kinne, O. 1964. The effects of temperature and salinity on marine and brackish water animals. II. Salinity and temperature salinity combinations. *Oceanog. Mar. Biol. Ann. Rev.,* 2:281–339.

Kinne, O., and E. M. Kinne. 1962. Rates of development in embryos of a cyprinodont fish exposed to different temperature-salinity-oxygen combinations. *Can. J. Zool.,* 40:231–253.

Kinne, O., and H. Rosenthal. 1967. Effects of sulfuric water pollutants on fertilization, embryonic development and larvae of the herring, *Clupea harengus. Mar. Biol.,* 1:65–83.

Korringa, P. 1957. Water temperature and breeding throughout the geographical range of *Ostrea edulis. Ann. Biol.,* 33:1–17.

Licht, P. 1966. Reproduction in lizards: influence of temperature on photoperiodism in testicular recrudescence. *Science,* 154:1668–1670.

Marler, P., and W. J. Hamilton, III. 1966. *Mechanisms of Animal Behavior.* Wiley, New York.

Marshall, A. J. 1960. Annual periodicity in the migration and reproduction of birds. *Cold Spr. Harb. Symp. Quant. Biol.,* 25:499–505.

Marshall, A. J. 1961. Breeding seasons and migration. *In* A. J. Marshall (ed.), *Biology and Comparative Physiology,* Vol. II, pp. 307–339. Academic Press, New York.

Marshall, A. J., and H. J. de S. Disney. 1957. Experimental induction of the breeding season in a xerophilous bird. *Nature,* 180:647–649.

Marshall, S. M., and A. P. Orr. 1957. The life history of the copepod *Calanus finmarchicus* in different latitudes. *Année Biol.,* 33:43–47.

Marshall, S. M., and A. P. Orr. 1964. Grazing by copepods in the sea. *In* D. J. Crisp (ed.), *Grazing in Terrestrial and Marine Environments,* pp. 227–238. Blackwell Oxford, England.

Mayr, E. 1965. *Animal Species and Evolution.* Belknap, Cambridge, Mass.

Moore, J. A. 1946. Incipient intraspecific isolating mechanisms in *Rana pipiens. Genetics,* 31:304–326.

Moore, J. A. 1947. Hybridization between *Rana pipiens* from Vermont and eastern Mexico. *Proc. Nat. Acad. Sci.,* 33:72–75.

Moore, J. A. 1949a. Geographic variation of adaptive characters in *Rana pipiens* Schreber. *Evolution,* 3:1–24.

Moore, J. A. 1949b. Patterns of evolution in the genus *Rana. In* G. L. Jespen, G. G. Simpson, and E. Mayr (eds.), *Genetics, Paleontology and Evolution.* Princeton University Press, Princeton, N.J.

Moore, J. A. 1950. Further studies on *Rana pipiens* racial hybrids. *Am. Nat.,* 84:247–254.

Orton, J. H. 1920. Sea-temperature, breeding and distribution in marine animals. *J. Mar. Biol. Assoc. U.K.,* 12:339–366.

Panikkar, N. K., and R. G. Aiyar. 1939. Observations on breeding in brackish-water animals of Madras. *Proc. Indian Acad. Sci.,* B9:343–364.

Ray, C. 1960. The application of Bergmann's and Allen's Rules to poikilotherms. *J. Morphol.,* 106:85–108.

Reish, D. J., and J. L. Barnard. 1960. Field toxicity tests in marine waters utilizing the polychaetous annelid *Capitella capitata* (Fabricius). *Pacific Nat.,* 1:1–8.

Remane, A., and C. Schlieper. 1958. *Die Biologie des Brackwassers.* E. Schweizerbart'sche Verlagsbuchnandlung, Stuttgart.

Ruibal, R. 1955. A study of altitudinal races in *Rana pipiens. Evolution,* 9:322–338.

Sastry, A. N. 1963. Reproduction of the bay scallop *Aequipecten irradians* Lamarck. Influence of temperature on maturation and spawning. *Biol. Bull.,* 125:146–153.

Sastry, A. N. 1966a. Temperature effects in reproduction of the bay scallop, *Aequipecten irradians* Lamarck. *Biol. Bull.,* 130:118–134.

Sastry, A. N. 1966b. Variation in reproduction of latitudinally separated populations of two marine invertebrates. *Amer. Zool.,* 6:325–326.

Sastry, A. N. 1968. The relationships among food, temperature, and gonad development of the bay scallop, *Aequipecten irradians* Lamarck, *Physiol. Zool.,* 41:44–53.

Schoener, A. 1968. Evidence of reproductive periodicity in the deep sea. *Ecology,* 49:81–87.

Scholander, P. F. 1955. Evolution of climatic adaptation in homeotherms. *Evolution,* 9:15–26.

Segerstråle, S. G. 1957. Baltic Sea. *In* J. W. Hedgpeth (ed.), *Treatise on Marine Ecology and Paleoecology,* Vol. I, pp. 751–800. Geol. Soc. Am. Mem. 67.

Smith, R. I. 1964. On the early development of *Nereis diversicolor* in different salinities. *J. Morphol.,* 114:437–464.

Sonneborn, T. M. 1957. Breeding systems, reproductive methods, and species problems in Protozoa. *In* E. Mayr (ed.), *The Species Problem,* pp. 155–324. A.A.A.S. Symposium, Washington, D.C.

Stauber, L. A. 1950. The problem of physiological species with special reference to oysters and oyster drills. *Ecology,* 31:107–118.

Taning, A. V. 1952. Experimental study of meristic characters in fishes. *Biol. Rev.,* 27:169–193.

Thorson, G. 1950. Ecology and larval development of marine invertebrates. *Biol. Rev.,* 25:1–45.

Vatovo, A. 1949. La Fauna Bentonica dell'Alto e Medio Adriatico. *Nova Thalassia,* 3:1–110.

Winn, H. E., and A. R. Picciolo. 1960. Communal spawning of the glassy darter *Etheostoma vitreum* (Cope). *Copeia,* 3:186–192.

9

Population Regulation

Although reproduction and development are clearly influenced by the interaction of environmental and endogenous factors, even under the most favorable circumstances, no animal population will increase in numbers indefinitely. Just why populations do not increase indefinitely has been the subject of vigorous debate. Biologists have disagreed not only about the way in which population size is limited, but also about the concept that animal populations are regulated (Ehrlich and Birch, 1967). The limitation of animal numbers generally is said to be either density-independent or density-dependent. Density-independent factors are extrinsic, acting independently of population density, and include climate and man-made changes in the environment such as those resulting from industrial pollutants. Density-dependent factors have been defined as intrinsic factors that enable a population to grow when population densities are low, and which limit growth when densities are high. Competition, predation, and disease are some of the density-dependent factors that can interact with the environment to limit population size.

DENSITY-INDEPENDENT FACTORS

The theory of limitation of populations through density-independent factors, particularly weather, is associated with Andrewartha and Birch (1954). This theory postulates that natural populations are inherently unstable and that both distribution and abundance of animals are constantly changing. Short-term effects on populations can be observed in seasonal changes, and long-term effects are associated with changes in the weather. For example, fluctuations in numbers of a natural population of Australian thrips, *Thrips imaginis,* were followed over a 14-year period. Nymph and adult stages of this insect live in roses and depend upon them for survival and reproduction. Annually, the peak of the thrips population occurs in the spring and coincides with the peak numbers of flowers that appear at this time of the year. The number of flowers increases more rapidly than the thrips so that the environment of these insects is unlimited. Following the spring, however, a drought season sets in, the flowers thin out, breeding places are farther apart, and the numbers of thrips decrease (Fig. 9-1). The decrease in numbers of thrips is not related to the abundance of food, however, since the small numbers remaining on the flowers would not be enough to consume all available food. Instead, the seasonal changes in numbers of thrips are due to the seasonal changes in the accessibility of food, and the variations in numbers of thrips from year to year are linked to the length of time that these animals have to multiply (Davidson and Andrewartha, 1948a,b). If the season is long, there is a greater number of thrips. Therefore, population size is said to be related to the weather, and limitation is said to occur regardless of the density of the population.

These conclusions have been critized by several ecologists (Solomon, 1957; Nicholson, 1958; Smith, 1961) who contend that without a density-dependent controlling factor, there would not have been such a close correlation between weather and fluctuation in numbers of thrips.

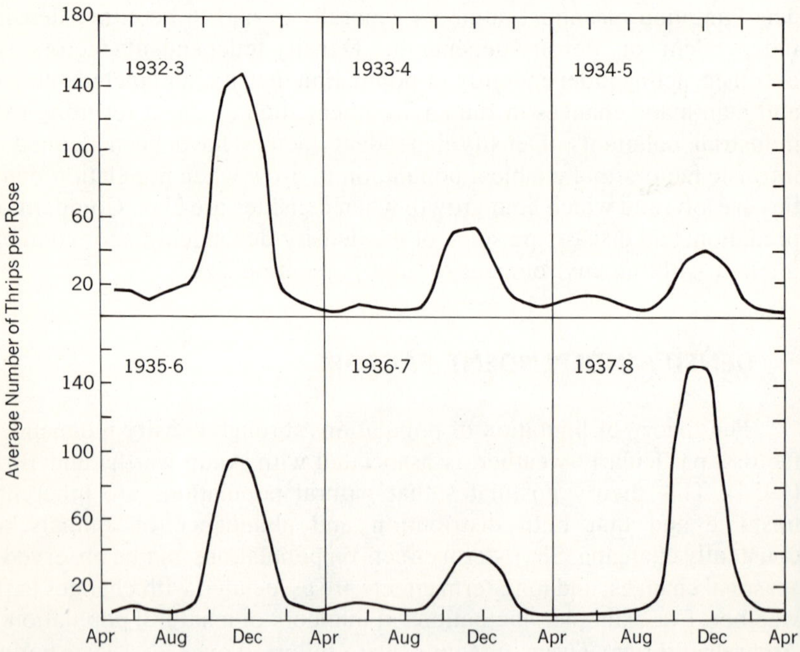

Figure 9-1. Seasonal fluctuations in a population of adult thrips living on roses. (From Andrewartha and Birch, 1954.)

DENSITY-DEPENDENT FACTORS

Regulation of populations through density-dependent mechanisms is a more widely accepted concept than that of population limitation independent of density. This theory implies that population density and the numbers which the environment can support are innately balanced through a homeostatic process; that is, population increases when the population density is low, and decreases when population size is high. Numerous studies indicate that the size of natural populations generally tends to be stable and regulated despite some fluctuation in numbers. The natural populations of birds in England, for example, were studied thoroughly by Lack (1954), who noted that most present-day species were present in the nineteenth century and that many occurred even in the Middle Ages. Furthermore, he concluded that fluctuations in number that have occurred are due mainly to man. Another study followed a natural population of brook trout for 11 generations, noting a decrease of numbers in some years and an increase in other years. The population fluctuations throughout the years, however, were within quite narrow limits (McFadden *et al.*, 1967).

Although population size generally is constant in nature, there are instances of startling explosive growth among animals that have been transplanted to a new ecological situation. A classic example is the great increase shown by striped bass after being transplanted from the Atlantic Ocean to the Pacific. Between 1879 and 1881 a total of 435 striped bass were transplanted from New Jersey to the San Francisco area. Within ten years they were being caught commercially on the west coast, and by 1899 the yearly commercial catch was 1,234,000 lb (Merriam, 1941). Such long-term population explosions are extremely rare, however.

Even in populations that undergo extremely rapid growth, the number is eventually more or less stabilized, and one of the major problems in population ecology is to determine just how population size is controlled. No one control mechanism is operative for all groups of animals, nor is the control mechanism necessarily the same at all stages in the life history of an organism. Often several mechanisms operating interdependently act to control natural populations, and these include competition, predation, and disease.

Competition

Generally, biologists hold the view that competition between organisms will occur whenever the same essential resources are in demand. Such competition may occur between individuals of the same species (intraspecies competition) or between two or more species (interspecies competition). Food and space are the common essential resources in competition.

Intraspecies Competition

The influence of competition on population has been illustrated by laboratory experiments on the sheep blowfly *Lucilia cuprina* (Nicholson, 1958). In one experiment, flies were cultured and maintained at 25°C in the laboratory under constant conditions. The larvae were fed 50 grams of liver every other day, and water and dry sugar were always present for the adults. Under these conditions the number of adults varied greatly and with a marked periodicity. When the supply of adults was plentiful, many eggs were produced. Once the eggs hatched, there were so many larvae that the food supply was exhausted before the larvae grew large enough to pupate. There were no adult offspring during these periods and the number of adults decreased progressively, with a corresponding decrease in the number of eggs produced. With fewer eggs, larval competition for food was reduced, and the hatched larvae had enough food to grow, pupate, and produce adults. About two weeks after the appearance of the first adults of the new generation, the adult population was again high, many eggs were produced, and the cycle was repeated (Fig. 9-2). Thus, despite oscillations in size, the population has a certain stability in that an indefinite increase or decrease does not occur. The underlying mechanism for this relative stability is competition for food among the larvae (Nicholson, 1958).

Competition for food among adults also can limit population size in these flies, as illustrated by an experiment in which the larvae were given a surplus of food and the adults were restricted to a constant amount of ground liver along with an excess of sugar and water. As in the preceding experiment, there were marked oscillations in population size, but in this situation, when numbers of adults were high, few or no eggs were produced because an adult fly requires a certain amount of protein before it can produce eggs. When numbers of adults reached a certain level, there was not enough protein for egg production in all flies, so that few adults were replaced with a new generation. Competition for the protein was thereby lessened, and some flies were able to obtain a sufficient amount for egg production. Since the larvae had an overabundance of food available, nearly all the hatched eggs were able to reach maturity. The final result was the same as when the larvae competed for food, that is, when the number of flies was very large, few adult offspring were produced.

The competition observed in these studies has been referred to as a "scramble" (Nicholson, 1958); that is, competition occurs between individuals with unlimited access to common nutrients, and the most efficient

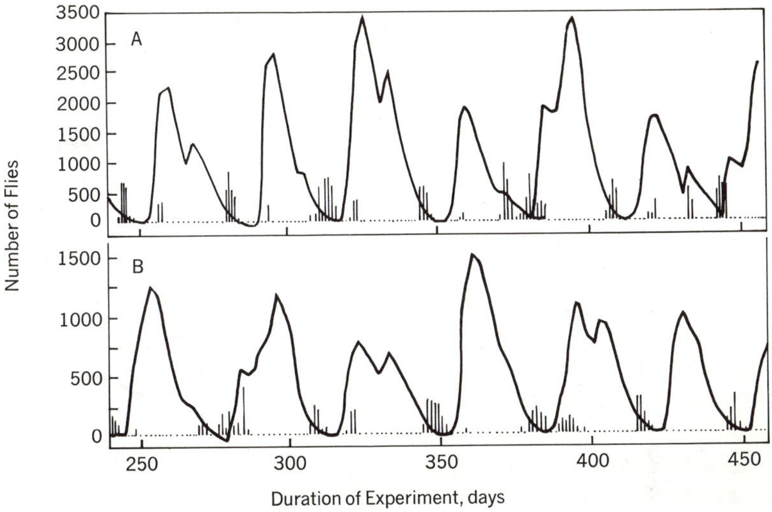

Figure 9-2. Changes in numbers of blowflies, *Lucilia cuprina,* over a period of time. The heavy line indicates the number of adults in the cage, vertical lines indicate numbers of adults eventually emerging from eggs laid on dates of plots. In both cultures conditions were identical except that the daily quota of larval food was 50 g in *A* and 25 g in *B*. (From Nicholson, 1958. Reproduced from *Cold Spring Harbor Symposia,* Vol. XXII.)

competitor is the one that can best exploit the available food. This form of competition, termed *exploitation* by some workers, is more characteristic of the lower metazoans among which physical environment is important and social interactions are not well developed. In higher animals, particularly some vertebrates, competition prevents a competitor, either directly or indirectly, from gaining access to a necessary resource, and this type of competition is known as *interference* (Miller, 1967).

Although the effect of competition in intraspecies population regulation is difficult to demonstrate, some evidence has indicated that competition for food influences the population size of some animals (Lack, 1966). In the wood pigeon *Columba palumbus,* for example, bird numbers and availability of food are directly related. These birds, which breed in farmlands, feed on grain during the summer months and through November, and on clover the rest of the year; eggs are laid during the season when grain is available. Each year that the birds were studied, a marked decrease in population was noted between the end of the breeding season and the beginning of December. The number of losses was linked to the availability of the grain. Further mortality occurred between early December and early March, and its extent during this period of time correlated closely with the clover supplies. Thus, Lack (1966) concluded that the relationship between available food and bird numbers probably was the result of direct competition for food combined with a peck-order, whereby weaker animals would give way and die quickly when stronger animals attacked.

Since animals seldom starve in nature, Wynne-Edwards (1962, 1965) has suggested that although availability of food is ultimately the limiting factor in population control, over population does not occur owing to the behavioral patterns of the animals. Consider, for example, the territoriality of various birds. Available food and breeding sites are parceled out to the stronger members, who then have enough food to survive and a place to breed; the weaker ones, however, have only enough food to survive, but not enough to breed.

Interspecies Competition

Darwin was one of the first to espouse the idea of competition between populations of animals. A number of years later, mathematical models were devised independently by Lotka (1934) and Volterra (1931) to describe relationships between two species utilizing the same food source. These formulas basically indicated that in such a competitive situation, only one species would survive. The formulas were first tested experimentally by Gause (1934) using Protozoa. When he placed two species of Paramecium, *P. aurelia* and *P. caudatum,* together with a fixed amount of bacterial food, *P. aurelia* always successfully eliminated *P. caudatum;* but when *P. bursaria* was placed in the same experimental situation with *P. caudatum,* these two species were able to coexist because *P. bursaria* fed only on the bacteria on the bottom, whereas *P. caudatum* fed on suspended bacteria. Thus, since these two species had different feeding requirements, both were able to survive. The concept that developed from this observation (that is, that no two species with identical niches can occupy the same environment) is often referred to as the *Gause principle* or axiom.

Gause was also one of the first to recognize the role of physiological factors in the growth of a population and to provide a direct demonstration of how the outcome of biological competition in that system can be altered if the physical environment is changed. For example, when Gause (1935) frequently changed the culture medium, the pattern of competition between the protozoan species, *Paramecium caudatum* and *Paramecium aurelia,* was altered. The accumulation of metabolic wastes had a greater effect on *P. caudatum* than on *P. aurelia,* so that the survival of *P. caudatum* was enhanced if metabolites were not allowed to accumulate.

The outcome of competition between species is often determined by only slight alterations in the environment. For example, temperature and humidity can determine the outcome of competition between two species of flour beetles, *Tribolium castaneum* and *T. confusum* (Park, 1954). In experiments made under conditions of high temperature and moisture, *T. confusum* was always eliminated, but under low temperature and humidity, *T. castaneum* was eliminated. One of the most interesting aspects of this study, however, was the intermediate outcome during intermediate conditions of temperature and moisture: Although *T. confusum* was able to eliminate *T. castaneum* in 71 percent of 28 replicates at 24°C and 70 percent relative humidity, the outcome was just the reverse in 29 percent of the replicates.

Undoubtedly the outcome of competition is due to the complex interaction of many factors, and competitive exclusion in the laboratory does not necessarily preclude coexistence in natural populations. In a study of whirligig beetles of the genus *Dineutes,* Istock (1966) noted that initial concentration, environment, and behavior all contributed to alteration of the competitive outcome. In laboratory experiments with two species of whirligig beetles, *D. nigrior* and *D. horni, D. horni* generally was more successful. At 30°C, *D. horni* very quickly overcame its competitor, whether or not it initially had the larger number of individuals (Fig. 9-3). At 25°C, however, *D. horni* successfully excluded *D. nigrior* only when *D. horni* was initially more abundant. If *D. nigrior* was initially more abundant, the species coexisted. Istock attributed the greater success of *D. horni* to two main factors: larval aggression, which is food-dependent and leads to cannibalism; and a lower age-specific survival and higher per larva demand in *D. horni* than in *D. nigrior*. As a result, when the two species are placed together, the amount of food becomes critical. Food supplies large enough to suppress aggression in *D. nigrior* are not high enough to suppress aggression in *D. horni,* and therefore individual *D. horni* generally are more aggressive than *D. nigrior* and most cannibalistic deaths fall to *D. nigrior*. In nature, despite this behavioral advantage of *D. horni, D. nigrior* does manage to coexist with *D. horni* in small ponds in Michigan, apparently because of the earlier onset of breeding. Since the ponds are small, high initial numbers of *D. nigrior* can develop before *D. horni* breeds so that a substantial population is maintained in more southerly ponds. *D. nigrior* has complete dominance in more northerly areas, where it has the advantage of a higher initial concentration and a more favorable temperature.

One of the first requirements of any study on competitive exclusion of one species by another is to demonstrate that when alone, each species investigated can survive indefinitely in the experimental environment. Al-

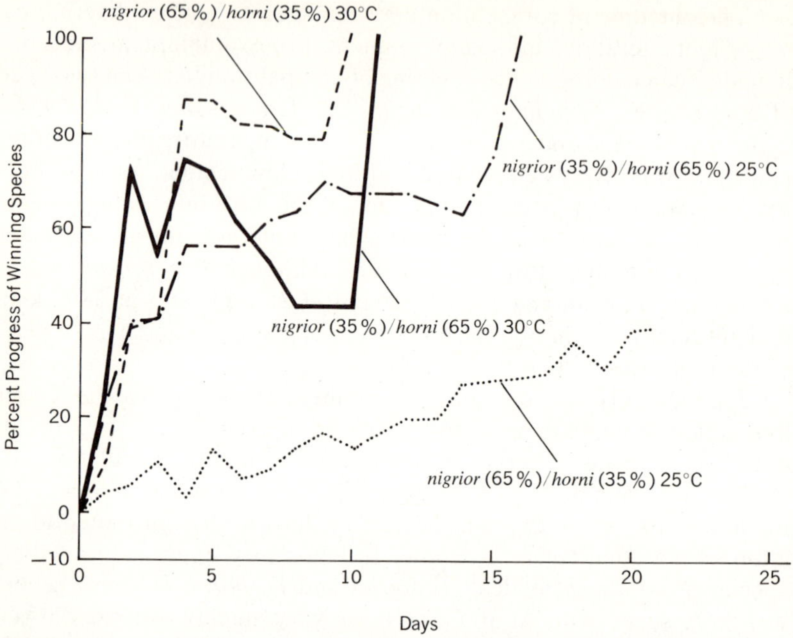

Figure 9-3. The outcome of interspecific competition between two species of whirligig beetles, *Dineutes nigrior* and *D. horni,* at different temperatures and at different initial concentrations. At 30°C *D. horni* eliminates *D. nigrior* quickly even when *D. horni* is less abundant initially. At 25°C *D. horni* excludes *D. nigrior* only when initial concentrations of *D. horni* are greatest.

though it is difficult to obtain this type of information in natural populations, the field study of Connell (1961) on intertidal barnacles does satisfy this requirement. In Scotland, the barnacle *Chthamalus stellatus* inhabits the upper part of the intertidal zone. Another barnacle, *Balanus balanoides,* occupies a slightly lower level of the intertidal zone. It was observed that the young *Chthamalus* often were attached to the rocks in the intertidal area occupied by *Balanus,* but no adult *Chthamalus* were ever observed there. Since barnacles are sessile and fixed in the adult stage, population histories can be accurately recorded by marking the settling plates where the barnacles first settle; using this system, Connell showed that *Balanus,* which lives in the lower zone, grows faster than *Chthamalus*. The *Balanus* would pry *Chthamalus* off the substrate or even grow over them. Because *Chthamalus* can live in the lower intertidal-zone level in the absence of *Balanus, Chthamalus* may be restricted to the higher zone in natural populations, owing to competitive exclusion by *Balanus.*

Predation

To maintain any animal population, organisms within the population must obtain a certain amount of energy from the total food supply. Without enough food, such maintenance is impossible, and a decline in population level is inevitable. Thus, predators face a dual problem: they must consume enough prey to maintain a population while not endangering survival by depleting the prey population.

The interaction between predator and prey is a very complex relationship, and many factors affect it. The size of the predator population is influenced by the quantity of available prey. Consequently, any fluctuations affecting the prey population will also affect the predator population. These two variables, predator and prey densities, form the two basic components of predation, a relationship in which the functional response is predator response to prey density and the numerical response concerns the density of the predator population (Holling, 1959, 1961). The functional response, however, is of central importance because it is measured in terms of food consumption. The numerical response is the outcome of the effects of food consumption on reproduction, immigration, emigration, and mortality (Holling, 1966).

Generally, predators will consume more prey as prey density increases up to a certain point; then prey consumption will level off. Three different types of functional response to increasing prey density have been described. In type I, predation increases in linear proportion to prey density until a plateau is reached (Fig. 9-4); this type of response has been observed in both crustaceans and arachnids. Type II, evident in many insects, is a negatively accelerated rise to a pleateau (Fig. 9-5). Type III is characteristic of most vertebrates that have been studied, and the functional response curve (Fig. 9-6) for these predators is an S-shaped rise to a plateau (Holling, 1965). These differences in response are caused by the behavior patterns of the predators: how they respond to hunger, how well they can hunt, how often they are exposed to prey, and how much time is spent in handling the prey.

That predation is not random can be well illustrated by the studies on deer mice, *Peromyscus leucopus* (Holling, 1959). In nature, these mice have a wide choice of food supplies, which vary widely in abundance and palatability; one of the most palatable to the deer mice is the sawfly cocoon, *Neodiprion sertifer*. In one series of experiments, deer mice were exposed to varying numbers of sawfly cocoons along with abundant supplies of alternate foods. One of these alternate foods, sunflower seeds, was highly palatable; the other, dog biscuits, was less so. When the cocoon densities were low, the deer mice relied primarily on alternate foods and took only a few cocoons. As the number of the sawfly cocoons

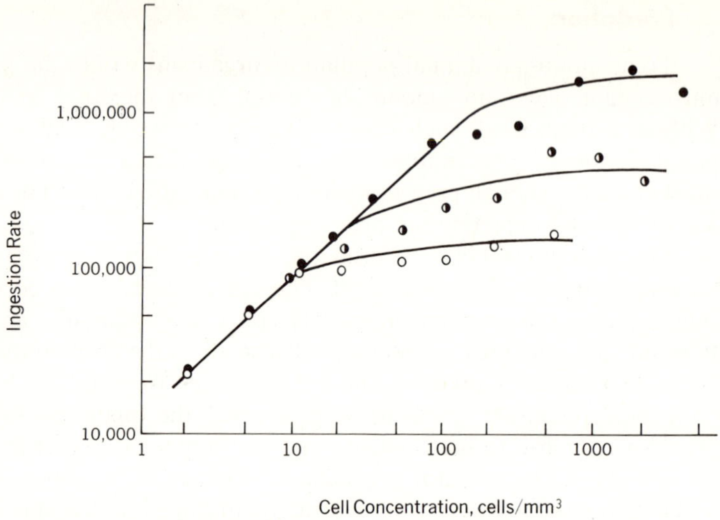

Figure 9-4. Type I functional response to increasing prey density, as illustrated by the response of *Artemia* to increasing cell concentration. ○ represents *Dunaliella;* ◑, *Phaeodactylum;* ●, Chlorella. (From Reeve, 1963.)

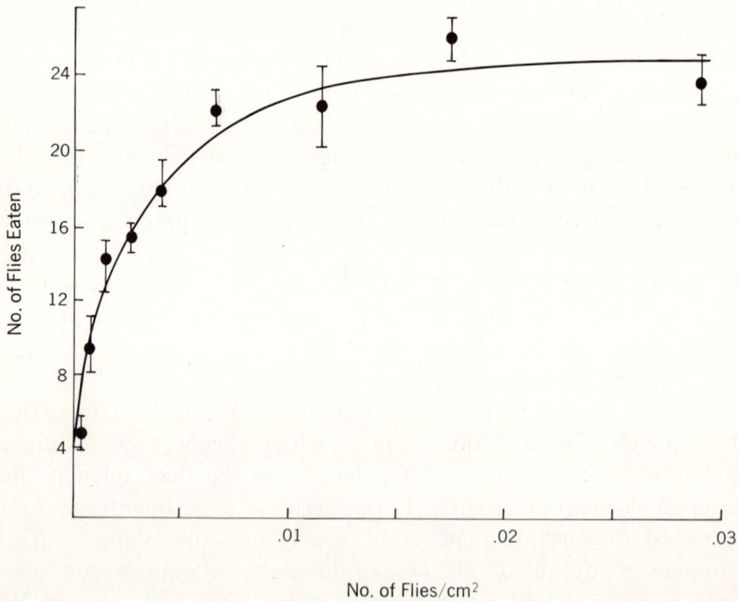

Figure 9-5. Type II functional response as illustrated by the response of individual mantids, *Hierodula crassa,* to density of adult female houseflies. (From Holling, 1965.)

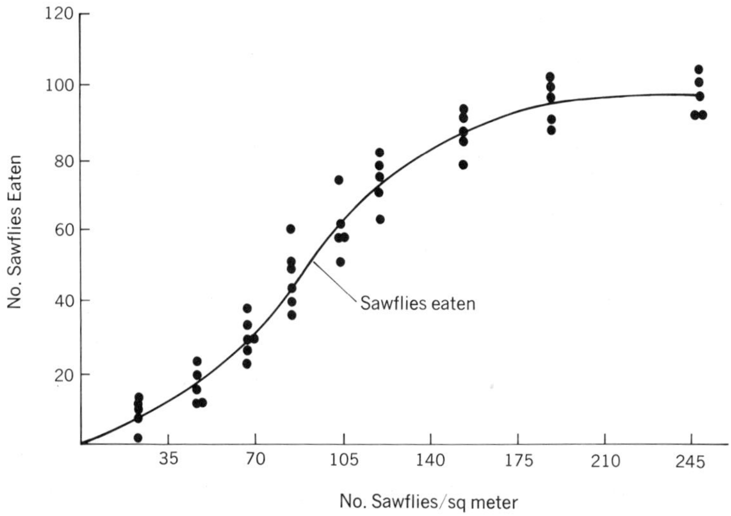

Figure 9-6. Type III functional responses as illustrated by the response of the deer mouse, *Peromyscus leucopus,* to density of sawfly cocoons, *Neodiprion sertifer.* (From Holling, 1965.)

increased, less of the alternate foods was eaten. The number of cocoons eaten, however, soon reached a plateau, and consequently the number of cocoons that were opened each day remained constant. The level of the plateau was higher with the dog biscuits than with the more palatable sunflower seeds. Even with dog biscuits as the only alternate food, however, the deer mice maintained a varied diet, and dog biscuit consumption never dropped to zero (Fig. 9-7).

Predators, such as birds described by Tinbergen (1960), may even search out food when supplies are abundant. The great tit, *Parus major,* feeds primarily on larvae of moths and sawflies found in pine trees during the breeding season. The density of the prey species fluctuates considerably. A study was made of the effect of these variations in numbers of prey that the birds gave to their nestlings as food. Tinbergen found that a new prey species was not eaten until a certain density was reached, but subsequently it was heavily exploited. In many cases the new prey necessitated a particular pattern of searching, and this behavior pattern, called the *specific searching image,* was adjusted to a particular species. In addition to prey density, other factors influence this searching: size, palatability, location, and alternate food species with the same characteristics. Since there can be only a limited number of specific searching images, the birds are able to concentrate on those prey that will give the largest return. Again, when densities rise, the number of any one species of prey taken reaches a plateau, and the birds always maintain a varied diet.

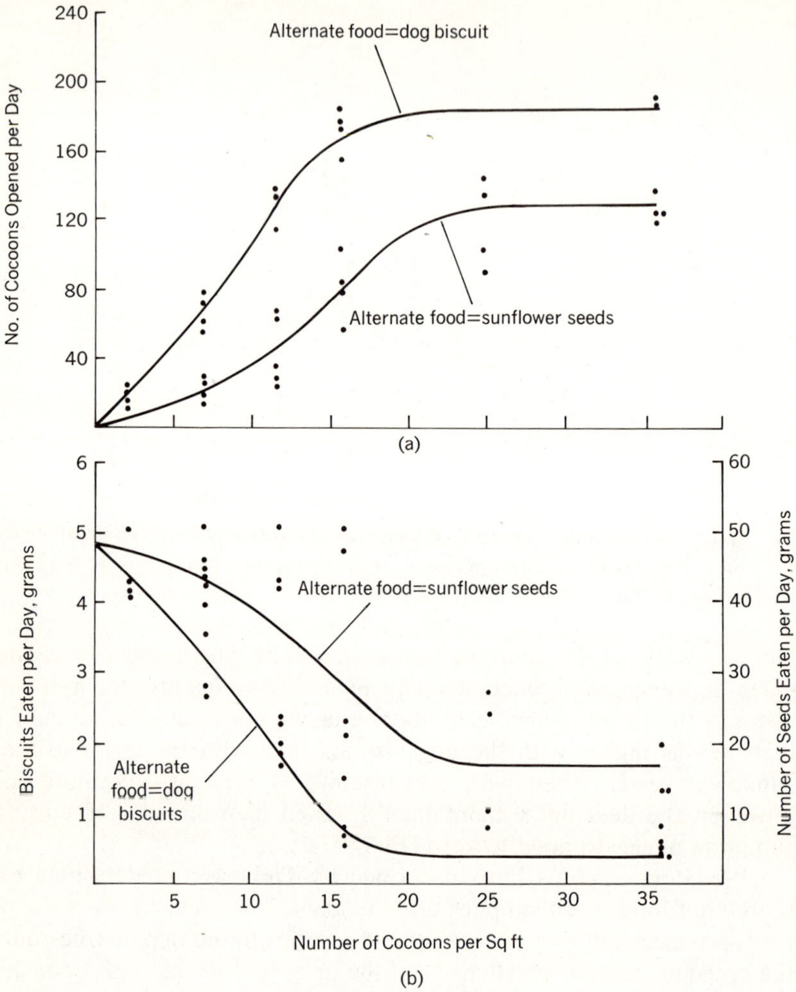

Figure 9-7. The relative acceptability of two alternate foods to a deer mouse, *Peromyscus leucopus* as influenced by the abundance of sawfly cocoons. (*a*) With alternate food present in abundance, the number of sawfly cocoons eaten is plotted against density. (*b*) Cocoon density is compared with amount of alternate food taken. (After Holling, 1959.)

The ability to respond to many stimuli and to screen out those without survival value is one of the chief characteristics of vertebrate predators. Consider this learning effect in deer mice, *P. leucopus*. The mice were raised in captivity on a diet of dog biscuits and then divided into two groups. In one group, sawfly cocoons with living prepupae, a highly palatable food, were buried in the sand in an experimental plot. In the other

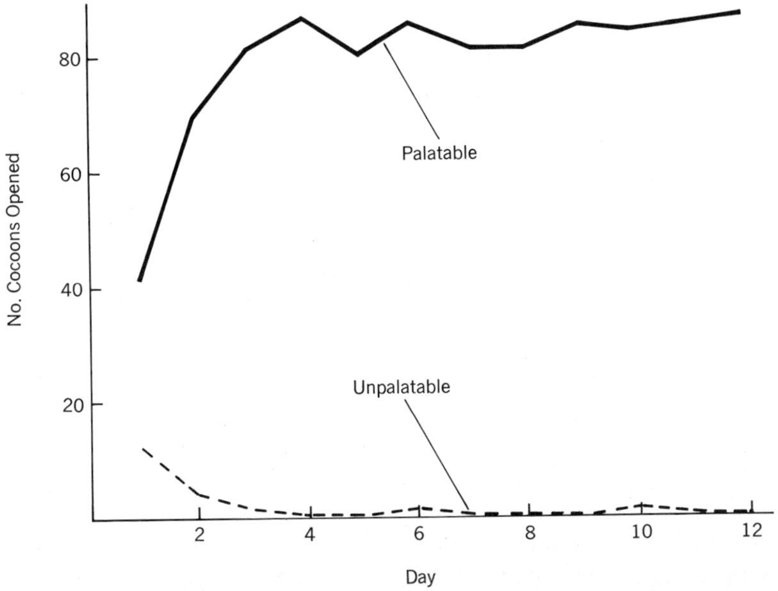

Figure 9-8. Learning effect in deer mice, *Peromyscus leucopus*. One group of mice was exposed to palatable, the other to unpalatable prey. (After Holling, 1965.)

group, cocoons with dead larvae were buired in the same manner. The initial response of both groups was to investigate the cocoons. The rate of attack on the live cocoons increased more than twice the initial rate within four days, whereas the rate of attack on the dead cocoons dropped close to zero (Fig. 9-8). If the mice were presented with both living and dead sawfly cocoons, a process of attraction and avoidance developed through two different channels, odor and taste. These channels were not perma-nent, however, and were not maintained without reinforcement. This learning process allows great flexibility in the exploration of the environ-ment (Holling, 1965).

Both predators and prey have evolved complex adjustments to each other. Many prey species blend into the environment and are difficult to see; their behavior patterns may be secretive and they may avoid ag-gregating in large numbers. Other prey species have well-developed de-fensive mechanisms such as bristles or the ability to procude noxious substances. These animals are conspicuous, as if to advertise their un-palatability, and often are observed together in numbers. The predator, on the other hand, has evolved the ability to find the prey species and to overcome the defensive mechanisms that are presented.

The other basic component of predation, the numerical response, is concerned with the density of the predator population. Cyclic oscillation in numbers is one characteristic common to both predator and prey populations, and often it has been noted that predators exhibit a cycle of abundance with the same frequency as that of the prey, but with a lag in the peaks and troughs of the predator population. The population changes that occur under predator-prey interactions have been observed experimentally in laboratory studies with mites (Huffaker *et al.,* 1963). One of the mites, *Eotetranychus sexmaculatus,* is the prey species, and *Typhlodromus occidentalis* is the predator species. Oranges served as food for the prey species, and the amount of surface exposed on each orange was varied by wrapping it with paper and cutting out circular areas for exposure. The oranges were placed on three shelves in a constant temperature cabinet; the shelves were of wire grid construction, which permitted a three-dimensional sphere of movement. At the start of the experiment, one mature female *E. sexmaculatus* of the prey species was introduced on each of 63 oranges selected at random. This number represented one-fourth of the oranges present. Twenty-eight days later, one mature female *T. occidentalis* of the predator species was added to each of one-third of the oranges, chosen at random from those previously stocked with the prey species. The density of the predator and prey populations oscillated in a characteristic manner (Fig. 9-9). As the number of prey increased, the predator population increased up to a certain point, and then the number of prey was cut back sharply. With the new low density of prey, the predator population declined. In each oscillation there was a lag in the recovery of the predator population until the prey began to increase once again.

Complexity and stability seem to be closely correlated in a predator-prey system. In an uncomplicated predator-prey experiment, the predator usually is so efficient in finding the prey that it soon consumes the entire population and then starves to death. When complications are introduced into the system, such as hiding places for the prey or alternate foods for the predator, the system will become more stable; spatial heterogeneity is also a factor in stability.

How important is predation in the control of prey populations? In the predator-prey relationship discussed above, the prey population, *E. sexmaculatus,* was definitely limited by the presence of the predator population, *T. occidentalis*. When another similar experiment was set up without the addition of predators, the prey population density was 2000 mites per whole-orange equivalent, but the predator density was only 500 mites (Huffaker *et al.,* 1963).

In nature, marked population increase of a prey population may occur once the predator is removed; a striking example is found among the deer of the Kaibab plateau of Arizona. In 1907, the population was only 4000, but after lions and wolves were destroyed to protect the deer, the deer

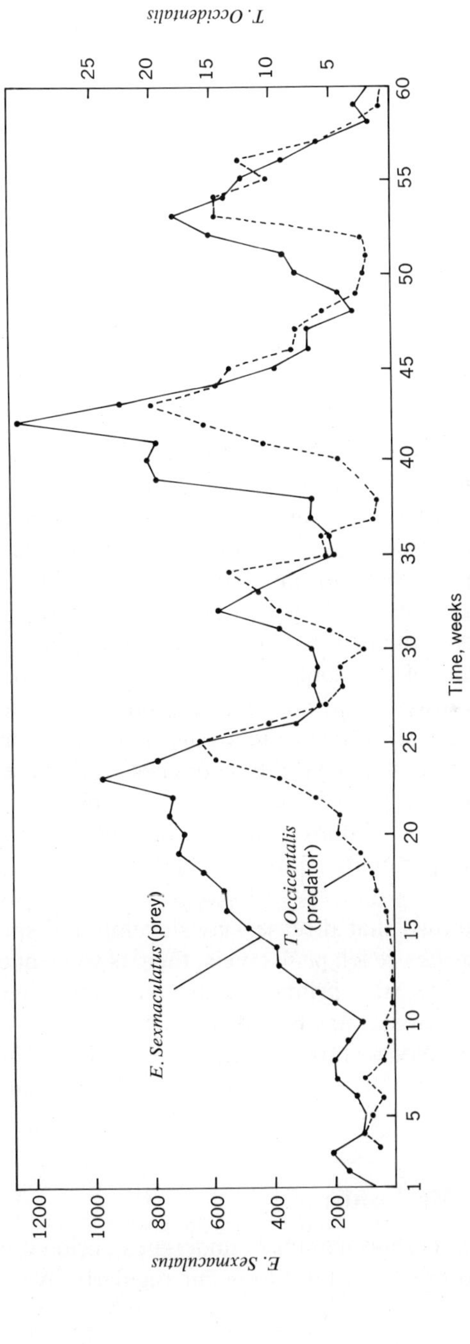

Figure 9-9. Predator-prey interaction between two species of mites, ***Eotetranychus sexmaculatus*** (prey) and ***Typhlodromus occidentalis*** (predator).

population rose to 100,000 by 1924. Since the range could not support such high numbers, it was overbrowsed within two years, and from 1924–1930 approximately 80,000 deer died from starvation. The population decline continued until it leveled off at around 10,000 animals (Rasmussen, 1941). On the other hand, Errington (1946) concluded that in most situations the only mammals that fall prey to predators are ill, old, or in some sense displaced from their natural environment. Perhaps Holling (1965) clarified the regulatory role of predation when he wrote: "So long as a reasonable number of the prey killed by predators would otherwise survive and reproduce, there are strong reasons to believe that vertebrate predation has the qualitative and quantitative features required to . . . affect population numbers."

Disease

At times, outbreaks of infectious diseases can greatly reduce numbers in natural populations. An outstanding example is the relationship between the myxomatosis virus and the European rabbit in Australia. Within 20 years after introduction into Australia in 1859, the rabbit population reached epidemic proportions. To combat the increasing rabbit population, the myxomatosis virus present in South American rabbits was introduced into the Australian population, and it spread so rapidly that approximately 98 percent of the rabbits were killed during the first epidemic. Virulence lessened, however, with each succeeding epidemic, so that by the sixth epidemic the virus caused a mortality of about 25 percent in the rabbits. Three factors lessened viral effectiveness: (1) the evolution by mutation of the virus to a less virulent strain; (2) a passive immunity to myxomatosis in offspring of does that survived the disease; and (3) a genetic change making rabbits more resistant to the virus (Pimentel, 1968).

Although it is known that disease may develop and spread rapidly when population densities reach peak levels, there is some question about whether or not disease is a primary cause of population reduction. Instead, the spread of disease may be linked to changes in host resistance, since laboratory experiments show a marked depression of inflammatory responses in animals from increasing populations (Christian and Davis, 1964).

Other Possible Mechanisms

Population size in certain mammals undergoes periodic fluctuations, and although these oscillations tend to occur regularly over a period of years, the increase and decrease in numbers may be spectacular among some mammals. Perhaps the most famous cyclic animals are the lemmings, which have a three- to four-year cycle. Among the European species, mass migrations may take place when the numbers become very large. The migration begins as a seasonal movement down the mountain, from one vegetation type to the next: from lichens, to willows, to birch forest, to conifers (Curry-Lindahl, 1962). As the lemmings move down the mountain, they breed, and new population highs are reached at the lower elevations. If they encounter a long body of water, the accumulated mass of lemmings panic and move off recklessly in all directions. This behavior pattern is not fully understood, but it may be due to endocrine malfunctioning. Lemming populations from Canada, Alaska, and Greenland, also have definite population cycles, but they seldom undergo the mass migrations reported for the European lemmings.

It is widely thought that periodic fluctuations in population size are more violent in the simpler ecosystems, such as the tundra and the desert, than in the more complex ones, but this probably is not a correct assumption. The vole *Microtus californicus* lives in a Mediterranean climate in a complex grassland ecosystem, and yet these animals not only have a three- to four-year cycle, but also have population fluctuations greater than those reported for northern rodents (Pearson, 1964; Krebs, 1966). Krebs (1966) therefore concluded that, "The idea that periodic fluctuations become more marked as one moves north must be abandoned as far as comparisons between species are concerned."

What are the causes of these cycles? Although climate, food, disease, and predation can cause population changes among these animals, none of these factors appears to control the cycles, and numerous studies substantiate this idea. During the four-year cycle of the brown lemming in northern Canada, there was no depletion of forage and no starvation (Krebs, 1963, 1964). Population cycles of *M. californicus* could not be linked to lack of food, disease, or predation (Krebs, 1966). Furthermore, in a population study of the vole *Microtus agrestis,* 80 percent of the adults were lost in the spring during a four-week period while inhabiting an apparently favorable environment (Chitty *et al.*, 1968).

Thus, none of the more conventional explanations can adequately account for the cyclical changes in population numbers. Krebs (1966) has suggested that the answer may lie in a self-regulatory mechanism wherein the quality or phenotypic fitness of the individual changes as the density of a population changes. There is evidence in some animals of self-regulation through behavioral physiological mechanisms. In dense populations, animals interact more often, and if such meetings are stressful, as indicated in laboratory studies, the final outcome will be glandular malfunction. With increasing population size, there is progressive adrenocortical hypertrophy and thymic involution, which result in a reduced birth rate or an increased death rate. Populations therefore decrease when populations are dense, and increase when sparse. The basic stimulus to endocrine change is a sociopsychological situation rather than actual physical contact such as fighting. Such endocrine changes have not yet been demonstrated in field animals, although it may be possible if the microscopic changes in the endocrine glands and the age and maturation of the animals are evaluated critically (Christian and Davis, 1964).

A genotypic hypothesis of self-regulation also has been suggested: animals born in stationary or declining populations are not the same as those born in expanding populations (Chitty, 1960, 1964; Chitty and Phipps, 1966). A comparison of *Microtus* born in peak and expanding populations showed that males born at peak populations were smaller than those from the expanding population. Females from peak populations had larger litters, but their offspring were smaller at weaning than those born to females from expanding populations. In addition to physical differences in offspring of such populations, Chitty and Phipps have suggested that quality of behavior might be related to genetic changes. This hypothesis is difficult to prove, although it is known that behavioral characteristics in mammals may be affected by genetic composition.

SUMMARY

The question of why and how—or even if—animal populations are regulated is one of the most intriguing in biology today. One school of thought holds that populations are limited in size by extrinsic density-independent factors such as weather. A more widely accepted concept is that populations are regulated by intrinsic density-dependent factors, although it is recognized that the intrinsic factors interact with environmental factors.

Although biologists generally agree that density-dependent factors are operative in population regulation, there is some question about which factors are involved. Some experimental evidence indicates that both intra- and interspecific competition for essential resources such as food and space can limit population size. The predator-prey relationship is another factor that has been implicated as a regulatory influence, as has disease. Some populations, however, show marked fluctuations in size which seemingly are not caused by any of these factors. Instead, behavioral physiological interactions have been suggested as a possible regulating mechanism in these populations. Thus, no single factor satisfactorily explains how population size is limited.

References

Andrewartha, H. G., and L. C. Birch. 1954. *The Distribution and Abundance of Animals*. University of Chicago Press, Chicago.

Chitty, D. 1960. Population processes in the vole and their relevance to general theory. *Can. J. Zool.,* 38:99–113.

Chitty, D. 1964. Animal numbers and behaviour. *In* J. R. Dymond (ed.), *Fish and Wildlife: A Memorial to W. J. K. Harkness,* pp. 41–53. Longmans, Toronto, Canada.

Chitty, D., and E. Phipps. 1966. Seasonal changes in survival in mixed populations of two species of vole. *J. Anim. Ecol.,* 35:313–331.

Chitty, D., D. Pimentel, and C. J. Krebs. 1968. Food supply of over-wintered voles. *J. Anim. Ecol.,* 37:113–120.

Christian, J. J., and D. E. Davis. 1964. Endocrines, behavior, and population. *Science,* 146:1550–1560.

Connell, J. H. 1961. The influence of interspecific competition and other factors on the distribution of the barnacle, *Chthamalus stellatus*. *Ecology,* 42:710–723.

Curry-Lindahl, K. 1962. The irruption of Norway lemmings in Sweden during 1960. *J. Mammal.,* 43:171–184.

Davidson, J., and H. G. Andrewartha. 1948a. Annual trends in a natural population of *Thrips imaginis* (*Thysanoptera*). *J. Anim. Ecol.,* 17:193–199.

Davidson, J., and H. G. Andrewartha. 1958b. The influence of rainfall, evaporation and atmospheric temperature on fluctuations in the size of a natural population of *Thrips imaginis* (*Thysanoptera*). *J. Anim. Ecol.,* 17:200–222.

Ehrlich, P. R., and L. C. Birch. 1967. The "balance of nature" and "population control." *Am. Nat.,* 101:97–107.

Errington, P. L. 1946. Predation and vertebrate populations. *Quart. Rev. Biol.,* 21:144–177; 221–245.

Gause, G. F. 1934. *The Struggle for Existence*. Williams & Wilkins, Baltimore.

Gause, G. F. 1935. La théorie mathématique de la lutte pour la vie. Hermann et Cie., Paris.

Holling, C. S. 1959. The components of predation as revealed by a study of small mammal predation of the European sawfly. *Can. Entomol.,* 91:293–320.

Holling, C. S. 1961. Principles of insect predation. *Ann. Rev. Entomol.,* 6:163–182.

Holling, C. S. 1965. The functional response of predators to prey density and its role in mimicry and population regulation. *Mem. Entomol. Soc. Can. No. 45.*

Holling, C. S. 1966. The functional response of invertebrate predators to prey density. *Mem. Entomol. Soc. Canada, No. 48.*

Huffaker, C. B., K. P. Shea, and S. G. Herman. 1963. Experimental studies on predation: complex dispersion and levels of food in an acarine predator-prey interaction. *Hilgardia,* 34:305–330.

Istock, C. A. 1966. Distribution, coexistence, and competition of whirligig beetles. *Evolution,* 20:211–234.

Krebs, C. J. 1963. Lemming cycle at Baker Lake, Canada during 1959–62. *Science,* 146:1559–1560.

Krebs, C. J. 1964. The lemming cycle at Baker Lake, Northwest Territories, during 1959–62. *Arctic Inst. N. Am. Tech. Pap. No. 15.*

Krebs, C. J. 1966. Demographic changes in fluctuating populations of *Microtus californicus. Ecol. Monographs,* 36:239–273.

Lack, D. 1954. *The Natural Regulation of Animal Numbers.* Oxford. University Press (Clarendon Press), New York.

Lack, D. 1966. *Population Studies of Birds.* Oxford University Press (Clarendon Press), New York.

Lotka, A. J. 1934. Théorie analytique des associations biologiques. *Actualités Scientifiques et Industrielles,* 187:1–45.

McFadden, J. T., G. R. Alexander, and D. S. Shetter. 1967. Numerical changes and population regulation in brook trout *Salvelinus fontinalis. J. Fish. Res. Bd. Canada,* 24:1425–1459.

Merriam, D. 1941. Studies of the striped bass (*Roccus saxatilis*) of the Atlantic coast. *Fish. Bull., U.S. Fish and Wildlife Serv.,* 50:1–77.

Miller, R. S. 1967. Pattern and process in competition. *In* J. B. Cragg (ed.), *Advances in Ecological Research,* Vol. IV, pp. 1–74. Academic Press, New York.

Nicholson, A. J. 1958. The self-adjustment of populations to change. In *Population Studies: Animal Ecology and Demography. Cold Spr. Harb. Symp. Quant. Biol.,* Vol. XXII, pp. 153–172. Cold Spring Harbor, L.I.

Park, T. 1954. Experimental studies of interspecies competition: II, temperature, humidity, and competition in two species of *Tribolium. Physiol. Zool.,* 27:177–238.

Pearson, O. P. 1964. Carnivore-mouse predation: an example of its intensity and bioenergetics. *J. Mammal.,* 45:177–188.

Pimentel, D. 1968. Population regulation and genetic feedback. *Science,* 159:1432–1437.

Rasmussen, D. I. 1941. Biotic communities of Kaibab Plateau, Arizona, *Ecol. Monographs,* 3:229–275.

Reeve, M. R. 1963. The filter-feeding of *Artemia.* I. In pure cultures of plant cells. *J. Exptl. Biol.,* 40:195–205.

Smith, F. E. 1961. Density dependence in the Australian thrips. *Ecology,* 42:403–407.

Solomon, M. E. 1957. Dynamics of insect populations. *Ann. Rev. Entomol.,* 2:121–142.

Tinbergen, L. 1960. The natural control of insects in pinewoods. I. Factors influencing the intensity of predation by songbirds. *Arch. Néerl. Zool.,* 13:265–343.

Volterra, V. 1931. Variation and fluctuations of the number of individuals in animal species living together. *In* R. N. Chapman (ed.), *Animal Ecology.* McGraw-Hill, New York.

Wynne-Edwards, V. C. 1962. *Animal Dispersion in Relation to Social Behaviour.* Hafner, New York.

Wynne-Edwards, V. C. 1965. Self-regulating systems in populations of animals. *Science,* 147:1543–1548.

10

Animal Associations

TYPES OF ASSOCIATIONS

Types of associations among different species of organisms are extremely numerous and varied, ranging from casual contact to absolute physiological dependence of one or both organisms involved. As a result of the many attempts to describe these various types of associations, a rather voluminous and complex terminology has evolved over the years, but unfortunately the various terms do not always have the same meaning to everyone. Here, the term *symbiosis* designates the relationship between two or more species living together, and embraces three main categories: mutualism, commensalism, and parasitism. It should be emphasized, however, that many associations do not exactly fit any of these three categories, but instead represent intergrades between categories.

Mutualism may be defined as a symbiotic relationship that is advantageous to both species involved. In many mutualistic associations, the two species are not together continuously and their dependency may be limited to one factor, as in the relationship between the "cleaner" shrimp and certain reef fishes. The shrimp seem to be responsible for removing the external parasites of the reef fishes, and in turn the shrimp depend on these ectoparasites for their food (Limbaugh *et al.,* 1961). In other mutualistic associations, all stages in the life cycle of the two species concerned may be almost completely interdependent. Mutualistic associations in which the metabolism of both partners is supplemented are distributed widely between insects and various species of protista. Insects house symbionts in their gut, in the Malpighian vessel, and in various regions of the reproductive tract; the consequences of symbiont loss in these insects may include marked retardation in development, stunting of growth, and eventual death in many. In some of these insects, the symbionts supply a necessary growth factor, such as vitamins, to the host animals.

In both commensalism and parasitism, one of the two species involved is largely dependent upon the other. *Commensalism* is commonly defined as a relationship between two organisms which is advantageous to only one of the organisms, but is not harmful to the host animal. In contrast, *parasitism* is described as an association in which the unilaterally dependent species is harmful to the host species. As noted by others, however, pathogenicity is not a valid criterion for judging parasitism because both physiological and environmental conditions may determine whether the dependent organism will have a detrimental effect on the host species. Some organisms classified as parasites will be extremely pathogenic in one host species, but relatively harmless in another host species. On the other hand, some dependent organisms classified as commensals do cause injury to the host, even though these injuries are referred to as inadvertent. For example, the limbs of the pea crab *Pinnotheres ostreum,* which lives in the mantle cavity of the American oyster *Crassostrea virginica,* often become entangled in the gills of the oyster so that when the crab tries to free itself, the gills of the host animal are lacerated. Therefore, Stauber (1945) has classified *P. ostreum* as a parasite, even though this crab is a filter feeder and does not feed off the host tissue. Thus, perhaps a more operational definition would describe parasitism as the metabolic dependence of one species on another, and commensalism as a dependence based on utilization of excess food or seeking of shelter in the burrow or tube of the host animal.

ORIGIN AND EVOLUTIONARY ASPECTS

Symbiotic relationships probably arose early in the origin of organismal existence, but this concept is difficult to prove, owing to a lack of substantiated fossil records. Some relict organisms, however, give clues to the antiquity of symbiotic relationships, as does the present-day but primitive termite *Mastotermes darwiniensis,* which has symbionts exactly comparable to those of modern blattids (Koch, 1938). This termite, a sole survivor of a termite family widely distributed during the Tertiary, has both bacterial and flagellate symbionts. As in blattids, the bacterial symbionts live in the region of the fat body of the termite and are transmitted from one generation to the next through ovarial infection. The flagellate symbiont is similar to the one in the American cockroach, and both are assumed to be descended from common ancestors, since cockroaches and termites are thought to have common progenitors. Thus it seems probable that similar symbiotic relationships occurred in their common ancestors during the Devonian period (Koch, 1967).

Just how symbiotic relationships arose is speculative, but in commensalism and parasitism the progenitors of the dependent organisms undoubtedly were free-living animals having the genotypic make-up that permitted invasions of habitats in or on other animals with which they had intimate or frequent contact. Fleas, for example, probably arose as winged scavenging flies whose larvae fed on the excrement in burrows of mammals (Rothschild, 1965). The larvae of the present-day flea still feed on debris in the nest or lair of the host, but somewhere in time the free-living adults took up residence among the fur of a passing mammal and became dependent on this animal for food. This development represents a direct transition from a free-living to a dependent existence. Not all associations can be interpreted so logically, for some symbionts (particularly the parasitic protozoa and helminths) have exceedingly complex life cycles with several ecologically distinct phases.

An interesting concept of the evolution of complex life cycles postulates that, initially, one part of the life cycle is transferred from the environmental resources at hand to a new set of supplementary resources (Istock, 1967). If this transfer is to occur, two conditions must be met. There must be enough genetic lability—or genetic preadaptation—to enable the organism to utilize a new set of supplementary resources, and this transfer to the supplemental resources must result in a net energy gain to the species. Once the transfer has been made, each of the ecologically distinct phases in the life cycle evolves more or less independently of the others. Thus a complex life cycle tends to be unstable over a long evolutionary time, for the balance is upset if either adult fertility and abundance decrease or the number of larvae passing to the adult stage is reduced to the level where the adult environment is not sufficiently exploited. The way that a complex life cycle evolves is the result of the genotypc of a particular organism. Although some genotypes would tend to complete the entire life cycle and utilize the maximum resources available from each phase, others would tend toward a reduction or loss of the adult phase, and still others toward a reduction of the larval stage(s).

The life cycles of many symbiotic organisms can be fitted into this concept, and an excellent example is the life cycle of the digenetic trematode, which usually involves a number of morphologically and ecologically distinct stages (Fig. 10-1). The adult generally is an intestinal parasite of vertebrates, releasing its eggs into the lumen of the host intestine, from which they are passed to the outside in the feces of the host. The eggs are deposited in the water, where they hatch into a free-swimming miracidium that penetrates the first intermediate host, nearly always

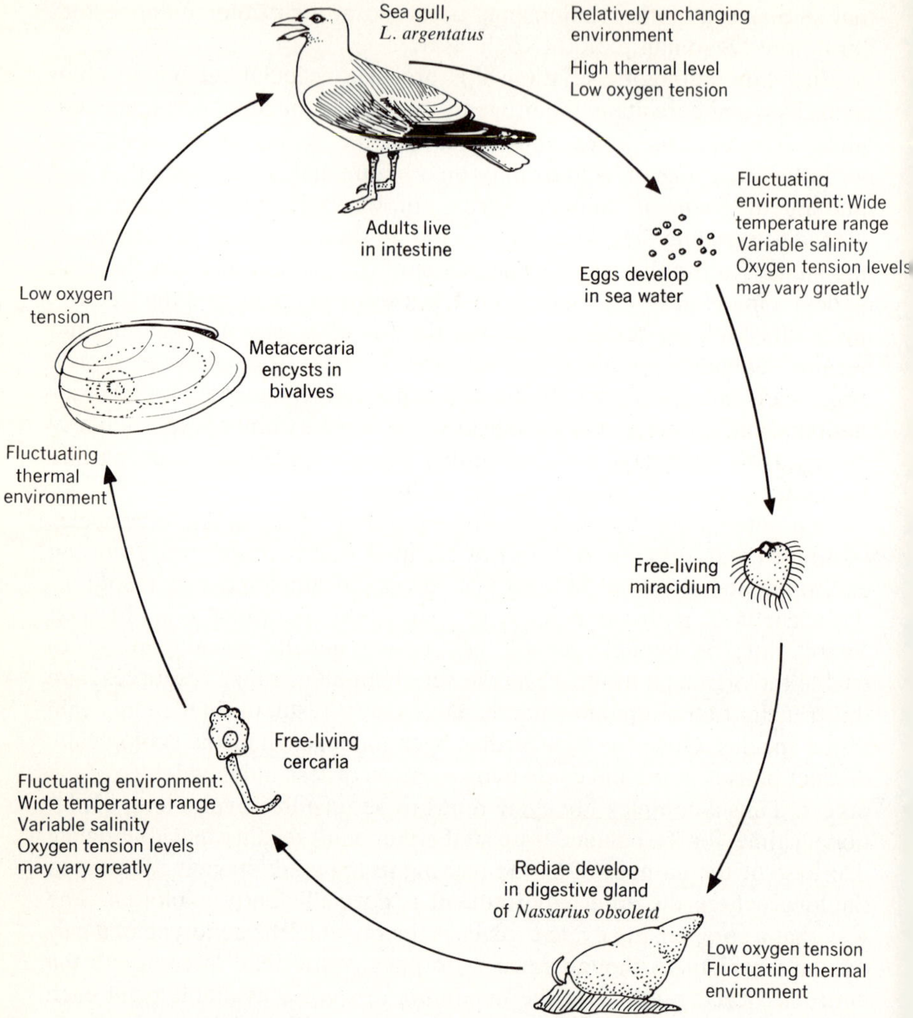

Figure 10-1. The life cycle of the trematode *Himasthla quissentensis.*

a mollusc. In the mollusc, the miracidium develops into a sporocyst stage having germinal cells that eventually differentiate into germ balls. The germ balls then differentiate into the next larval stage, the rediae, which in turn give rise to the free-living cercarial stage. The free-swimming cercaria comes in contact with a suitable host, penetrates and encysts; this stage is called the *metacercarial stage,* and its host may be either an invertebrate or poikilothermic vertebrate host. When this second intermediate host is eaten by a vertebrate (the definitive host), the metacercaria excysts and develops into the adult stage. Not only do stages in the life cycle differ morphologically and ecologically, but also physiologically many responses are quite distinct for each stage of the life cycle. For example, in all species studied to date, the free-living cercarial stage is susceptible to oxygen lack and can survive anaerobic conditions for only a short time, whereas the adult stages are highly resistant to anaerobic conditions. The rediae of *Himasthla quissetensis,* which occur in the digestive gland of the mud-flat snail *Nassarius obsoleta,* are relatively independent of oxygen tension metabolically, but the metabolic rate of the free-living cercarial stage is entirely dependent on oxygen tension (Vernberg, 1963). These differences in response to oxygen deprivation and to lowered oxygen tensions can be correlated readily with environmental conditions of each stage in nature. The free-living cercarial stage has access to a much greater oxygen supply compared with that available to a parasitic stage living in the digestive tract or tissue of another organism.

Different stages in the life cycle have other distinctly different metabolic requirements. The redial stage of *H. quissetensis* readily utilizes amino acids *in vitro,* but cannot utilize sugars. The free-living cercarial stage does not utilize either sugars or amino acids. This stage is a short-lived one, apparently nonfeeding, and relies on stored reserves to carry it through until contact is made with a compatible host, which in this particular cercariae may be any one of several bivalve molluscs. Almost immediately after encystment in the molluscan host, the metacercaria is infective to the definitive host (herring gulls). Once in the definitive host, it can utilize sugars readily, as demonstrated by *in vitro* experiments (Vernberg and Hunter, 1963). Thus, in this complex life cycle, each stage apparently has evolved more or less independently, utilizing a different set of resources and having its own set of limiting factors; yet, it has maintained morphological, physiological, and behavioral characteristics necessary to make the transition from one stage to another.

HOST-SYMBIONT CONTACT

The method of transition from one stage to another is one of the most intriguing aspects of the host-symbiont relationship. For some symbionts the transfer from one host (or habitat) to another depends on the feeding habits of the host species. Cestodes, for example, nearly always have one or more intermediate hosts, and the transfer from one host to another usually is accomplished when the larvae or their host are eaten by another organism. In most symbiotic relationships, however, contact with a suitable host is not accomplished through passive transfer. As a result, it has been speculated that symbionts either may have mechanisms for seeking a host or may search randomly. On the basis of host-parasite interaction in insects, Thompson (1939) discounted the theory of random searching, concluding that host selection by the parasite is generally "both specific and adaptive." Considerable evidence supports this viewpoint, since many symbionts and hosts show specific and adaptive responses toward gaining and maintaining the association.

In the mutualistic relationships between insects and their symbionts, the physiological and behavioral patterns of the host guarantee the continuity of the symbiotic relationships. Insect symbionts may be transferred in one of three ways, depending on their location in the host (Koch, 1967). If they are housed in the region of the gut, the female may ensure that the larvae take the symbionts orally by discharging a few symbionts where the young are sure to feed, or the female may smear the developing eggs with the symbionts, or if the symbionts are housed in the germinal glands, infection with the symbiont takes place before the eggs are laid.

One of the most interesting interdependent symbiotic relationships is the one between hermit crabs and commensal sea anemones. In some of these associations the behavior patterns of the crabs evoke responses from the sea anemones which enable a crab to detach and place anemones on its shell. Experiments with the anemone *Calliactis tricolor* and the crab *Dardanus venosus* in the Caribbean, and *C. polypus* and *D. gemmatus* at Hawaii, showed that when the crabs moved their appendages on the surface of the anemones, the anemones responded by relaxing completely, whereupon the crab could lift the anemone easily (Ross and Sutton, 1968). Further, the investigators could simulate crab movement by gently prodding the column of the anemones with pipe cleaners, and when an electric current was applied to *C. polypus* at a frequency of one stimulus per second, within 20–30 sec the anemone opened and relaxed in the same manner as it did in response to tactile stimuli. Ross and Sutton concluded that the behavior of the crab stimulates a basic motor mechanism of the anemone, producing first a total general inhibition that is then followed by a local contraction at the base, which allows the anemone to be transferred from one substratum or host to another.

In some associations among sea anemones and hermit crabs, however, the crabs show no distinctive behavioral pattern in relation to the anemone. In the relationship between *Calliactis parasitica* and *Eupagurus bernhardus*, Ross (1960) observed that the anemones did not discriminate between empty shells in which no crabs had lived, shells of living molluscs, or shells with crabs living in them. Instead, *C. parasitica* responded to some organic constituent in the periostracum of the molluscan shell; thus, both crab and anemone are bound together by their independent responses to molluscan shells, not by any attraction between these two animals.

In other symbionts, morphological and behavioral patterns have evolved to ensure that the organism will be eaten by a suitable host. To complete the life cycle of the azygid trematode *Proterometra marcrostoma,* the cercarial stage must be eaten by any one of several species of fish belonging to the family Centrarchidae. This cercaria is large, yellow in color, and swims with a vertical movement, and its conspicuous appearance, which is attractive to many fish, ensures that some cercaria will be eaten by centrarchid hosts (Horsfall, 1934).

One of the first experiments to demonstrate symbiont interaction among environmental factors, behavioral patterns, and physiological response to a diffusable substance was conducted by Welsh (1930) with the mite *Unionicola ypsilophorus* var. *haldemani,* which lives in the mantle cavity and on the gills of the fresh-water clam *Anodonta cataracta.* When the mite is under stimulation by host material, it is negatively phototrophic, but the negative phototrophism is reversed when there is no chemical sign from the host animal. This type of response signifies strong adaptive capability of the free-living larval mites; out of contact with host stimulation, they tend to be dispersed and are photopositive to light, but upon contact with the host material or host factor, they become photonegative and fall to a lower level, where they can be ingested more easily by the in-current siphons of their hosts.

A study by Lees (1948) on the sheep tick *Ixodes ricinus L.* also illustrates the interrelationship between ecology and behavioral response among symbionts. This tick is found on rough moorland grazings that have a thick vegetative cover. Near the roots of the vegetation, the humidity is always very high, whereas near the tips of the vegetation the humidity is generally lower, although it may be variable. The unfed tick has an upward-turning response that guides it to the tips of the vegetation, a favorable position for encountering the host; the ticks usually climb from the base of the vegetation to the tip and assume a questing position with the forelegs extended. If the quest for a host is unsuccessful, the tick tends to become desiccated after a period of time; then it walks downward to rest at the roots, absorbs water in this high humidity region, and again moves up to the tips of the vegetation. The unfed ticks respond readily to tactile stimuli, such as any unusual movements of the vegetation or vibrations, and to any light shadows, which serve as a warning of the approach of a possible host. When the host is near, the ticks orient toward warmth and the odor of the sheep. The response to temperature and odor are interrelated, for below 20°C the ticks do not respond to the odor of sheep wool. After the tick is on the sheep and becomes engorged, it drops off; guided by an avoidance reaction to light, it seeks the deep layers of vegetation while it digests its blood meal.

Host recognition may be observed even among the protozoans. In studying the relationships between the parasitic protozoan *Hydramoeba hydroxena* and the hydra *Chlorohydra viridissima,* Stiven (1964) reported three observed methods that could be utilized in the transfer of the protozoan from host to host: (1) The infection could be passed directly from one hydra to another by tentacular contact; (2) since the free amoebae are sticky, they could be picked up when the tentacles of the hydra were trailed over the substrates; and (3) the hydramoebae could migrate up the stalk of the host animal from the substrate. By the first two methods, infection obviously would occur by accidental contact. Because it was not clear, however, whether the third method (that is, the symbiont's climbing up the stalk of the host) was a random or directed movement, uninfected *C. viridissima* were placed in the vicinity of hydra heavily infected with *H. hydroxena.* Chi square analyses between control and experimental dishes after 48 hr indicated that the movements of the protozoan were not random, but rather that the hydramoebae moved differentially into the areas where uninfected hosts were present.

Growing evidence suggests that the specific recognition factor mediating many symbiotic associations is chemical in nature, particularly in aquatic partnerships. One of the most common experimental procedures for studying host-finding behavior is to use some type of "choice apparatus"; the methods described by Sastry and Menzel (1962) serve as an illustration of the general procedures involved in such studies. These investigators were interested in the attraction of the commensal crab *Pinnotheres maculatus* to its host, the bay scallop *Aequipecten irradians* and the penshell *Atrina rigida.* The circular "choice apparatus" used in these experiments was designed and described by Bartel and Davenport (1956). See Fig. 10-2. A host was introduced into one of the radial chambers, and after a period of equilibration the crabs were placed into the central chamber. The behavior of the crabs and their subsequent distribution were observed over a period of several hours, and then the distribution of the crabs in the radial chambers was noted and analyzed statistically. Results of these experiments showed that the crab is capable of recognizing its host; the crabs were equally attracted to both scallops and pen shells, but only when the hosts were living, for empty shells did not attract them. It is noteworthy, however, that when water from an aquarium containing a host was siphoned into one of the radial chambers,

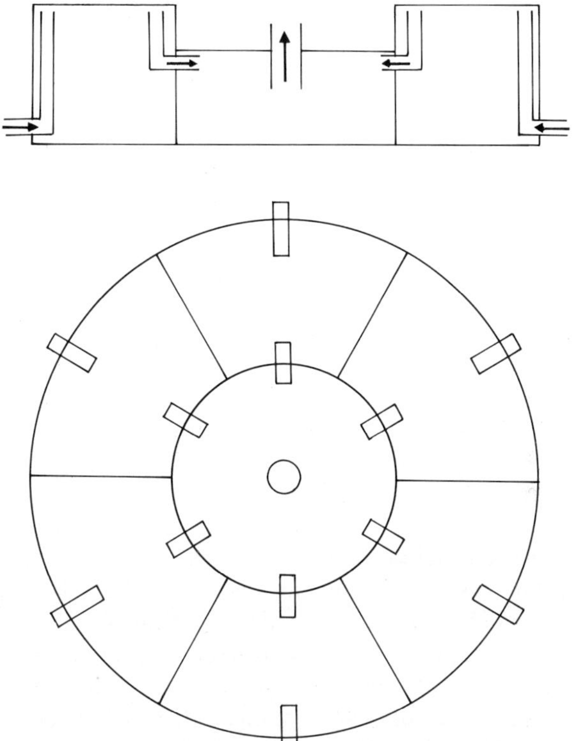

Figure 10-2. Choice apparatus for studying host-finding behavior. The upper diagram shows the apparatus in section. See text for details. (After Bartel and Davenport, 1956.)

attraction to the crab was lost. The authors suggested that "perhaps a spatial proximity of hosts to commensals is necessary for demonstration of attraction under experimental conditions. The decreased attraction would have resulted either from a gradient or a highly diffusible nature of the attractant." Thus, although the host does produce an attractant to the crab, this attractant does not permeate over long distances.

In studying host-finding by trematode miracidia, Wright (1959, 1960) suggested that three stages are involved in host selection. During the first stage, the miracidia locate the habitat of the host by response to physical stimuli. Once in the area of the host habitat, the larvae move around randomly until they come in contact with the host attractant. It seems probable that similar behavioral patterns occur in other symbionts. For example, many commensal polychaetes living in burrows or tubes of other animals have certain characteristics in common that lead them into burrows and crevices, which are areas where potential hosts are likely to be encountered. Most of these worms move away from light, many are positively skototaxic (moving toward dark areas), and others are positively geotactic (Dales, 1967). Since many are relatively host-specific, it seems likely that once they reach the host's environment, the symbiont will recognize some sign from the potential host.

Despite the difficulty of identifying the host attractant in any particular organism, some studies have done so successfully, as demonstrated by a study of the host-symbiont relationships between the monogenetic trematode *Entobdella soleae* and its fish host *Solea solea* (Kearn, 1967). After hatching from the egg, the free-living ciliated oncomiracidium of *E. soleae* has about 24 hr in which to find a suitable host; if a host is not available within this time period, the larvae will die. To determine the degree of host specificity, freshly hatched larvae were offered a choice between scales still attached to pieces of skin of *S. solea,* and those of several other species of flat fish. It was found that the larvae could identify *S. solea* scales in total darkness, and did not settle randomly on any fish scale in their vicinity. Since the larvae could pick out the desired host even in total darkness, visual clues were ruled out. To eliminate sense of touch as a mechanism for host identification, the larvae were offered a choice between agar jelly that had been in contact with host tissue and control jelly that had not been in contact with the host tissue. The oncomiracidia were able to discriminate readily between the experimental and control jellies (Fig. 10-3).

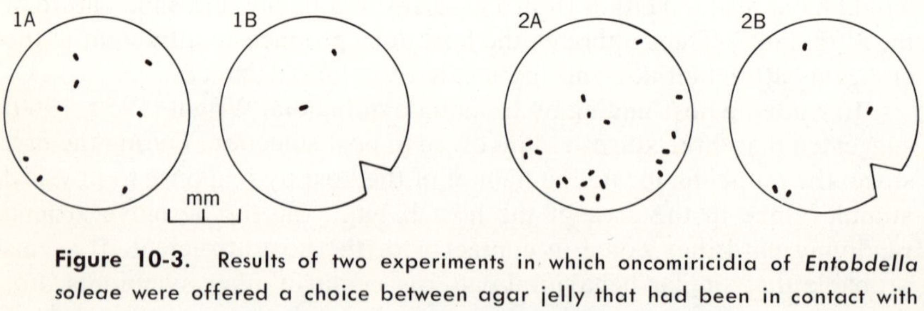

Figure 10-3. Results of two experiments in which oncomiricidia of *Entobdella soleae* were offered a choice between agar jelly that had been in contact with *Solea solea* skin (*1A, 2A*) and control jellies (*1B, 2B*). Few larvae chose the control jellies. (After Kearn, 1967.)

Further experiments traced the positive response of these larvae to the epidermal mucus cells of the host. When whole scales of *S. solea* were boiled to destroy the chemical attraction, the scales no longer attracted the oncomiracidia. The cornea, which lacks a mucus layer, was also ineffective as an attractant. Kearn has speculated that the host specificity exhibited here could be the result of host-finding by chemoreception. A species of monogenetic trematode is attracted to a specific substance "a" secreted by a particular fish A. At some point in time, the population of fish A is somehow isolated into two populations, which ultimately produce two allopatric fish species, B and C. Over a long period of time, the larvae of the monogeneans become attracted to chemicals "b" and "c," respectively. Then, even if the ecological barrier is removed, allowing fish B and C to freely intermingle, the parasite of fish B will be attracted only to substance "b," and that of fish C only to substance "c." In effect, each monogenetic trematode is isolated on a single host species because only one specific host emits the odor that attracts the parasite.

DEVELOPMENTAL STIMULI

Symbionts having both a free-living stage in the life cycle and a stage that is dependent on a host organism develop only so far in the free-living stage and then stop until they reach the proper host. Rogers (1963) suggested that the stimulus for further development in nematodes comes from the host. He postulated that, as an adaptation to parasitism, these organisms have lost that part of the system controlling development and are now dependent upon the host to trigger further development. Two examples among the nematodes, which may serve as illustrations, are the eggs of the parasite of man, *Ascaris lumbricoides,* and the larvae of the stomach nematode of sheep, *Haemonchus contortus*. The eggs of *A. lumbricoides* pass out in the host's feces into the soil, where they develop, and subsequently a larva is formed within each egg shell. The larva then molts into a second-stage larva, but no further development takes place until it is eaten by the definitive host. In *H. contortus,* the eggs pass to the exterior in the host's feces, where, under favorable climatic conditions, they hatch into first-stage larvae which are free-living in the soil. Since two more molts occur, this nematode has three nonparasitic stages. The third-stage larva, however, retains the "skin" or "sheath" after the second molt, and can no longer feed. No further development occurs until the larva is eaten by the sheep host. Thus, the host supplies some intrinsic mechanism within the worm which allows development to continue.

Stimuli that have been implicated in the hatching of infective eggs and the exsheathment of infective larvae include dissolved, gaseous carbon dioxide or a related substance in conjunction with a temperature similar to that of the host. Other factors, such as pH and the presence of a reducing agent, have also been implicated. In *in vitro* experiments such factors have caused infective eggs of *A. lumbricoides* to secrete a "hatching fluid" (Fairbairn, 1961), whereas in infective larvae of *H. contortus,* these stimuli caused the secretion of an exsheathing fluid containing an enzyme that was, at least in part, responsible for the initial breakdown of the sheath. Rogers (1966) has presented a possible scheme to show how the stimulus from the host relates to the early development of parasitic nematodes that infect the host by ingestion (Fig. 10-4).

Studies of excystment stimuli in tapeworms demonstrate that several factors are involved in the process of excystment (Read and Simmons, 1963). Working with cysticercoid larvae of *Hymenolepis diminuta,* Read (1955) observed that a temperature approximating that of the host (rodents) was necessary before any of the media was effective. In bile salts at a temperature of 37°C, there was movement of the scolex and excystment of some larvae, and when trypsin was added to the bile-salt solution, most larvae had excysted in about an hour. When pretreated with acid pepsin, all larvae excysted within 15 min after being placed in a bile-salt trypsin solution. From these effects *in vitro,* Read hypothesized five physiological effects of secretions in the host gut: pepsin causes a "priming effect" on the outer cyst, which makes it swell; bile salts produce an "irritability effect" as evidenced by movement of the scolex; a "protein alteration effect" is brought about by bile salts and demonstrated by the effects of trypsin when added to larvae treated with bile-salt solution; a "proteolytic effect" of trypsin is manifested by the excystment of the larvae after they have been exposed to acid pepsin and bile salts; and finally, the "temperature effect" induced movement of the larvae.

Host stimulation of development has been observed in other symbiotic relationships. Kearn (1967) has speculated that with the larvae of the monogenetic trematode *Entobdella soleae,* the host fish *Solea solea* produces a substance that stimulates the oncomiracidia to become attached to the fish skin and lose their ciliated cells. Some evidence indicates that molluscan hosts contribute a stimulus that initiates the shedding of the ciliated epidermis in some of the miracidia of digenetic trematodes (Cheng, 1967). Further evidence shows that other stages in the life cycle of digenetic trematodes also depend on host stimulation to develop into the next stage, as demonstrated in *Gorgodera amplicava,* the common bladder fluke of frogs (Cheng, 1963). During its life cycle this species utilizes the fresh-water mollusc *Muscilium partumeium* as the

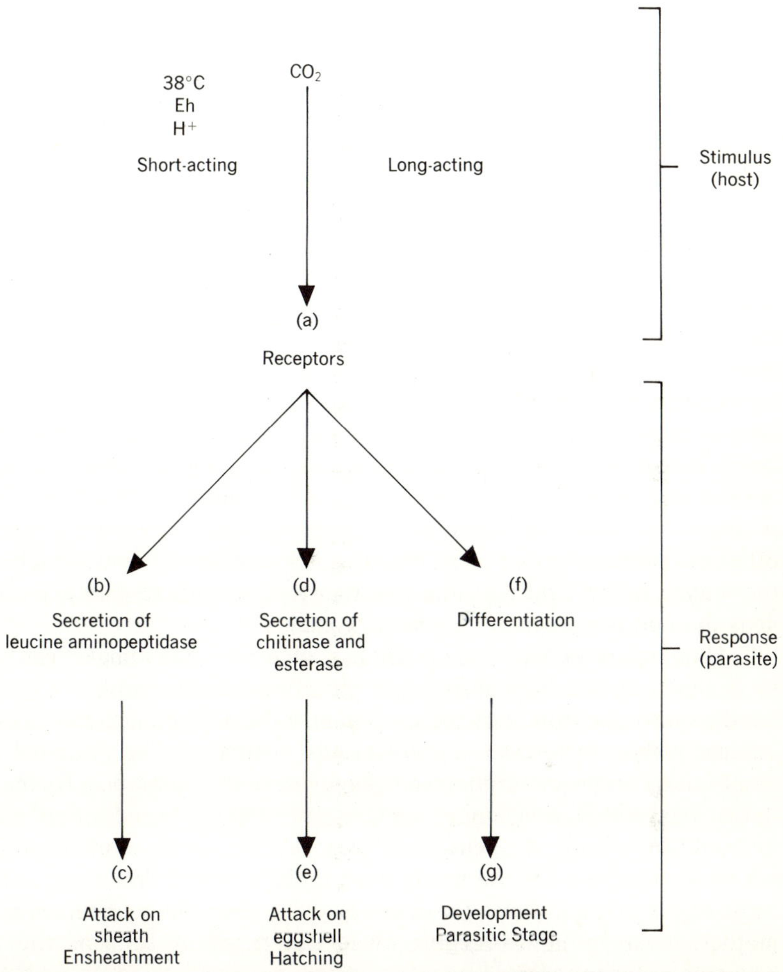

Figure 10-4. Possible scheme showing how the stimulus from the host relates to the early development of parasitic nematodes that infect the host *per os*. (From Rogers, 1966.)

first intermediate host; subsequently, the cercariae escape from this mollusc and are eaten by one of several species of mollusc, where they encyst as metacercariae; this second intermediate host is then eaten by one of several species of frogs, where the trematode develops into the adult stage. The cercarial stage is inactive until it is ingested by the second intermediate host. Cheng noted that it was the sera (plasma or hemolymph) from the host—which in this study included any of five species of fresh-water molluscs—that stimulated encystment into the metacercarial stage. Activation of the cercariae is not due to the pH of the plasma, but apparently is caused by some component fraction of the plasma. Cheng has suggested that once the quiescent cercaria is eaten by the molluscan second-intermediate host, plasma seeps into the digestive tract, activating the cercaria. In this manner the cercaria is able to migrate to a site favorable for encystment as a metacercaria.

The effect of host hormones on symbionts is another aspect of growth and developmental stimuli, and is well illustrated in the study by Rothschild (1965) of the interrelationships between the rabbit flea and its host. The breeding cycle of this parasite is controlled by the sex-hormone cycle of the vertebrate host. These fleas attach to the ears of the rabbit, and unless their host becomes pregnant or they are able to transfer to an expectant rabbit, the fleas cannot breed. The breeding cycle of the flea is thus controlled by the secretion of the sex hormones concentrated in the host's blood. As soon as ovulation occurs in the female rabbit, sex hormones are released by the anterior lobe of the pituitary gland, and these pituitary hormones stimulate the ovaries and adrenal glands to secrete other hormones. At this point the fleas become more firmly attached to the female rabbit, and the future mother rabbit tends to gain even more fleas than do nonpregnant females and males.

About ten days before the rabbit is to give birth, the female fleas begin to develop under the influence of the adrenocorticotrophic hormone, another hormone from the anterior pituitary. This hormone stimulates the adrenal glands to release corticosteroids, which are the principal hormones that control maturation and egg laying in the rabbit flea. By the last day of the rabbit's pregnancy, the fleas are ready to breed. Shortly after the rabbits are born, the fleas move from the ears of the mother rabbit to her face, passing on to the newly born rabbits, where they mate and lay eggs. Egg laying continues for about 12 days; then the fleas return to the mother, where a new breeding cycle can begin as soon as the host becomes pregnant again.

GROWTH AND POPULATION REGULATION

Once the symbiont has become established on or in a host, the rate of population growth depends on many of the same factors that are important in population growth among free-living organisms. In addition, the symbiont encounters yet another environmental parameter, its host. Ectosymbionts relate directly to the same environmental stresses that the host does. A copepod living on the skin of marine fish must contend with the same environmental fluctuations in temperature, salinity, and oxygen tensions that the host does, along with possible host defense mechanisms that the parasite must overcome. In endosymbionts the host is the external environment. Although the endosymbiont is inextricably tied to the host and must overcome any defense mechanism thrown up by the host, the physiological reactions to environmental stress may not necessarily be the same in both symbiont and host. For example, endoparasitic trematode larvae in molluscs do not necessarily respond to temperature in the same way metabolically as does the host animal. Further, two species of trematode larvae found in the mud-flat snail *Nassarius obsoleta* have characteristic and different metabolic responses to temperature, neither of which is the same as that of the host.

Metabolic thermal-acclimation patterns were determined for the intramolluscan redial stage of *Himasthla quissetensis* and the intramolluscan sporocyst stage of *Zoögonus lasius,* as well as for the snail host. The adult stage of *H. quissetensis* is found in birds, that of *Z. lasius,* in fish. As can be seen in Fig. 10-5, the thermal-acclimation patterns are quite distinct for each of these species (Vernberg and Vernberg, 1965). Such differences occur not only at the whole-animal level of organization, but also at the enzymatic level. When cytochrome oxidase activity patterns were assayed over a wide thermal range in the sporocysts of *Z. lasius* and in the digestive gland of the snail host, it was found that both level of activity as well as activity patterns were completely dissimilar in the host and its symbiont (Vernberg and Vernberg, 1968). Thus, although the parasites are metabolically dependent upon the host, they have maintained some measure of physiological integrity.

In contrast to the number of experimental studies on factors governing population growth among free-living organisms, there have been relatively few studies of symbiotic organisms. Among them is one based on the simple and easily maintained host-parasite relationship between two species of hydra and a protozoan parasite (Stiven, 1964). The protozoan parasite *Hydramoeba hydroxena* is an ecto- and endoparasite of hydra. It starts on the tentacles of the hydra and migrates inward, and then finally becomes established in the gastrovascular cavity of the host animal, where it feeds on the gastrodermal cells. Stiven found that the pattern of population growth was altered by the host species, by the number of hydramoebae initially introduced onto the host, and by temperature. On one of the hosts, *Hydra pseudoligactus,* the parasite tends to grow exponentially and always has a lethal effect on the host. With an initial infection of one hydramoeba at a temperature of 20°C, a peak density of 30–40 hydramoebae was reached within a few hours after the death of the host and about seven or eight days after infection. At 25°C, the rate of population increase was much greater. The peak density

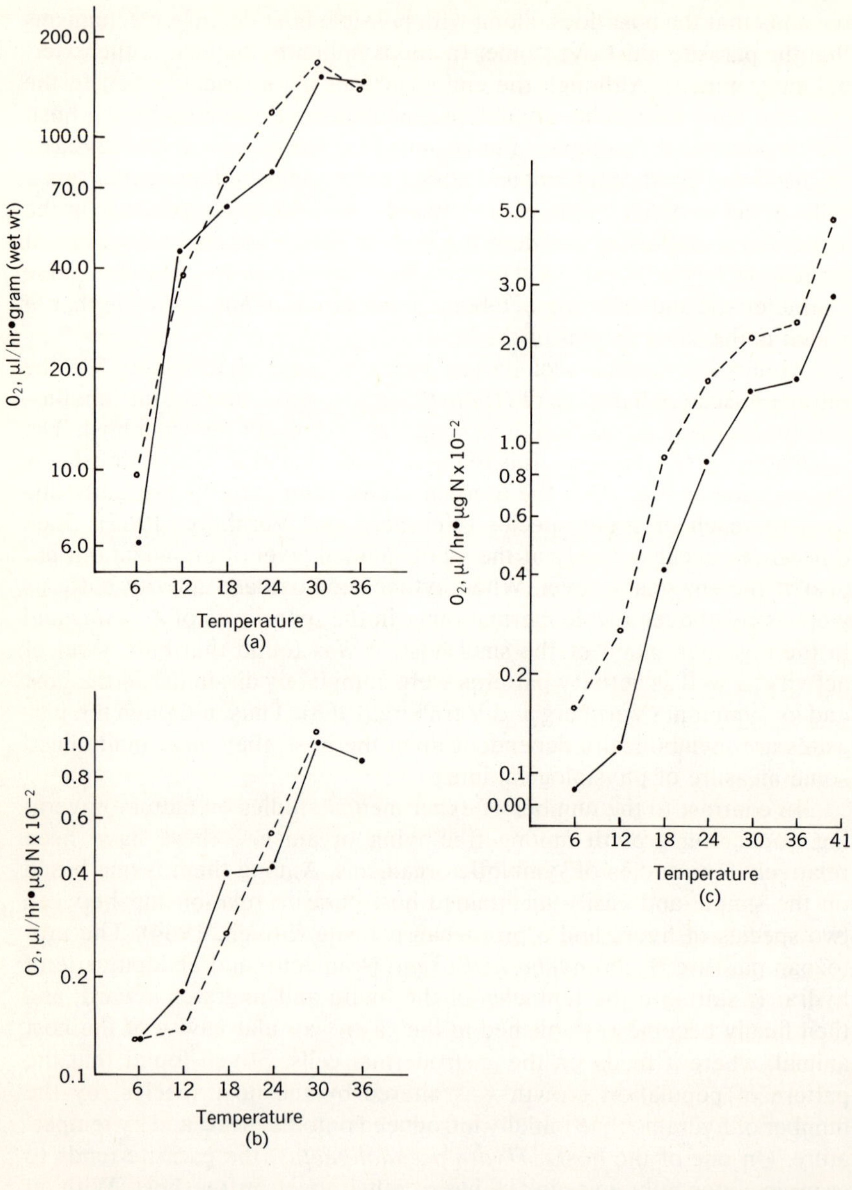

Figure 10-5. Metabolic-temperature acclimation patterns of uninfected *Nassarius obsoleta* (*a*) and two species of trematode larvae, *Zoogonus lasius* (*b*) and Himasthla quissetensis (c) that utilize *N. obsoleta* as the first intermediate host. Broken line: cold-acclimated; Solid line: warm-acclimated. (From Vernberg and Vernberg, 1965.)

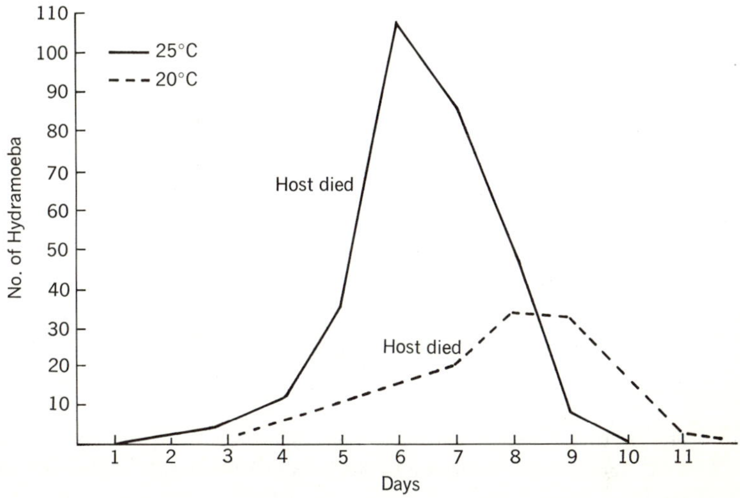

Figure 10-6. Comparison of rate of increase in parasite populations on *Hydra pseudoligactus* at 20°C and 25°C with initial infection of one hydramoeba per hydra. Peak density and death of host occurred about day 6. (Stiven, 1964.)

totaled 100–175 parasites and was reached by about the sixth day, at which time the host died from undetermined causes (Fig. 10-6). At 20°C, the parasites multiplied 1.67 times per day; at 25°C, the population multiplied 2.57 times per day.

On another host, *Chlorohydra viridissima,* a different pattern emerged. At 25°C and with an initial infection of one, the population increased to a peak density of 14 by the sixth day. This represents a rate of increase of about 1.86 times per day. After reaching the peak density, the numbers decrease and fluctuate for several days, but then a major increase occurs, reaching a peak density of about 28 by the sixteenth day, followed by the death of the host. With an initial infection of four parasites, the rate of increase is not exponential. Again there are fluctuations in numbers of parasites, followed by an eventual second rise in numbers leading to the death of the host (Fig. 10-7). At 20°C, the pattern with initial infections of 1:4 is similar at first to that at 25°C. After the initial peak density is reached, numbers again fluctuate, but there is no major rise in numbers followed by the death of the host; instead, at this temperature, the numbers of parasites decrease gradually until the host is free of hydramoebae and survives (Fig. 10-8). Thus, population growth in this species involves a combination of the parasite genotype, environmental factors, and host reaction.

The regulation of symbiont populations is accomplished by many of the same means that are operative in free-living organisms. Environ-

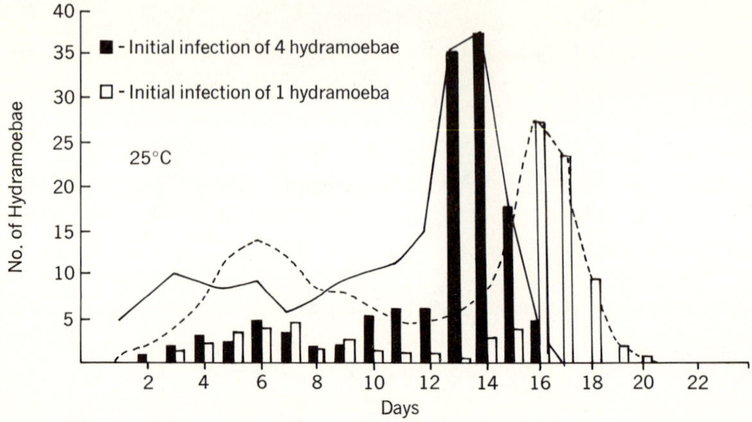

Figure 10-7. Population fluctuations of hydramoebae on host C. *viridissima* at 25°C. Total number indicated by the lines and the free or unattached parasites by the bars. (Stiven, 1964.)

mental conditions outside the hosts tend to limit populations among some symbionts. Climatic conditions have been shown to limit abundance of parasitic helminths in their hosts because an unfavorable microclimate reduces survival of free-living stages to such a degree that the host does not become heavily infected. As reported by Schad (1966a), climatic conditions aid in regulating hookworm infection in rural India. The eggs of human hookworm, a nematode parasite, pass out with the host's feces and develop in the soil; larval development continues up to a point, but then the larvae must come in contact with a host before developing further. Following contact with a host, the larvae penetrate the skin and migrate through the body, ending up in the small intestine. During the free-living stage, hookworm larvae are subject to predation and the same environmental fluctuations as other free-living organisms. Thus, Schad observed that during long dry periods, development and survival of hookworm

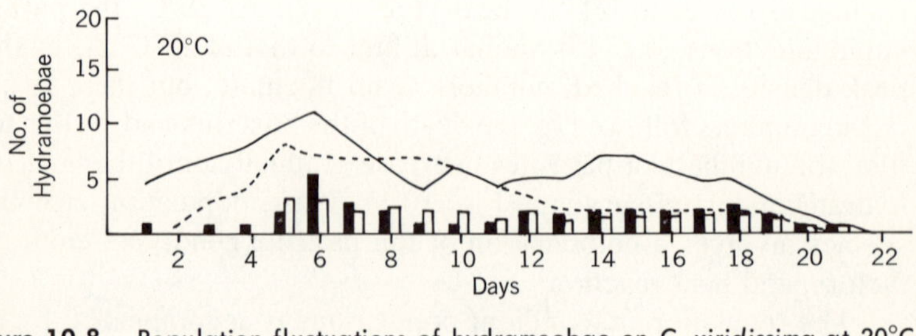

Figure 10-8. Population fluctuations of hydramoebae on C. *viridissima* at 20°C. (Stiven, 1964.)

larvae were severely reduced, and during the monsoon season when the soil became waterlogged, it was difficult for the larvae to move. Hookworm larvae also need relatively high oxygen tensions in order to develop, and during this period of heavy rain the oxygen tensions were reduced in the muddy soil. After the monsoons, optimal conditions prevail for the free-living larvae, making intense transmission possible for a brief period of time. For most of the year, however, transmission of hookworm is negligible.

Once entrance has been gained into a host, regulation of symbiont populations is as much of a density-dependent process as it is among free-living organisms. In cestode infections, a number of workers have observed an inverse relationship between the number of worms and individual size, which has been termed the *crowding effect* (Read, 1951, 1959; Read and Simmons, 1963), and has been attributed to competition for utilizable carbohydrates. The work of Roberts (1961) has demonstrated intraspecific competition in populations of the adult cestode *Hymenolepis diminuta*. Rats were fed a known number of this species of larval tapeworms, then sacrificed, and the developing worms removed at specified intervals following ingestion of the larval tapeworms. The length of the developing worms was used as an index of growth rate. Up to 24 hr after infection, there was a short lag phase, followed by a period of rapid acceleration. By the second day, the rate of growth became exponential and continued at this rate until the sixth or seventh day of postinfection. A period of retardation followed until patency at 16 to 17 days. Although the time scale of maturation was not affected by population density, high population density did cause a lower growth rate so that mature proglottids and gravid proglottids were markedly reduced in numbers (Fig. 10-9). Thus the population density had a direct influence on the germinative region which served to reduce the reproductive potential of the worms. Maximum size was attained when there were only three to five worms or less per host. In further studies of *H. diminuta,* Roberts (1966) concluded that intraspecific competition for host carbohydrates is a primary factor in limiting the size of these worms in large populations.

Holmes (1961) has suggested that interspecific competition for host carbohydrate also occurs. When rats were fed larvae of *Hymenolepis diminuta* and the acanthocephalan *Moniliformis dubius,* both species showed alteration of size and general body form when compared to single infections. The same number of worms of the heterologous species had a greater effect on each species than did an equivalent number of the homologous species.

Competition between two species of parasites for host carbohydrates leads to the question of whether two species with similar habits and the same ecological relationships may actually coexist. Such coexistence

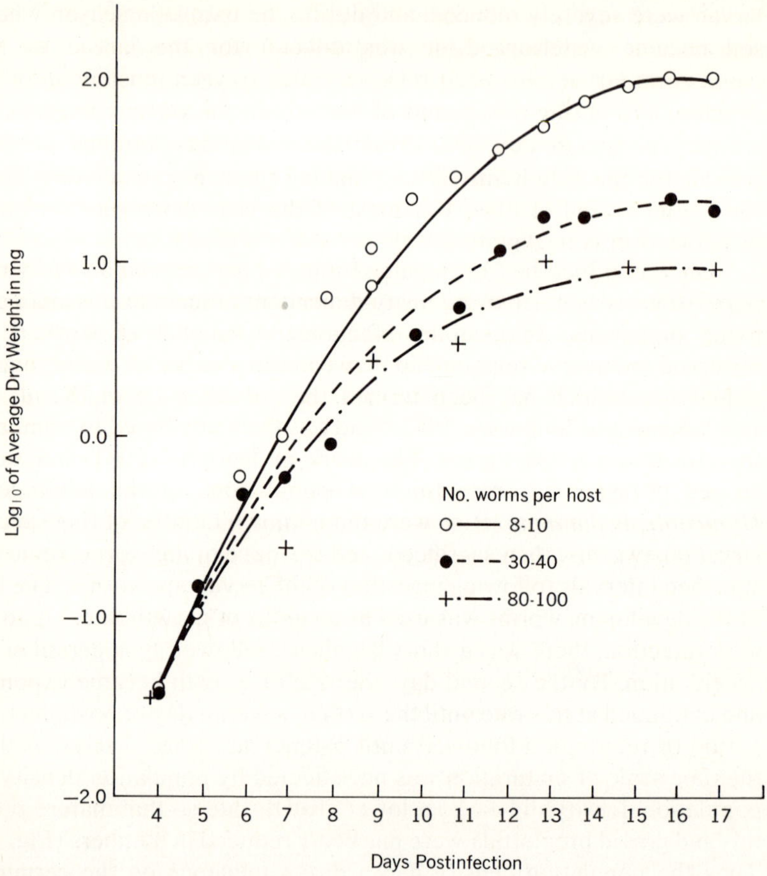

Figure 10-9. Growth rate patterns of the cestode *Hymenolepis diminuta* and their relation to population density. (Roberts, 1961.)

would, of course, be contrary to one of the basic tenets of ecological thought. Although experimental demonstrations of this premise are more extensive among free-living organisms, a few published studies indicate that symbionts may also coexist, as shown by a study on the tapeworm *H. diminuta* and the acanthocephalan *Moniliformis dubius* (Holmes, 1961). In single infections, *H. diminuta* extends throughout the anterior three-fourths of the intestine, although about 74 percent of the worms are within the anterior quarter of the host intestine. In the presence of *M. dubius,* however, no *H. diminuta* were found in the anterior quarter of the intestine. Instead, 98 percent were in the second and third quarters. In single infections, *M. dubius* is much more limited in distribution than is *H. diminuta,* with 84 percent of the worms in the second and third tenths of the intestines. In double infections, the position of both species was similar, but the acanthocephala shifted slightly anteriad (Fig. 10-10). Holmes also noted only slight distributional overlap in concurrent infections of individual rats, with *M. dubius* always attached anteriorly to *H. diminuta*.

An even more complicated situation was investigated by Schad (1963) in the European tortoise *Testudo graeca*. The colon of this particular tortoise serves as the habitat for ten or more species of nematodes of the genus *Tachygonetria* (Oxyuroidea). Much interspecific distributional overlap was observed along the longitudinal or linear axis of the colon, but the interspecific overlaps were greatly reduced when investigated on a radial basis. Two categories were recognized: those living in the paramucosal region (the area ringing the mucosal layer of the intestine) and those distributed throughout the lumen of the intestine. Schad noted very different food preferences of two of the more broadly overlapping species: *T. robusta* is an indiscriminate eater of the lumen contents, whereas *T. stylosa* feeds almost entirely on fine particulate matter. Although they occupy the same areas, their food habits are entirely different. The feeding habits of the two worms could be correlated readily with differences in oral morphology.

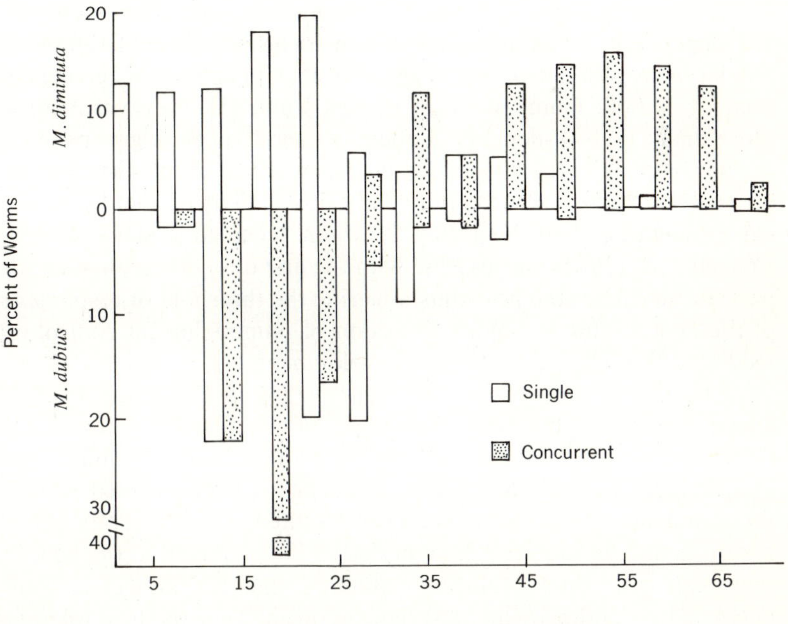

Figure 10-10. Distribution effects during concurrent infection of *Hymenolepis diminuta* and *Moniliformis dubius*. (Holmes, 1961.)

Population regulation also may be mediated through the immune response of the host, as demonstrated by Dineen *et al.* (1965) and Donald *et al.* (1964) in studies of the trichostrongylid sheep nematode *Nematodirus spathiger.* The experimental animals were divided into three groups and given known dosages of infective larvae of *N. spathiger.* Group I was given a single dose of 50,000 infective larvae, group II a single dose of 130,000 larvae, and group III received 2000 larvae daily for 25 days. Thus groups I and III received the same total dosage. The total egg production of the worms was lowest in group III, where larvae were administered in small daily doses, greatest in group I and intermediate in group II. These results suggest that population control of the parasites is mediated around a certain threshold level of abundance and host responsiveness. That is, the host will tolerate the parasite up to a particular level of abundance, after which it exerts an effective immunological response that limits the population. This immunological control was expressed by several mechanisms: the elimination of worms at the infective stage; arrested development before attainment of the adult stage; reduction in egg production in the adult female; elimination of adult worms; and discrimination against the female in inhibition of the development in the larval stages; and greater elimination of adults. The lowered egg count reflects a combination of these various mechanisms.

Differences in groups I and II indicate that population control is density-dependent; for example, infection with a single dose of 130,000 infective larvae resulted in a lower egg count than in animals receiving a single dose of 50,000. Comparison of groups I and III, however, shows that more than a density-dependent effect is operative. Both groups of experimental animals received the same total number of infective larvae, yet group III, which received small daily doses, showed more limitation of egg production than did group I, which received a single large dose. Dineen *et al.* (1965) suggest that with the first dose of infective larvae, the parasite sensitizes the host thus lowering the threshold of responsiveness of the host so that a major expression of immunological control is then mediated through retardation of development.

The effectiveness of immune response in regulating population abundance in parasites has been the subject of considerable interest. Dineen (1963a,b) hypothesized that during the evolution of a helminth parasite and vertebrate host relationship, mechanisms have evolved which limit the abundance of a parasite to a level that the host can tolerate. Since helminths generally require a long period of development in the host before they can reproduce, Dineen contends that an antigenic similarity between host and parasite must have evolved. Thus the host tolerates the parasite up to a certain critical threshold before reacting immunologically to limit the parasite. Damian (1964) has suggested that the parasites produce antigenic determinants resembling those of their hosts to such an extent that they do not cause the formation of host antibodies, and he has termed this phenomenon *molecular mimicry*.

Schad (1966b) has suggested yet another method of population regulation among helminth parasites; he has hypothesized that nonreciprocal cross immunity has evolved in the adaptation of a parasite to its environment so that it can limit the populations of competing parasitic species. In such an interaction the abundance of one species is limited by a previous infection with or by the presence of another species. As demonstrated between species of tapeworms by Heyneman (1962), infections with *Hymenolepis nana* in white mice limited the population of another species of cestodes, *Hymenolepis diminuta*, when both species were present simultaneously and also when a previous infection with *H. nana* had been lost. In other words, *H. nana* elicited a greater immune response against *H. diminuta* than it did against itself. Hypothetically, then, during the course of development of a host-parasite relationship, if species A elicits an impotent immunogenic response against A, but one that is potent against species B, then genotypes of A permitting this expression of antigenicity would be favored. Schad viewed this type of response as "a special kind of competitive interaction between two species, which is mediated by a third species, the host."

There are instances in which neither immunity nor direct competition seem to be involved in population regulation. The mud-flat snail *Nassarius obsoleta* is known to serve as the first intermediate host for at least ten species of parasites, yet the incidence of double infections is extremely low. In enzyme activity studies on tissue from infected snails, it was noted that each parasite distinctly alters the activity patterns of the host tissue. This would suggest that the snail tissue is biochemically altered to such an extent that other species no longer find it a suitable habitat (Vernberg and Vernberg, 1968). Lie *et al.* (1965) have reported that echinostome rediae are actual predators on any sporocysts that try to coexist in the same snail host.

All these interactions, then, indicate that populations of symbionts are subject to natural population regulation just as are free-living hosts.

SUMMARY

Many problems that symbionts must solve are basically the same as those of free-living animals, but there are also some additional ones their unencumbered cousins do not encounter. Habitat selection, for example, requires making the right host contact, and since contacts are not entirely due to chance, the symbiont must have evolved methods for gaining entrance into the host's environment in order to make contact. Once in the proper host, the symbiont does not need to seek food actively, but it must be physiologically adapted to live on food made available by the host. Population growth and regulation in symbionts also are governed by many of the same factors as those in free-living animals: environmental conditions, competition, predation. There is, however, an additional factor—that of the resistance of the host species.

Thus, although symbionts may be said to be taking a "free ride," almost invariably they must pay for it in some other coin. Below a certain population level, host resistance is minimal; at a greater level of population growth, the host may build up an immune response, but unregulated growth of the symbiont population will kill the host and in turn destroy the symbiont population.

References

Bartel, A. H., and D. Davenport. 1956. A technique for the investigation of chemical responses in aquatic animals. *Brit. J. Anim. Behav.*, 4:117–119.

Cheng, T. C. 1963. Activation of *Gorgodera amplicava* cercariae by molluscan sera. *Exptl. Parasitol.*, 13:342–347.

Cheng, T. C. 1967. Marine molluscs as hosts for symbioses with a review of known parasites of commercially important species. *In* F. S. Russell (ed.), *Advances in Marine Biology,* Vol. V. Academic Press, New York.

Dales, R. P. 1967. Symbiosis in marine organisms. *In* S. M. Henry (ed.), *Symbiosis,* Vol. I, pp. 299–326. Academic Press, New York.

Damian, R. T. 1964. Molecular mimicry: antigen sharing by parasite and host and its consequences. *Am. Nat.*, 98:129–149.

Dineen, J. K. 1963a. Immunological aspects of parasitism. *Nature,* 197:268–269.

Dineen, J. K. 1963b. Antigenic relationship between host and parasite. *Nature,* 197:471–472.

Dineen, J. K., A. D. Donald, B. M. Wagland, and J. H. Turner. 1965. The dynamics of the host-parasite relationship. II. The response of sheep to primary and secondary infection with *Nematodirus spathiger. Parasitology,* 55:163–171.

Donald, A. D., J. K. Dineen, J. H. Turner, and B. M. Wagland. 1964. The dynamics of the host-parasite relationship. I. *Nematodirus spathiger* infection in sheep. *Parasitology,* 54:527–544.

Fairbairn, D. 1961. The *in vitro* hatching of *Ascaris lumbricoides* eggs. *Can. J. Zool.*, 39:153–162.

Heyneman, D. 1962. Studies on helminth immunity. II. Influence of *Hymenolepis nana* (Cestoda: Hymenolepididae) in dual infections with *H. diminuta* in white mice and rats. *Exptl. Parasitol.,* 12:7–18.

Holmes, J. C. 1961. Effects of concurrent infections of *Hymenolepis diminuta* (Cestoda) and *Moniliformis dubius* (Acanthocephala). I. General effects and comparison with crowding. *J. Parasitol.,* 47:209–216.

Horsfall, M. W. 1934. Studies on the life history and morphology of the cystocercous cercariae. *Trans. Am. Micro. Soc.,* 53:311–347.

Istock, C. A. 1967. The evolution of complex life cycle phenomena: an ecological perspective. *Evolution,* 21:592–605.

Kearn, G. C. 1967. Experiments on host-finding and host-specificity in the monogenean skin parasite *Entobdella soleae. Parasitology,* 57:585–605.

Koch, A. 1938. Die intrazellulare symbiose von *Mastotermes darwiniensis* Froggatt. *Z. Morphol. Oekol. Tiere,* 34:584–609.

Koch, A. 1967. Insects and their endosymbionts. *In* S. M. Henry (ed.), *Symbiosis,* Vol. II, pp. 1–106. Academic Press, New York.

Lees, A. D. 1948. The sensory physiology of the sheep tick, *Ixodes ricinus* L. *J. Exptl. Biol.,* 25:145–207.

Lie, K. J., P. F. Basch, and T. Umathevy. 1965. Antagonism between two species of trematodes in the same snail. *Nature,* 206:422–423.

Limbaugh, C., H. Pederson, and F. A. Chace, Jr. 1961. Shrimps that clean fishes. *Bull. Mar. Sci. Gulf Caribbean,* 11:237–257.

Read, C. P. 1951. The "crowding effect" in tapeworm infections. *J. Parasitol.,* 37:174–178.

Read, C. P. 1955. Intestinal physiology and the host-parasite relationship. *In* W. H. Cole (ed.), *Some Physiological Aspects and Consequences of Parasitism,* pp. 27–43. Rutgers University Press, New Brunswick, N.J.

Read, C. P. 1959. The role of carbohydrates in the biology of cestodes. VIII. Some conclusions and hypotheses. *Exptl. Parasitol.,* 8:365–382.

Read, C. P., and J. E. Simmons, Jr. 1963. Biochemistry and physiology of tapeworms. *Physiol. Rev.,* 43:263–305.

Roberts, L. S. 1961. The influence of population density on patterns and physiology of growth in *Hymenolepis diminuta* (Cestoda: Cyclophyllidea) in the definitive host. *Exptl. Parasitol.,* 11:332–371.

Roberts, L. S. 1966. Developmental physiology of cestodes. I. Host dietary carbohydrate and the "crowding effect" in *Hymenolepis diminuta. Exptl. Parasitol.,* 18:305–310.

Rogers, W. P. 1963. Physiology of infection with nematodes: some effects of the host stimulus on infective stages. *Ann. N.Y. Acad. Sci.,* 113:208–216.

Rogers, W. P. 1966. Reversible inhibition of a receptor governing infection with some nematodes. *Exptl. Parasitol.,* 19:15–20.

Ross, D. M. 1960. The association between the hermit crab *Eupagurus bernhardus* (L.) and the sea anemone *Calliactus parasitica* (Couch). *Proc. Zool. Soc. London,* 134:43–57.

Ross, D. M., and L. Sutton. 1968. Detachment of sea anemones by commensal hermit crabs and by mechanical and electrical stimuli. *Nature,* 217:380–381.

Rothschild, M. 1965. The rabbit flea and hormones. *Endeavour,* 24:162–168.

Sastry, A. N., and R. W. Menzel. 1962. Influence of hosts on the behavior of the commensal crab *Pinnotheres maculatus* Say. *Biol. Bull.,* 123:388–395.

Schad, G. A. 1963. Niche diversification in a parasitic species flock. *Nature,* 198:404–406.

Schad, G. A. 1966a. Natural control of the abundance of parasitic helminths. *Bull. Ind. Soc. Mal. Com. Dis.,* 3:18–27.

Schad, G. A. 1966b. Immunity, competition, and natural regulation of helminth populations. *Am. Nat.,* 100:359–364.

Stauber, L. A. 1945. *Pinnotheres ostreum,* parasitic on the American oyster, *Ostrea* (*Gryphaea*) *virginica. Biol. Bull. Mar. Biol. Lab., Woods Hole,* 88:269–291.

Stiven, A. E. 1964. Experimental studies on the epidemiology of the host-parasite system, *Hydra* and *Hydramoeba hydroxena.* II. The components of a simple epidemic. *Ecol. Monographs,* 34:119–142.

Thompson, W. R. 1939. Biological control and the theories of the interaction of populations. *Parasitology,* 31:299–388.

Vernberg, W. B. 1963. Respiration of digenetic trematodes. *Ann. N.Y. Acad. Sci.,* 113:261–271.

Vernberg, W. B., and W. S. Hunter. 1963. The utilization of certain substrates by larval and adult stages of *Himasthla quissetensis. Exptl. Parasitol.,* 14:311–315.

Vernberg, W. B., and F. J. Vernberg. 1965. Interrelationships between parasites and their hosts. I. Comparative metabolic patterns of thermal acclimation of larval trematodes with that of their host. *Comp. Biochem. Physiol.,* 14:557–566.

Vernberg, W. B., and F. J. Vernberg. 1968. Interrelationships between parasites and their hosts. IV. Cytochrome *c* oxidase thermal acclimation patterns in a larval trematode and its host. *Exptl. Parasitol.,* 23:347–354.

Vernberg, W. B., F. J. Vernberg, and F. W. Beckerdite. 1969. Larval trematodes: Double infections in the common mud-flat snail. *Science,* 164:1287–1288.

Welsh, J. H. 1930. Reversal of phototropism in a parasitic water mite. *Biol. Bull. Mar. Biol. Lab., Woods Hole,* 59:165–169.

Wright, C. A. 1959. Host-location by trematode miracidia. *Ann. Trop. Med. Parasitol.,* 53:288–292.

Wright, C. A. 1960. Relationships between trematodes and molluscs. *Ann. Trop. Med. Parasitol.,* 54:1–7.

11

Ecosystems

INTRODUCTION

The various levels of biological organization recognized and emphasized by biologists range from the molecular through the organismic to the aggregation of organisms forming a distinct ecological unit that is called a *community*. Defined broadly, the community is "a natural assemblage of organisms which, together with its habitat, has reached a survival level such that it is relatively independent of adjacent assemblages of equal rank; to this extent, given radiant energy, it is self-sustaining" (Allee *et al.*, 1949). Thus the assemblage of organisms found in the desert, for example, is distinct from that in the grassland region. The community concept may be illustrated in any given environmental complex where certain species of plants or animals commonly occur together, as in the landscape where certain obvious plants are not haphazardly distributed but rather exhibit distinct patterns. Although this association of certain plant species is more readily observed, animals also tend to occur together in definite patterns, and these aggregations of plants or animals living together constitute an initial approximation of a community.

The community concept is not new; certain ideas, such as the spatial grouping of specific species, can be traced to the ancient Greeks. Many early naturalists recognized that diverse species which live together in an ordered fashion form natural groups rather than being distributed randomly. One of the important contributions to the study of community ecology was made in 1877 by Möbius, who reported than an oyster bed represented a community of living organisms that normally had available all the substances required for survival and reproduction. He proposed the word *biocoenosis* to designate this ecologic relationship, but it was later changed to *biocoenose*. Although Möbius defined the biocoenose in static descriptive terms, he was clearly aware of its dynamic nature; for example, he stated that if one of the external factors varied for a long time, the community would be dramatically transformed. Since these early formulations of the community concept, the major lines of research have dealt with both detailed descriptions (the anatomy) of specific communities and the quantitative and dynamic aspects (the physiology) of communities. It has been suggested that the community is analogous to a superorganism in which the various component organisms may represent the structural and functional units in the same manner that different types of cells make up an organism. As such, a community exhibits all those life processes of an individual organism that were stressed in earlier chapters: that is, metabolism, reproduction, growth, and adaptive responses to environmental fluctuations.

Communities differ in their complexity and can range from a community of one prominent plant species occupying a single layer of a habitat, such as the desert salt-grass community, to a multispecies, multilayered system such as the beech-maple forest. Since a community is a natural assemblage of organisms that have evolved together to a high collective persistence level, its structure is based on its relatively distinctive species composition—which may be either sharply different from that of an adjacent community or may show degrees of overlap. Some species are found in only one community (an exclusive species), whereas others live in many communities (an ubiquitous species). Typically, in the aggregation of species that make up a community, only a few species are abundant in either number of individuals or biomass (weight), and the majority of species are relatively rare (Sanders, 1968).

The community may be named after the dominant species, whether plant or animal, such as an oak-hickory community or the barnacle-mussel community. Other communities are named by ecologists after distinct habitats, such as the sandy-beach, intertidal-zone community. In widely separated examples of the same type of community, a composition of different species may be found with different species occupying the equivalent niche. Obviously, the first method of nomenclature is more taxonomically oriented, although the correct identification of the constituent species is important in all community studies. In the second method, more emphasis is placed on the physical characteristics of a community. Another classification of communities is based on community metabolism (Odum, 1956). Regardless of the basis for its naming, the community consists of an anatomical (structural) and a functional component. Both must be studied thoroughly if the total community is to be understood.

The boundary lines between communities are generally indistinct, but a gradual blending in species composition may even form a *continuum*, which can result when organisms characterized as members of one community also have significant interaction with organisms or with a physicochemical environmental factor in one or more designated commu-

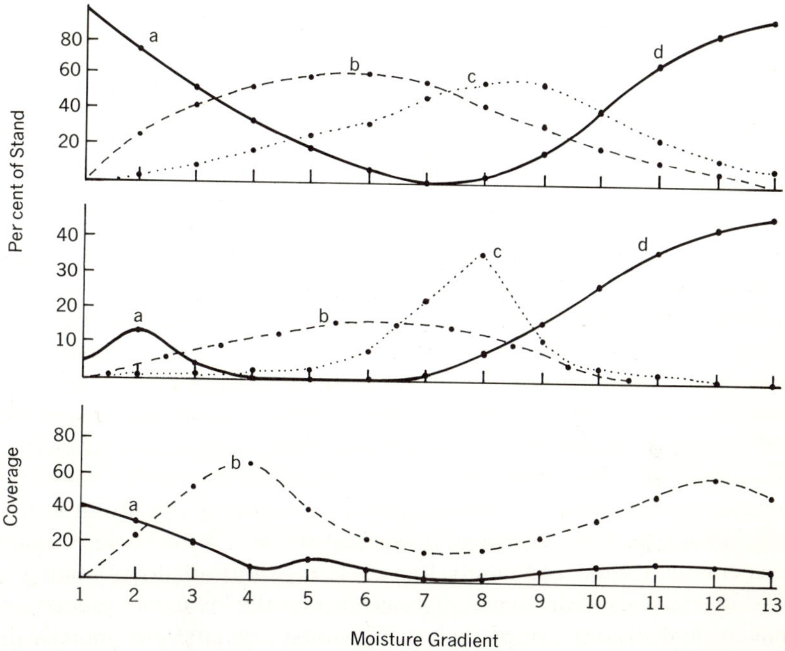

Figure 11-1. Progressive change in structure of vegetation in the Great Smoky Mountains along a moisture gradient. Top curves represent tree types: *a*: mesic; *b*: submesic; c: subxeric; *d*: xeric. Middle curves are for tree species: *a*: birch; *b*: dogwood; c: oak; *d*: pine. Lower curves depict: *a*: herbs; *b*: shrubs. (From Kormondy, 1969; based on data of Whittaker, 1956.)

nities. For example, the distribution of birds does not conform to the three major biogeographical approaches to the classification of communities (biomes, life zones, and biotic provinces), although each approach expresses generalizations about avian distributional relations (Miller, 1951). In another example, plants in the Great Smoky Mountains occurred in abundance at different points to follow a moisture gradient. For each species, the distributional curve was bell-shaped, with considerable overlap of the different species-abundance curves (Fig. 11-1).

Ecologists are in a healthy state of debate over the nature of communities, with the various views being polarized around two major positions. One contends that a community is a discrete entity having distinct boundaries. Certain dominant species are significant only in this one community. To some ecologists representing this school of thought, particularly botanists, the community exists as an organism (Becking, 1957). At the opposite pole is the "continuum" school described above, which emphasizes the individuality of species (Curtis, 1959; Whittaker, 1965). Although the concept of a community as defined by Allee *et al.* (1949) stresses natural interrelationships between organisms, all classifications of communities are somewhat arbitrary, and a community depends in part on the descriptive ability of the scientific investigator and the level of sophistication he uses in quantifying the data.

Within a community, the distribution (dispersion) of a single species is determined by its physiological capabilities and the effects of various physical and biotic environmental factors. Three principal patterns of spatial dispersion have been recognized: (1) regular dispersion in which individuals are uniformly found throughout the community, as reported for the lamellibranch *Tellina* (Holme, 1950); (2) clumped dispersion in which individuals are aggregated in one place, such as the schooling behavior of fish; (3) random dispersion in which individuals are haphazardly scattered throughout the community in a manner less uniformly distributed than type 1, as illustrated by spiders (Cole, 1946). To date, clumped dispersion is most commonly found. The pattern of distribution varies within a species with such factors as stage of the life cycle, species abundance, and availability of niches. Obviously, plants and animals differ because of locomotor ability. Knowledge of dispersion patterns is important to the ecologist in determining the methods of quantitative sampling of populations. Without concern for this variable, erroneous conclusions about community structure and function could result.

Each community has an outer border contiguous with the border of an adjoining community. Since these borders generally are indistinct, a transition area known as an *ecotone* is formed, which represents competing components of each community. The size and character of an ecotone depend on the nature of the adjacent communities; for instance, it may be very sharp and extend for only a few feet, as seen between loblolly pine and post-oak–blackjack-oak communities, or it may extend for miles, as in the post-oak–short-grass prairie ecotone. A basic aspect of the ecotone concept is competition between the adjoining communities. Organisms living in an ecotone may be faced with physical and biotic conditions strikingly different from those more centrally located in the community. Some species, which have been called *ecotone constituents,* thrive best in ecotone regions, whereas others are barely able to survive these environmental stresses. Excellent earlier examples of ecotone conditions are those outlined by Van Deventer (1936) and Kendeigh (1944). However, knowledge of the physiological ecology of ecotone inhabitants is very limited.

CHANGES IN A COMMUNITY

Various factors of the environmental complex may constantly change; a community reflects this fluctuation by some alteration of its structure and function. A moderate change of brief duration seldom disturbs community dynamics. If the changes are of longer duration, occurring daily, seasonally, or yearly, relatively minor alterations occur in the community (Moreau, 1944). However, catastrophic events, such as fire or flooding, are of short duration, but may result in the complete destruction of a community. Long-term subtle changes resulting from either physical or biotic causes also will influence a community. For example, new species may replace the resident species if their range of physiological tolerances is in better equilibrium with the altered environment. Many of these general environmental changes will produce marked microenvironmental alterations, which will favor the competitive advantage of physiologically adapted species. If the dominant species are replaced or the habitat is altered, the integrity of the original community may be disrupted and a new community will replace the old. This process of replacement is called *ecological succession.*

Since the environment is constantly changing, long-term studies of communities are necessary in order to interpret whether changes are oscillations around a mean, such as seasonal change, or whether they represent a consistent directional change in community structure. Many communities are transitory and are replaced by others. In the view of some ecologists (e.g., Tansley, 1935), this replacement of communities typically occurs in an orderly fashion until a community attains a certain degree of stability and persists unless disrupted by unusual circumstances. This "final" successional stage is called a *climax community,* and each intervening stage is called a *seral stage.* Two major types of successional patterns have been described. If succession starts from a barren area, such as bare rock or open water, it is called *primary succession.* But if an advanced successional stage retrogresses or is destroyed, the recovering successive stages leading toward a climax community represent *secondary succession* (Fig. 11-2). The general theory underlying the concept of ecological succession defines a successional series from pioneer stage to the climax as a *xerosere,* if it starts from bare rock or soil, or as a *hydrosere,* if it begins from open water (Figs. 11-3 and 11-4). Varying numbers of seral stages are formed, and any relatively stable stage below the climax represents a *subclimax.* Different types of climaxes have been described on the basis of edaphic (soil) or topographic features. In contrast to this view of succession, other ecologists (Dice, 1952) have mentioned that the replacement of one community by an-

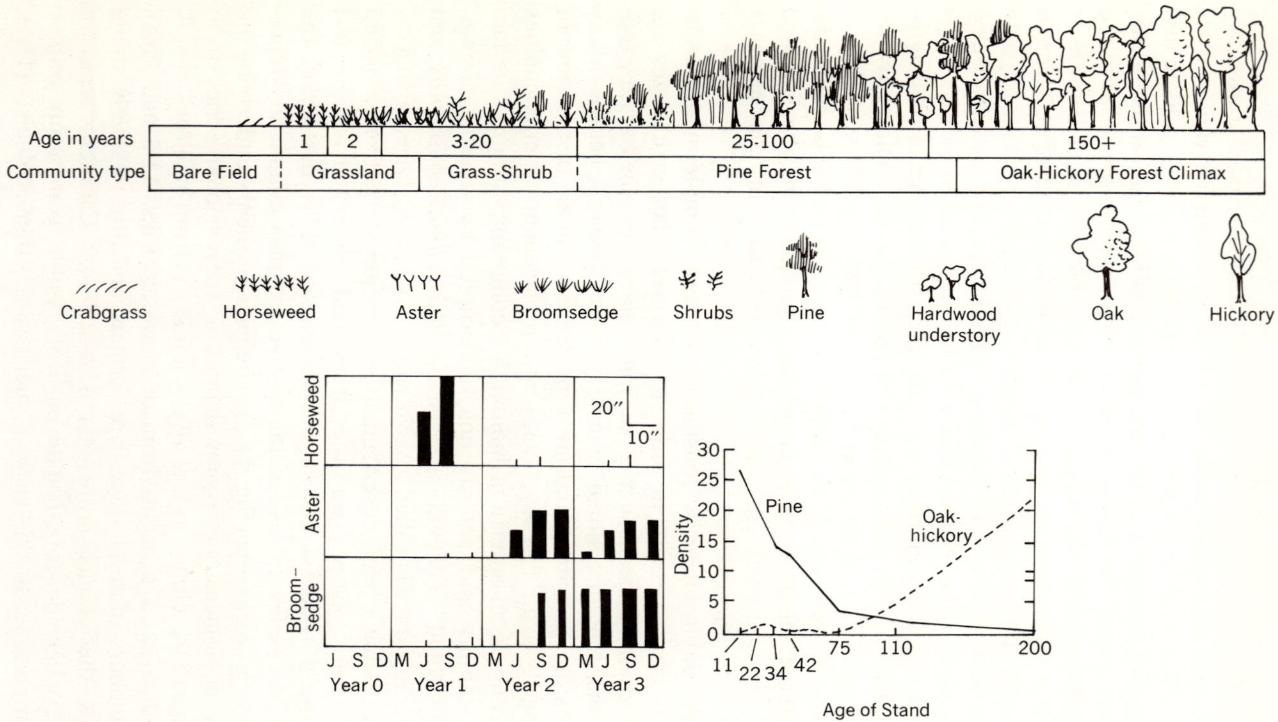

Figure 11-2. Secondary succession. The upper diagram shows principal dominant plants which progressively occupy an abandoned crop field. Lower left diagram depicts succession of three pioneer species with time. The height of the columns represents average height of plants in inches; the width represents relative diameter of stems. Lower right diagram shows progressive change from pine to hardwood trees with time. (From Odum, 1959.)

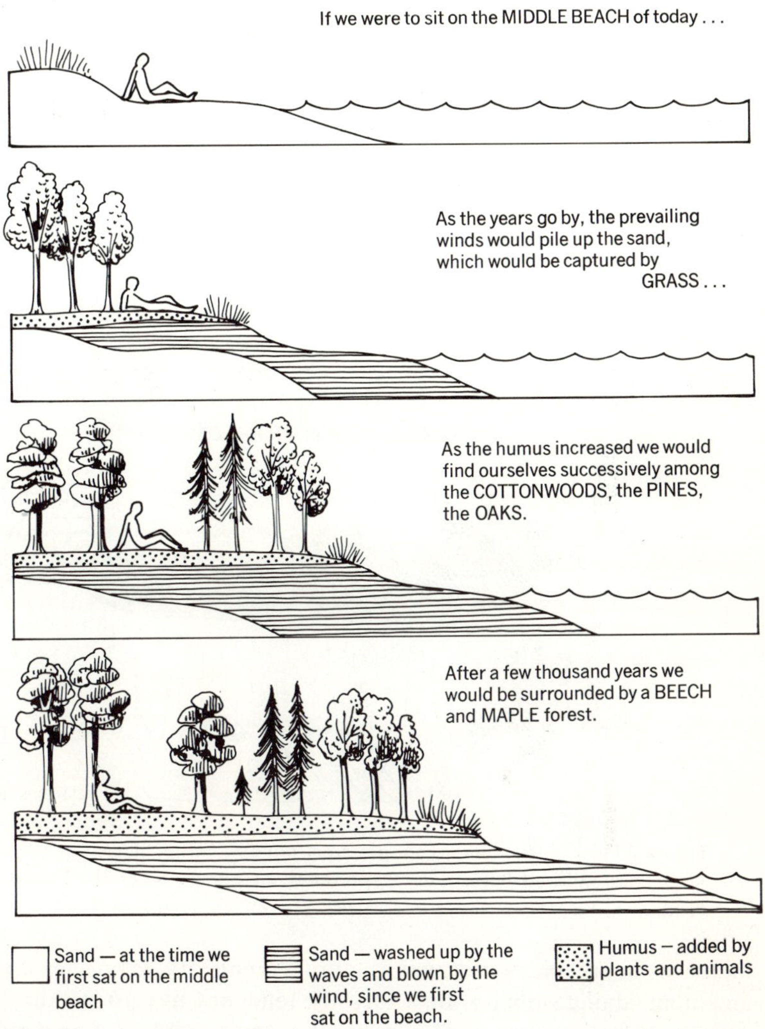

Figure 11-3. Succession and development operating through time in the forest-on-sand sere in the north Indiana dunes. (From *Basic Ecology* by Ralph and Mildred Buchsbaum. Reproduced by permission of The Boxwood Press, Pittsburgh, Pa.)

other represents succession, whether or not it leads toward a climax community.

Certain attributes are inherent in the ecological succession leading to a stable climax community. The initial seral stage represents a harsh environment with few niches which can support few species. As this and subsequent stages change, due to biotic and physical influences, there are

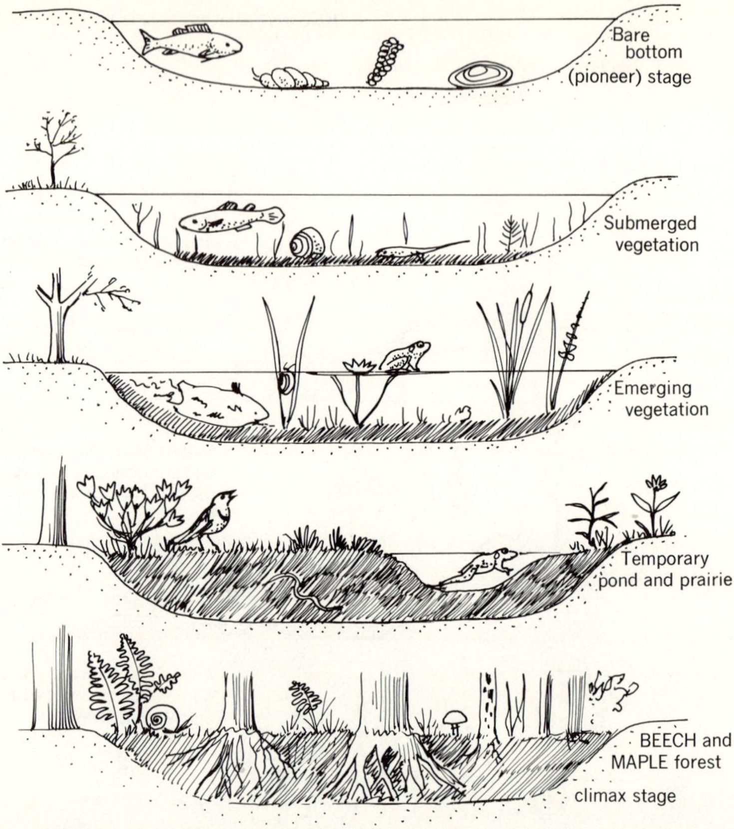

Figure 11-4. Successional changes from pond to forest. (From *Basic Ecology* by Ralph and Mildred Buchsbaum. Reproduced by permission of The Boxwood Press, Pittsburgh, Pa.)

increases in organic matter, in structural complexity, in species diversity, and in metabolic stability. Basically, the tendency toward stability, which appears to be inherent in all communities, may result in a community in which all the possible niches are fully occupied (Lindeman, 1942). With this increased complexity, an adverse happening in one segment of the community is less likely to disrupt the community as a whole.

ECOSYSTEMS

One fundamental problem in understanding ecology on a broad biological basis is that the study of animal communities frequently proceeds along lines different from those in investigations of plant communities.

This divergence is due in part to the presence of separate departments of zoology and botany and also to differences in the relative influence of environmental factors on the behavior of plants and animals. Since each plant and animal species may have a different range of tolerance along any environmental gradient, boundary lines between communities often are indistinct and delineation of specific communities is arbitrary. Furthermore, when a plant or an animal community is considered independently, the view of nature as an integrated whole is lost, and consequently the interaction of biotic and physical factors cannot be adequately appreciated. Technically, it is very difficult to study the complex dynamics of a biotic community. However, recent advances in computer technology and systems analysis should permit better approximations of what communities are and what they do. Therefore, some ecologists have proposed that a more accurate view of nature may be derived from the study of the species as the basic ecological unit, rather than the community, and that it is in fact the adaptive capacity of individual species which permits the coexistence of plant and animal species in that dynamic equilibrium called a community.

When the biotic community and its physical environment are considered together, this intricate interacting unit is called an *ecosystem*. Although this term was proposed earlier by Tansley (1935) to emphasize that both the abiotic environment and the biota together form the basic ecologic unit of study, it has not been until relatively recent years that ecosystem analysis has become a dominant line of ecological study. In broad terms, one of the basic goals in analyzing ecosystems is to understand the incorporation of energy from solar radiation into organic material by green plants and the subsequent transfer of energy to other organisms. Energy is the coinage that permits the economy of a complex ecosystem to exist.

The size of an ecosystem is not fixed. Although the biosphere is an ecosystem, for purposes of study the ecologist will select a smaller unit as an ecosystem, such as a pond, an estuary, or a balanced aquarium. To be suitable for investigation, an area should be large enough to contain all basic factors operative in an ecosystem. Ecosystem analysis has been based on the study of systems, both in the field and in the laboratory. Results of field studies have not proved to be entirely adequate, not because of the lack of intensive work of competent investigators but because of the complexity of the problem and the limitations of techniques (Whittaker, 1961). Artificial ecosystems have therefore often been used to examine the dynamics of simplified systems under controlled conditions, despite the difficulty of correlating laboratory and field data (McIntire and Phinney, 1965).

One method of studying an ecosystem is to subdivide it into two principal component parts: the environment and the biota. As emphasized earlier, an organism does not exist in the absence of a physical environment, but has a complex interaction with it. The influence or effect of the environment on a biological unit is referred to as *action,* whether this unit is an organism or another level of biological organization. The influences of the biological interplay between different species are grouped under the ecological heading of *coaction.* And the alteration of the environment by an organism is called *reaction.*

It is convenient to think of the environmental component as consisting of two units. One includes those physical and chemical factors that act on organisms, but are not depleted, such as temperature, gravity, and light. The second includes resources that can be utilized, such as nutrients.

Included in the biotic component are all interactions between species as well as the vital phenomena of a single species. The species interaction subcomponent includes predator-prey relationships and competition. The intraspecific subcomponent encompasses such functions as growth, reproduction, metabolism, and homeostatic mechanisms.

A simplified ecosystem, consisting of certain functional and structural units, is graphically presented in Fig. 11-5. The radiant energy from the sun is received by green plants (*autotrophs*), which bind it into organic material. This process is called *primary production* and constitutes the first nutritional or trophic level. The next higher trophic level is reached when this plant material is utilized and incorporated into new organic compounds by consumer organisms called *herbivores.* This process is referred to as *secondary production.* Numerous other trophic levels of consumers may be involved:

1. Carnivores feed on herbivores or on other carnivores.
2. Upon death, each organism is fed upon by *saprophytic* organisms, which may in turn provide energy for consumers.
3. The breakdown of complex organic compounds from dead organisms to inorganic material is accomplished by *decomposers.*
4. Certain organisms, called *transformers,* convert inorganic compounds into chemicals which can be used by primary producers.

Energy for the support of each higher trophic level obviously is derived from a lower one until the basic level of primary production is reached. This interpretation of the trophic dynamics of an ecologic system is one of the unique contributions of the ecosystem approach.

The structural anatomy of an ecosystem is based on descriptions of

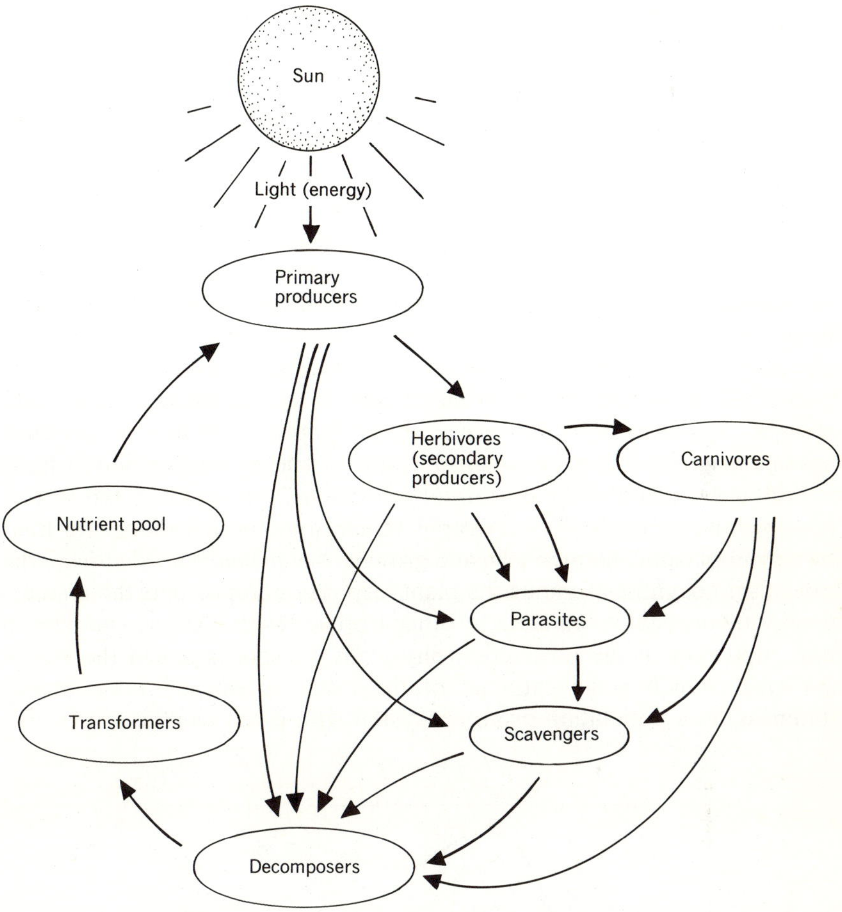

Figure 11-5. A generalized scheme of an ecosystem.

both biotic and abiotic components. Descriptions of the biota include such basic data as species composition, biomass, seasonal changes, and the distribution of species in time and space. The abiotic component includes both chemical and physical analyses. These analyses should include data from qualitative and quantitative studies of chemical composition of species in time and space, and the physical parameters of the environment, such as temperature and salinity, expressed as the range of fluctuations with time. In contrast, the functional view of an ecosystem emphasizes: (1) the rate of various ecosystem functions, such as primary production, energy flow through various trophic levels, and biogeochemical cycling; and (2) regulation of ecosystem processes.

TECHNIQUES FOR MEASUREMENT OF PRODUCTIVITY

To evaluate the energy relationships in ecosystems, ecologists have focused their attention on estimating the production of organic matter. Although most emphasis has been on primary production, some information on secondary production is available. *Productivity* is the total amount of organic matter produced in a given space during a given period. Various units have been proposed to express the rate of production. This rate may be expressed on the basis of calories (or kilocalories) produced per unit time per unit area, or of the dry or living weight of organisms per unit time per unit area. *Primary productivity* is a function of the ability of autotrophic organisms to fix solar energy and to manufacture organic substances. In some special cases, chemosynthetic organisms produce organic material, using a geological source of energy not previously fixed by biological activity. *Gross primary productivity* is equal to the rate of photosynthesis; however, autotrophs utilize some of this energy for their own protoplasmic maintenance and growth. *Net primary productivity,* the rate of photosynthesis minus the plant respiration, represents the residual energy theoretically available to other trophic levels. At any moment in time, that part of the production physically present is called the *standing crop,* which is an estimate of the total amount of living matter (biomass) in a population or an ecosystem at a given trophic level.

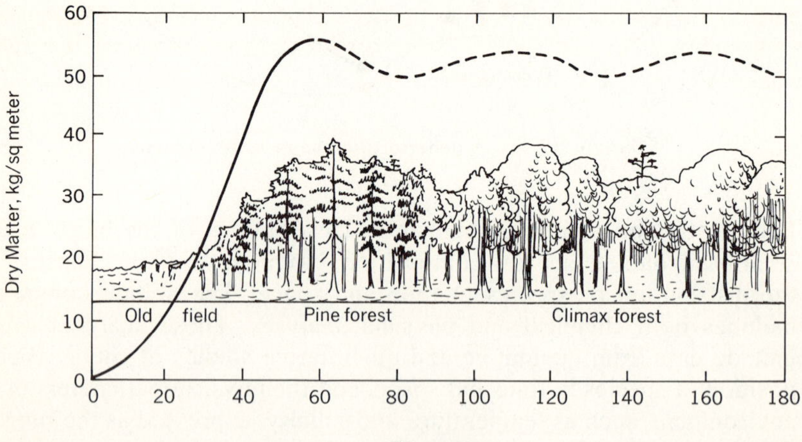

Figure 11-6. Biomass changes during field-to-forest succession. The dotted line indicates that some oscillations are assumed in the climax forest. (From Woodwell, 1967.)

Table 11-1. Standing Crop Estimates.

Type of Vegetation and Geographical Location	*Dry-Weight Biomass*	*References*
1. Above-ground plant parts from mature temperate-zone forests in favorable environments	30–50 kg/sq meter	Woodwell and Whittaker, 1968
2. Above-ground plant parts, open forests and woodlands	5–20 kg/sq meter	Woodwell and Whittaker, 1968
3. Above-ground plant parts, cove forests in the Great Smoky Mountains	60 kg/sq meter	Woodwell and Whittaker, 1968
4. Tropical rain forest, excluding roots, litter, and dead branches	56.1–301.8 (1×10^3 kg/$\frac{1}{2}$ acre); av 164.1 for 8 stands	Ovington, 1965
5. Temperate coniferous forest, excluding roots, litter, and dead branches	37.5–249.6 (1×10^3 kg/$\frac{1}{2}$ acre); av 133 for 15 stands	Ovington, 1965
6. Temperate hardwood forest, excluding roots, litter, and dead branches	46.5–223.7 (1×10^3 kg/$\frac{1}{2}$ acre); av 114.4 for 13 stands	Ovington, 1965

Earlier, primary production was estimated indirectly as long-term changes of the standing crop, using two general criteria: (1) harvesting methods; and (2) the concentration of inorganic nutrients or dissolved gases.

1. Using harvesting methods, the standing crop is sampled periodically and the rate of production is calculated. This technique has been used in studies of various terrestrial and aquatic ecosystems (Westlake, 1965; Whittaker and Woodwell, 1968), but it can be very time consuming and the sampling techniques are complex. In addition, the resultant data do not always reflect the energetics of the ecosystem; for instance, all the organic material may not be digestible, or an inestimable amount of plant material may have been consumed by animals or decomposed during the period of observation. A further limitation in these methods is due to the variations in the biomass between different ecosystems (Table 11-1). In a field-to-forest successional series, a biomass increase has been reported (Fig. 11-6), but during the climax stage there is no increase in biomass (Woodwell, 1967).

2. Indirect estimates of production also have been made by measur-

ing the rate at which organisms alter the concentration of significant materials in the ambient environments. Using these data, the amount of biotic activity needed to produce this change can be calculated and expressed as production. Specific studies have measured such factors as a decrease in CO_2, oxygen, phosphate, phosphorous, and nitrate (see review of Goldman, 1968). Results based on these indirect techniques represent only rough approximations of production, however, because of the uncertain interrelationships of plants and animals in modifying the concentration of these materials. Nevertheless, these estimates have stimulated newer attempts to achieve more precise measurements.

A more direct technique for measuring primary productivity has been to estimate photosynthetic rates. In aquatic ecosystems, this is done by the so-called light-and-dark bottle method. This technique is as follows. Three equal samples of water are collected: one is placed in a clear bottle, the second in a bottle that is darkened to prevent light from influencing the water sample, and the third sample is placed in a bottle that has been analyzed for oxygen content at the beginning of the experiment. The light and dark bottles are suspended side by side in the water for a standard period of time, usually 24 hr, and then they are analyzed for oxygen. The reduction in oxygen in the dark bottle indicates the amount of respiration of the sample. Changes in oxygen concentration in the light bottle reflect the rate of photosynthetic activity and respiration. These total values give an estimate of the total gross food production for this period of time. This method has also been modified for studies of terrestrial ecosystems (see review of Woodwell and Whittaker, 1968), in which biotic samples are placed in either transparent or darkened chambers and the uptake or output of carbon dioxide or oxygen is measured by means of sensitive instruments, such as the infrared gas analyzer for CO_2 detection. Although these methods have been used extensively, their reliability has been questioned, particularly when used in regions of low primary productivity.

In 1952, Steemann-Nielsen introduced a more sensitive method of measuring photosynthesis in phytoplankton, using radioactive carbon (C^{14}). In brief, this technique consists of the introduction of a known amount of C^{14} (in the form of bicarbonate) into the water sample, which is then exposed to light for a prescribed period of time, usually 6 hr. Subsequently, the water sample is filtered and the amount of C^{14} fixed in the plankton is determined (for specifics of technique see Steemann-Nielsen, 1952; Doty, 1961; and Strickland and Parsons, 1965). In certain lakes and inshore marine waters, the samples may be positioned *in situ* for the experimental time period, but this method is difficult at sea. On board ship, experimental bottles are placed in a tank on the deck. The temperature and the amount and type of light to which the samples are exposed are controlled to simulate various depths. This technique is used widely, although numerous factors affect the interpretation of the C^{14} method, such as daily productivity of photosynthesis rate and the excretion of assimilated C^{14} (Thomas, 1961; and Goldman, 1968). Further, it has been questioned whether this technique measures gross production, net production, or some process in between. Although Bunt (1965) maintained that it represents net production, he also stated that results depend on experimental conditions.

Another measure of primary production initially proposed by Manning and Juday (1941) is based on the amount and type of chlorophyll present plus data on light intensity at different depths. Some but not all investigators have obtained similar production values, using the C^{14} and chlorophyll methods (Ryther and Yentsch, 1957, and Holmes *et al.*, 1957). Talling and Driver (1961) discuss the problems in estimating chlorophyll.

Productivity Measurements

Primary Productivity

The fundamental objective in studying primary production is to obtain data on the quantity of energy potentially available to support other segments of an ecosystem. Although data on primary productivity are available for many types of ecosystems, Goldman (1968) recently emphasized the need for cautious consideration of energy flow through an ecosystem in order to avoid excessive reliance on published values of primary productivity. Investigators active in this phase of ecology are continuing concentrated studies to determine more precisely what is being measured. Although differences in techniques, for example, make strict comparisons of values inadvisable, certain generalizations do appear valid (Woodwell and Whittaker, 1968). For most mature temperate-zone forests of favorable environments, net primary production is estimated to be between 1200–1500 grams/sq/meter and about 400–1000 grams/sq meter/yr for woodlands and smaller forests. The range of values from 1200–1500 grams/sq meter/yr for mature forests has been considered a norm for climax communities of favorable environments.

Some ecosystems are more productive because of very favorable environmental conditions. Marshes, tropical grasslands, flood plain forests, and redwood flats may produce 1500–3000 grams/sq meter/yr or higher. Special crops such as sugar cane, rice, and maize also may be extremely productive. Shrublands and grasslands vary considerably: heath balds and high-grass prairie have values about 100 grams/sq meter/yr; desert

Table 11-2. Productivity of Benthic Marine Plants.

Population	*Location*	*Production in Terms Reported*	*gram C/sq meter/day*[b]	*Method*
Laminaria	Scotland and Calif.	66 grams dry matter/m^2/day	9.	O_2 bottles
Pelvetia	Calif.	35 grams dry matter/m^2/day	5.	O_2 bottles
Fucus	Calif.	19–42 grams dry matter/m^2/day	2.–5.	O_2 bottles
Alaria	Calif.	14 grams dry matter/m^2/day	2.	O_2 bottles
Egregia	Calif.	25 grams dry matter/m^2/day	3.	O_2 bottles
Iridophycus	Calif.	19 grams dry matter/m^2/day	2.	O_2 bottles
Gigartina	Calif.	54 grams dry matter/m^2/day	7.	O_2 bottles
Porphyra	Calif.	11–21 grams dry matter/m^2/day	1.–3.	O_2 bottles
Ulva	Calif.	3–7 grams dry matter/m^2/day	0.4–1.0	O_2 bottles
Zostera marina	Denmark	277 grams dry matter/m^2/year	0.3?	Harvest
Ruppia maritima	Denmark	140 grams dry matter/m^2/year	0.2?	Harvest
Chara spp.	Denmark	283 grams dry matter/m^2/year	0.3?	Harvest
Zostera marina	Denmark	500–2000 grams dry matter/m^2/year	0.6–2.5?	Harvest
Zostera + *Ruppia*	Denmark	184 grams dry matter/m^2/summer	0.2?	Harvest
Microbenthos	Denmark	0.11–0.22 grams C[a]	0.6–1.1	C^{14} bottles
Microbenthos	Georgia	0.4–1.0 grams C/sq meter/day	0.4–1.0	O_2 bell jars + CO_2 gas
Symbionts in *Pocillopora*	Rongelap Atoll	0.042 ml O_2/grams dry wt/hr		O_2 bottles
Symbionts in *Acropora*	Rongelap Atoll	0.035–0.075 ml O_2/grams dry wt/hr		O_2 bottles
Porolithon	Rongelap Atoll	0.046 ml O_2/grams dry wt/hr		O_2 bottles

Table 11-2. Productivity of Benthic Marine Plants. (*Continued*)

Population	*Location*	*Production in Terms Reported*	*gram C/sq meter/day*[b]	*Method*
Coral reef community	Rongelap Atoll	12 grams glucose/sq meter/day[c]	4.[c]	Diurnal O_2 curve
Coral reef community	Eniwetok Atoll	24 grams glucose/sq meter/day[c]	7.[c]	Diurnal O_2 curve
Coral reef community	Hawaii	7.7–8.3 grams C/sq meter/day[c]	7.7–8.3[c]	Diurnal O_2 curve
Thalassia testudinum	Long Key, Fla.	34 grams O_2/sq meter/day[c]	10.[c]	Diurnal O_2 curve
T. test. + microbenthos	Boca Ciega Bay, Fla.	5[c] grams O_2/sq meter/day	2.[c]	Diurnal O_2 curve
Macrocystis pyrifera	Calif.	15 grams dry matter/sq meter/day	2.	Diurnal O_2 curve
Spartina alterniflora	Georgia	4248 kcal/sq meter/yr	5.	Harvest

SOURCE: Pomeroy (1961).

[a] Expressed as "potential productivity." To estimate daily production in grams C/sq meter/day, use the factor "hours of daylight/2."

[b] To make the findings more readily comparable, all have been converted to grams of carbon per square meter per day. The conversions are necessarily approximate in some cases.

[c] Gross production. All other values are net production.

Table 11-3. Productivity of Various Ecosystems.

GROSS AND NET ORGANIC PRODUCTION OF VARIOUS NATURAL AND CULTIVATED SYSTEMS (GRAMS DRY WT/SQ METER/DAY)[a]

System	*Gross*	*Net*
A. *Theoretical potential*		
Av radiation (200–400 gram cal/cm^2 day)	23–32	8–19
Max radiation (750 gram cal/cm^2 day)	38	27
B. *Mass outdoor* Chlorella *culture*		
Mean		12.4
Max		28.0
C. *Land (max for entire growing seasons)*		
Sugar cane		18.4
Rice		9.1
Wheat		4.6
Spartina marsh		9.0
Pine forest (best growing years)		6.0
Tall prairie		3.0
Short prairie		0.5
Desert		0.2
D. *Marine (max for single days)*		
Coral reef	24	(9.6)
Turtle grass flat	20.5	(11.3)
Polluted estuary	11.0	(8.0)
Grand Banks (Apr.)	10.8	(6.5)
Walvis Bay	7.6	
Continental Shelf (May)	6.1	(3.7)
Sargasso Sea (Apr.)	4.0	(2.8)
E. *Marine (annual av)*		
Long Island Sound	2.1	0.9
Continental Shelf	0.74	(0.40)
Sargasso Sea	0.88	0.40

ANNUAL NET PRIMARY PRODUCTION OF AQUATIC COMMUNITIES ON FERTILE SITES (METRIC TON DRY ORGANIC MATTER/HALF-ACRE)[b]

Marine phytoplankton	1–4.5
Lake phytoplankton	1–9
Fresh-water submerged macrophytes (water weeds)	4–20
Marine submerged macrophytes (seaweeds)	25–40
Marine emergent macrophytes (salt marsh)	25–85?
Fresh-water emergent macrophytes (reed swamp)	30–85

SOURCES: [a] Ryther, 1959; [b] Westlake, 1965.

grassland probably ranges from 200–300 grams/sq meter/yr to very low values. Certain parts of the ocean are more productive than others; for example, where upwelling occurs and nutrients are brought from the deeper waters to the surface, the rate of primary production is high. Holmes (1961) reported that in the north equatorial Pacific Ocean (latitudes of 0–25°N) the *in situ* measurement of surface productivity of inshore waters ranged from 0.605–140 mg C/cu meter/day, while offshore waters ranged from 0.13–414 mg C/cu meter/day. The median values were 5.5 and 6.1, respectively. In addition to geographical differences, productivity at one location varies from month to month: the lowest median value was 6.3 in September; the highest value was 43.0 in May.

Lakes vary considerably in their energy production (Goldman, 1968). Lake Vanda (near McMurdo Sound, Antarctica), for example, is very oligotrophic, with a mean productivity of only about 1 mg C/sq meter/hr. The mean for Castle Lake, California, is 70 mg C/sq meter/hr, and it is classified as a mesotrophic lake. A high rate of production (300 mg C/sq meter/hr) during the growing season occurs in Clear Lake, California, a shallow lake which is termed eutrophic. In contrast, however, another California shallow lake, Cedar Lake, which has been nearly filled with sediment and invaded by terrestrial and high aquatic plants, has a low productivity value of 6.0 mg C/sq meter/hr. This lake is well on its way to extinction and hence is classified as dystrophic.

Much of the earlier data on productivity of benthic marine plants as reviewed by Pomeroy (1961) is found in Table 11-2; productivity values of other ecosystems are presented in Table 11-3 (after Ryther, 1959, and Westlake, 1965).

Secondary Productivity

The potential energy resulting from primary production furnishes the energy required by other trophic levels in an ecosystem. Some energy, in the form of food, is consumed by herbivores or omnivores who may be eaten by carnivores, who in turn may be eaten by other carnivores. However, much of this ingested food is not assimilated: herbivores may assimilate only 10 percent of the ingested food, whereas the assimilation coefficiency of carnivores is typically higher. For example, various species of fish absorbed between 86 and 96 percent of the ingested food, but the efficiency absorption of different classes of food substances varied: about 92–98 percent of the ingested protein was absorbed, while only 80 percent of the fat and about 63–79 percent of carbohydrates was absorbed by the fishes (Pandian, 1967). The unassimilated material leaves the animal's body as waste material to serve as an energy source for other organisms. Assimilated energy is used by these consumer species for various metabolic processes, such as respiration, excretion, and secretion (see discussion in Chapter 8). The resultant amount of energy stored in the tissues of heterotrophs is termed *net secondary production.*

Estimates of secondary production have great relevance to both theoretical considerations of energy flow through an ecosystem and applied aspects of ecological energetics such as animal husbandry and wildlife management. The need for man to manipulate an ecosystem for the purpose of increased secondary productivity is of increasing importance in the face of the almost overwhelming population growth.

To determine the level of secondary production, expressed as an increase in biomass, requires an analysis of the energy budget of a given species. A simplified expression of secondary productivity may be obtained by merely weighing organisms over their entire life cycle. If the population is in equilibrium with its environment and is neither increasing nor decreasing in size, secondary production may be estimated by determining the caloric content of all individuals dying during a prescribed period of time. However, because secondary production is extremely variable and subject to environmental fluctuation, the following nonequilibrium expression of secondary productivity has been proposed by Golley (1968):

$$\text{Net production} = \text{yield (dying individuals)} \pm \text{caloric equivalent of changes in standing crop}$$

Obviously, if the net production decreases during the observational period, the standing crop declines; if the net production increases, the standing crop increases.

Energy Flow

Although estimates of primary and secondary production are biologically important in themselves, ecologists are especially concerned with the overall behavior of energy in a total ecosystem. By adopting the energy-flow diagram technique of engineers, the directional flow of energy through an ecological system can be simulated. Many representations of energy-flow diagrams or "energetic" balance sheets in the literature represent the energy consideration of a single species, the bioenergetics of a "simple" ecosystem, or the representation of more complex ecosystems. Recently, E. P. Odum (1968) published a historic review of energy flow in ecosystems, in which he proposed a "universal" model of energy flow (Fig. 11-7). This model can be used to depict the energetics of individuals, populations, or various species representing a trophic group, and food chains can be represented if several models are linked together.

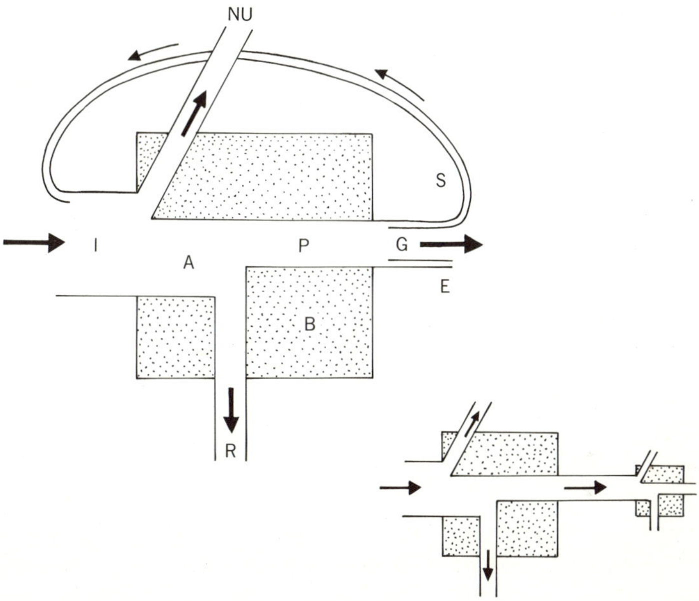

Figure 11-7. A "universal" model of energy flow through biological systems. (From Odum, 1968.)

It should be remembered that since there is no universal agreement on terminology, the symbols of Odum may not be comparable to those of other investigators. In Odum's model, *B* represents the biomass of the unit of structure, *I* is the total energy input or intake. *NU* is that part of the energy that is not used, *A* represents utilized or assimilated energy, *R* is that part of *A* that is used in respiration, and *P* is that part of energy transformed to new or different organic matter. Three subdivisions of production are suggested: *G* is addition to the biomass or growth; *E* refers to organic matter that has been assimilated but is excreted or secreted; and *S* is stored energy such as fat, which may be mobilized at a later time. Odum represents *S* by a loop back to *I*, to indicate stored energy necessary to ensure that new energy is obtained (that is, reserve energy used in search of food). Acutally, this function does not add energy to *I* from *S* because *S* is utilized. The energy available to the next trophic level is represented by *P*, whereas the energy of *NU* is available at the same trophic level. Earlier studies, as reviewed by Odum (1968), stressed metabolism of whole communities without having detailed information on all component populations, whereas one of the current investigative trends concentrates on detailed energy-flow studies of the individual organisms comprising a community (Darnell, 1968).

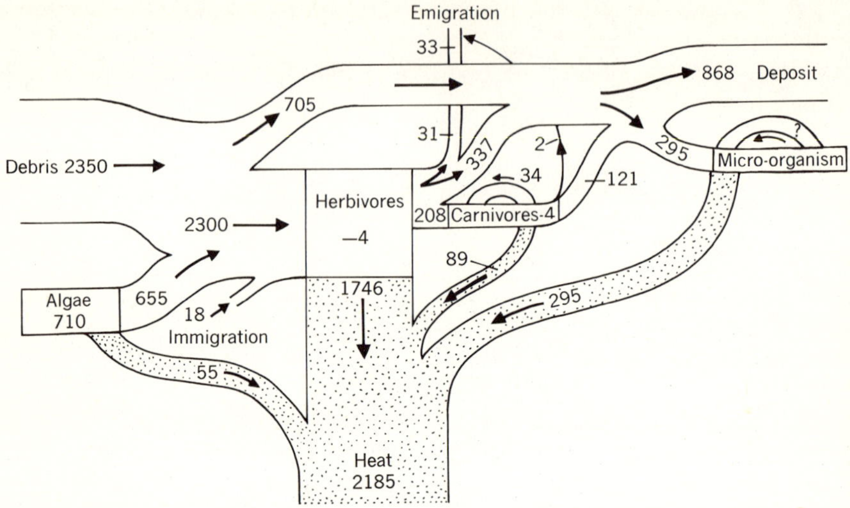

Figure 11-8. Energy flow diagram for a small spring. Figures in kcal/sq. meter/year; numbers inside boxes indicate changes in standing crop. (From Teal, 1957.)

An early attempt to analyze energy flow in a natural ecosystem was described by Teal (1957), whose work is graphically presented in Fig. 11-8. He studied a small spring, which was not much larger than an artificial laboratory ecosystem. Debris washing into the spring provided most of the energy input, although some came from algae. The standing crops of both herbivores and carnivores were slightly reduced at the end of the year's study. The herbivores consumed 2300 kcal/sq meter/yr, and the carnivores consumed 208 kcal./sq meter/yr. An *ecological efficiency* for herbivores was estimated to be about 9 percent. Ecological efficiency is defined according to Phillipson (1966) as

$$\frac{\text{Calories of prey consumed by predator}}{\text{Calories of food consumed by prey}} \times 100$$

Odum (1957) reported an efficiency value of 11.4 percent for herbivores in Silver Springs, Florida, when measured over a short period of time. Ecological efficiencies were reported to be from about 7 percent for *Hydra* to about 13 percent for *Daphnia* under relatively controlled laboratory conditions (Slobodkin, 1962).

In contrast to studies based on all trophic levels in an ecosystem, the annual energy-balance sheet for one group of terrestrial animals, oribatid mites, estimated (Fig. 11-9) that most of the energy is egested, and most of the assimilated energy is dissipated in respiration (Englemann, 1961,

1968). The ecological efficiency values for mites and several other soil arthropods ranged from 8.5–45.4 percent but most values were from 14–28 percent, which are not unlike values for aquatic organisms (Englemann, 1968).

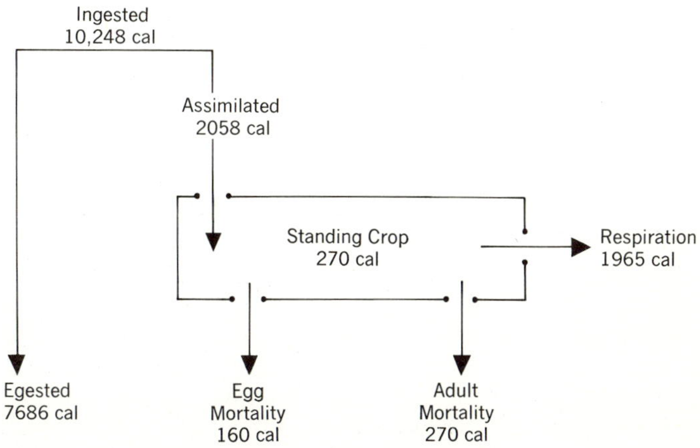

7686 + 2058=9744 (504 cal unaccounted for)

1965 + 270 + 160=2395 (337 cal overestimated)

Figure 11-9. Annual energy budget for oribated mites found in soil. (From Engelmann, 1961.)

After reviewing the literature dealing with ecological energetics, Slobodkin (1962) concluded that there is no apparent systematic pattern of taxonomical, ecological, or geographical variation in ecological efficiency. Although ecological efficiency may tend to be maximized through evolution, no theoretical reason has been found to suggest why any one value should be preferred by all ecological systems.

As in many young areas of investigation, precise terminology has not yet been established in ecological energetics. Recently, Kozlovsky (1968) reviewed the various definitions of ecological efficiencies and suggested a scheme for equating the terminology of various investigators. At the same time, he strongly refuted the generally expressed idea that consumers at higher trophic levels are progressively more efficient in the use of their food supply (Odum, 1959).

Studies of energy flow should continue because they are central to any interpretative, generalized, or predictive attempts in ecosystem analysis. Odum (1968) listed a number of important problems that can now be attacked by means of more sophisticated instrumentation and more advanced computer technology:

1. Systems analysis of ecosystems—an applied mathematic and computer method of assessing the entire ecosystem
2. Energetics of laboratory populations—with particular emphasis on self-sustaining microecosystems, which would have practical value to problems of space travel
3. Energy flow in populations—applying new techniques of field census and respirometry, which do not require confinement in cages or respiration flasks.

Food Chains and Webs

As emphasized in Chapter 9, food is of primary ecological importance to an individual organism. At the community-ecosystem level, this need is also obvious, but the intricacies of supply and demand of food are very complex. Food requirements involving different trophic levels may be expressed in various ways. In the earlier literature, food relationships were described in terms of *food chains* wherein one species (generally a holophytic plant) was devoured by another species (an herbivorous organism), which was consumed by a flesh-eating species (a carnivore), which in turn might be eaten by still another carnivore. The number of links in the chain varied, the simplest structure being a two-link chain with a herbivore feeding on a plant; in other cases, there were as many as five animal links in the chain. Elton (1927) pointed out that, typically, each new predator is slightly larger and less abundant than its prey. The net result of this relationship can be graphically represented in the form of ecological pyramids which may be based on numbers, biomass, or energy.

The pyramid of numbers indicates the total number of individuals at each trophic level in a food chain (Fig. 11-10). The higher the position on a food chain, the smaller the number of individuals, while great numbers of food producers are found at the base of the pyramid. This method, however, does not always adequately represent the energy interchanged

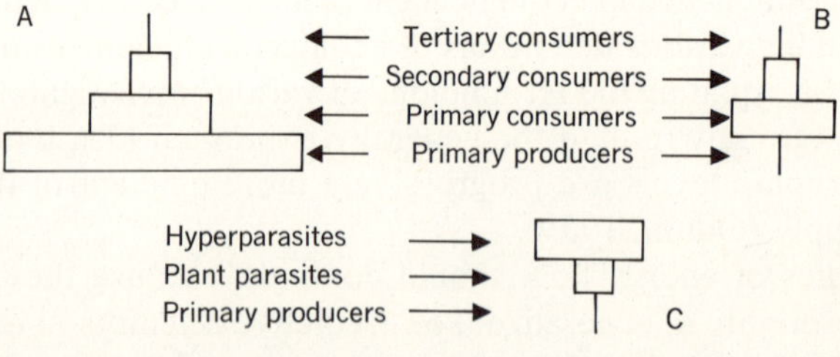

Figure 11-10. Pyramids of numbers. (A) The primary producers are small sized. (B) The primary producers are large sized. (C) A plant-parasite food web. (From Phillipson, 1966.)

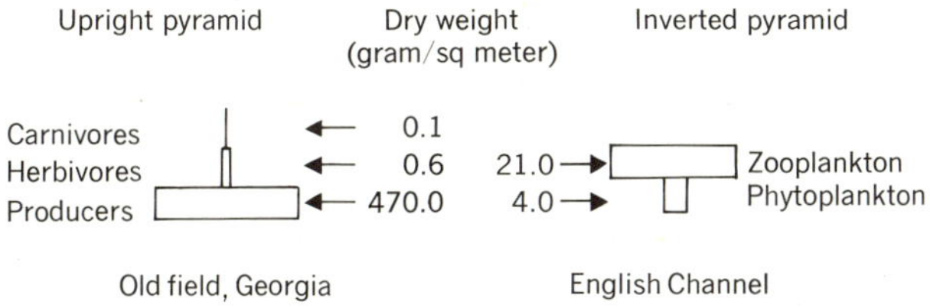

Figure 11-11. An upright and an inverted pyramid of biomass. (Adapted from Odum, 1959.)

between trophic levels, since differences in such factors as body size and differential rate of energy utilization by different species are not reflected. Although a better indication of the total amount of protoplasm is provided when the biomass at each trophic level is determined (Fig. 11-11), care must be taken in evaluating biomass estimates. These data may not be accurate representations of the amount of active protoplasm, owing to great interspecific differences in the amount of relatively inert material included in the total weight of a species; for example, the total living weight of an adult barnacle includes more shell and less protoplasm than that in a polychaete worm. To overcome this objection, some ecologists base their calculation of biomass on a more significant index of productive protoplasm, such as the amount of carbon or nitrogen. Not all species, however, utilize an equal amount of energy when equated on the basis of the same unit of weight and time (see Chapter 4). Even within a species, the same individual may utilize energy at different rates, depending upon fluctuations in the environment. As a further complication, the size of the population may vary seasonally, with an accompanying change in the size class of individuals represented. These and other reasons have prompted ecological energeticists to build pyramids based on energy, expressed in units such as a caloric (Fig. 11-12). Since energy utilization may vary with body size and other factors, an estimate of the energy budget for a species or a trophic level is very complex and is meaningful only when carefully determined over a long period of time.

Some inverted pyramids have been reported, but inverted energy pyramids are temporary and cannot be sustained. An inverted pyramid of numbers results if the primary producer is large (a tree) and the second trophic level consists of many small-sized organisms (plant parasites). Either an inverted biomass pyramid or an inverted energy pyramid could result in a geographical region where a primary producer (phytoplankton) could be introduced from an adjacent water mass to feed on zooplankton. Other causes could be a rapid rate of turnover by the lower trophic level or a transitory but marked decrease in biomass resulting from such factors as disease or overgrazing.

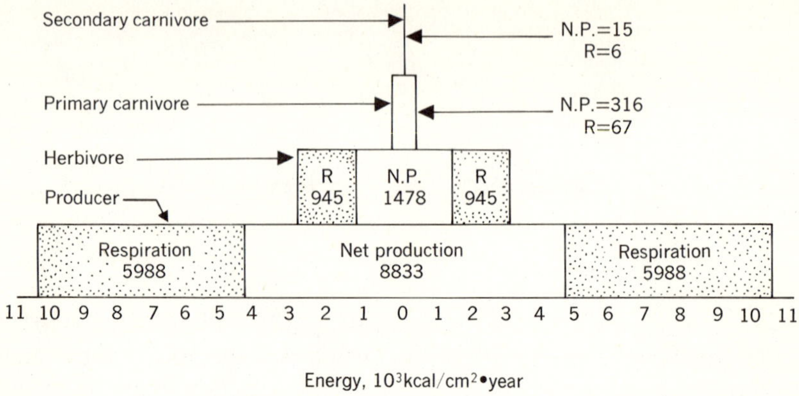

Figure 11-12. Energy pyramid. *N.P.* is net production and *R* is respiration. (From Kormondy, 1969; based on data of Odum, 1957.)

In summary, pyramids may offer a graphic method for representing trophic level interaction in a general way, but various technical problems reduce their value for critical quantitative comparisons.

In nature, predator-prey relationships are not so simple as a single chain. Since most organisms do not feed exclusively on a certain species, but are type feeders (such as filter feeders), some links in a chain may feed on a number of prey species, and thus a number of chains will be intertwined. The result is an interlocking of food chains, which has been called most appropriately a *food web* (Fig. 11-13). Thus, species in a food web may not be restricted to a single trophic level, but actually may participate at several trophic levels. This intertrophic interaction obviously influences and complicates the energy-flow diagram described earlier in this chapter.

Darnell's (1961) data on trophic studies of Lake Pontchartrain illustrates this point. He summarized his major findings as follows:

1. Several sources of food are utilized by a given size group of a species; for example, the adult blue catfish consumed molluscs (12 percent), macrocrustaceans (17 percent), fishes (30 percent), vegetation (26 percent), and miscellaneous material (15 percent).

2. Depending upon availability, alternative foods may be utilized by one species from time to time and from place to place. These food substitutions do not necessarily occur within the limits of a given trophic level, but may represent different trophic levels.

3. During the growth and development of a species, the food habits of an animal change in a regular progession. The direction of change in terms of trophic level during ontogeny varies with the species. For

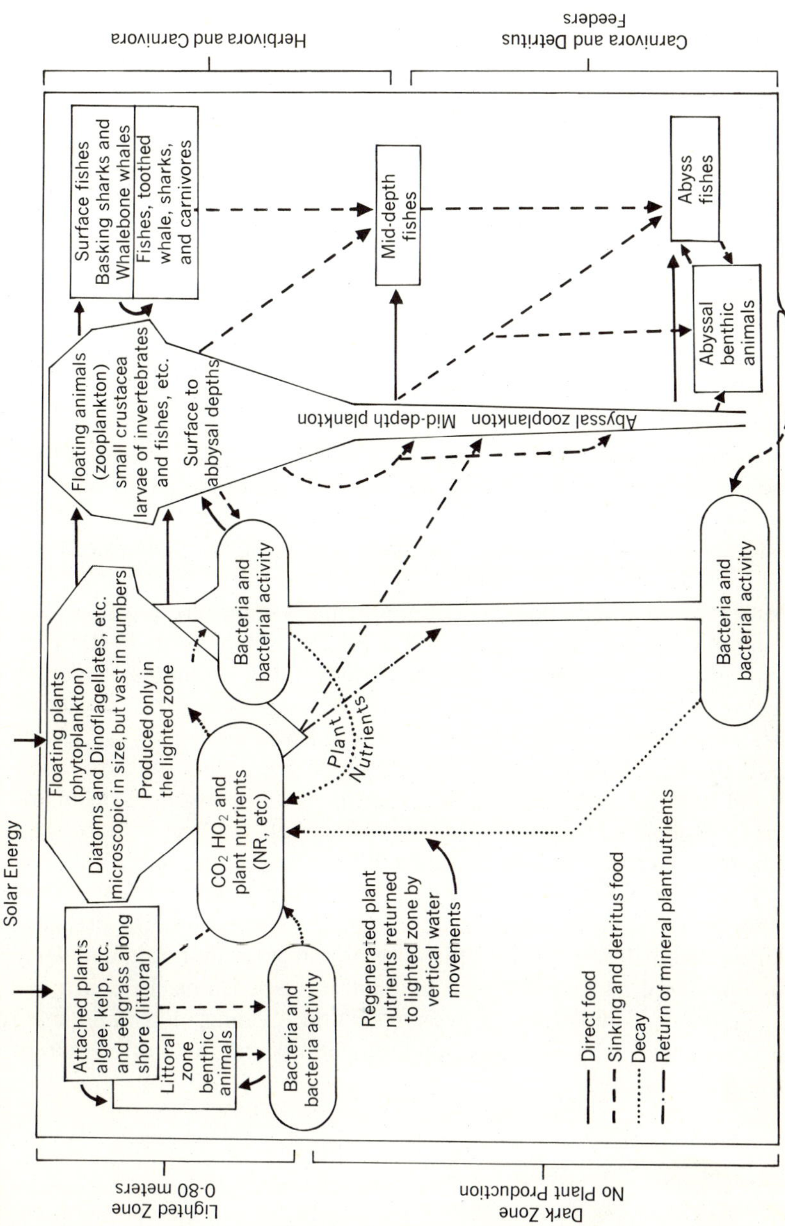

Figure 11-13. Generalized food web in the sea. (After Sverdrup *et al.*, 1942.)

example, the food preference of the Atlantic croaker, a fish, changes from small-sized zooplankton to small bottom animals to detritus to larger crustaceans and fishes, while in the pinfish the change is from small crustaceans to detritus to vegetation.

Various techniques which have been used in studying food webs are of two general types: (1) direct observation or (2) indirect methods. As the term implies, direct observational techniques consist of watching the feeding habits of the organism under study. There are obvious drawbacks and limitations to this method, especially when it is applied to nocturnal and/or burrowing forms. However, the direct observational approach has been successfully used in studies of various marine, rocky, intertidal-zone habitats (Connell, 1961; Paine, 1966). Indirect methods are varied. For example, the gut content of various species of animals from one habitat may be analyzed, and although valuable information on the source of food can be obtained by using this method, accurate identification of ingested material is often difficult, even when based on hard skeletal parts, and is almost impossible if the organism feeds on soft-bodied organisms or body fluids.

Recently, an immunological technique, the precipitin test, was used to demonstrate that certain predators fed regularly on the broom beetle *Phytodecta olivacda* (Dempster, 1960). Briefly, this technique involved producing an antiserum to the broom beetle and testing this antiserum against an extract of a suspected predator. Formation of a precipitate suggests the ingestion of beetles. Another technique employs radioactive tracers in both laboratory and field experiments. For example, Rice and Smith (1958) labeled four different sizes of planktonic algae with radioactive phosphorus and fed them to clams in both single and mixed cultures. The rate of filtering was higher in suspension of the larger-sized algae than in suspension of smaller-sized species. In mixed suspension, the presence of a larger-sized algae increased the filtering rate over that obtained when the smaller-sized algae were used in unialgal suspensions. In controlled pond enclosures simulating an ecosystem, the cycling of a radioactive tracer can be traced by sampling pond inhabitants and the biotic environment at different times.

CHEMICAL CYCLING

A transfer of chemical compounds or elements from one trophic level to another is concomitant with the flow of energy in an ecosystem, since the protoplasm that organisms feed on contains at least 40 elements. This transfer of chemical compounds or elements is cyclic, since it involves passage through a biotic phase (the various trophic levels) and subsequent release from an organism, either as waste products or upon death, to an abiotic phase when the chemical compounds or elements then become available to other organisms, especially primary producers. The abiotic phase consists of either an atmospheric (gaseous) component or a substrate (sedimentary) component. Since a cycle involves a biotic phase and a geologic (abiotic) phase, it is termed a *biogeochemical cycle.*

Cycling of chemical substances has been recognized for years, but with the recent development of newer techniques (especially radioactive isotopes) a greater emphasis has been placed on their quantitative and qualitative relationships. Certain cycles have been known and studied longer than others. For example, the carbon, nitrogen, oxygen, and phosphorous cycles were known earlier than were cycles of other substances such as radioactive isotopes, which are man-made products. Since basically all substances must be in solution to be biologically useful, water is essential for the functioning of other cycles as well as being important in itself as a cycling compound. Not all chemical elements are cycled at the same rate, nor is the rate the same for one element in various environments or in different parts of the cycle.

Specific examples representing different types of cycles—gaseous, sedimentary, and radioactive—will furnish a general background for understanding other cycles.

Carbon Cycle

To describe a cyclic phenomenon, obviously it is necessary to enter the circle at one point and follow the process around to the beginning point. One relatively simple but very important cycle concerns carbon (Fig. 11-14). During photosynthesis, carbon from atmospheric CO_2 is incorporated into the production of carbohydrates that subsequently may be converted to other organic compounds. Carbon utilized by plants may be stored as food and passed on to the trophic level of herbivores or parasites, or retained by the plant until it serves as food for decay organisms. Some of the carbon is returned to the atmosphere (or the enveloping aqueous medium) in the form of CO_2, a by-product of plant respiration. Decomposing microorganisms are important in breaking down dead material with the release of carbon back into the cycle.

Carbon taken up by herbivores or parasites may travel a number of routes. It may be incorporated into protoplasm and stored until the organism dies, whereupon it is utilized by decomposers; it may serve as live food for other organisms; or finally it may be stored in the environment as CO_2. Similar fates await carbon at the carnivore trophic levels. Some

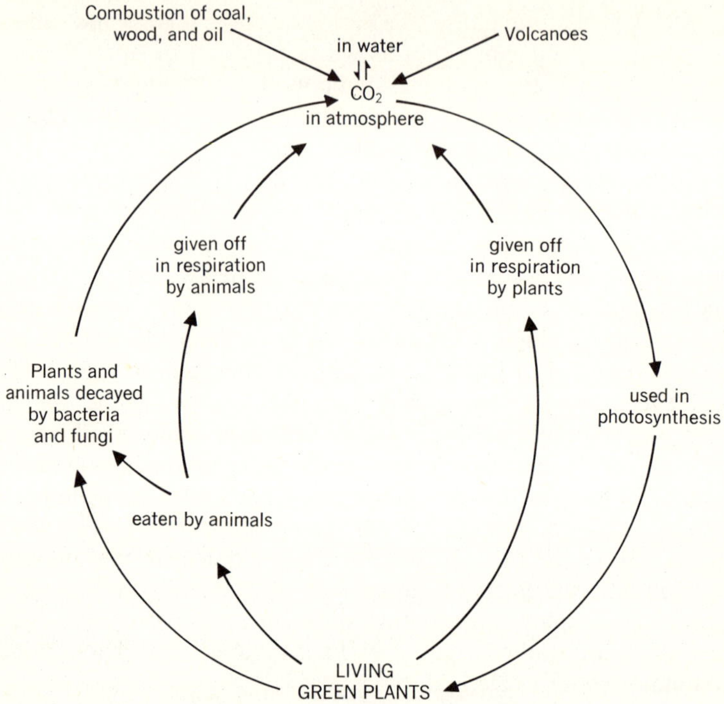

Figure 11-14. The carbon cycle.

animals are capable of fixing CO_2, while others immobilize CO_2 in the formation of skeletons and shells. These phases of the carbon cycle involve living entities. However, some of the carbon is incorporated into inert parts of the earth's surface in the form of coal, gas, oil, and limestone. Generally, the turnover rate of carbon is much faster when it passes through the biotic phase, whereas carbon trapped by the abiotic processes may be indefinitely separated from the cycle until some intermediary releases it. For example, weathering of carboniferous formations liberates some carbon, while man's use of coal and oil has greatly hastened the return of carbon, chiefly in the form of CO_2, for cyclic dynamics.

Certain control mechanisms are inherent in the carbon cycle, although the experimental demonstration of their existence is technically difficult (refer to C^{14} in the previous section, "Primary Productivity"). The rate of carbon utilization is dependent on its availability. If excessive amounts are taken up in any one phase of the cycle, other phases of activity may be inhibited or slowed down. For example, if the pH of water is alkaline, more carbon is tied up in a carbonate and less is in solution. This removal of carbon in solution would upset the equilibrium established between the atmospheric and the dissolved CO_2, and the net effect would be a movement of CO_2 into solution until equilibrium was reached.

Phosphorous Cycle

The mineral phosphorous, which is one of the main nutrients needed to sustain an ecosystem, is an illustration of a sedimentary cycle. Phosphorous, which is mined primarily from apatite rocks and phosphate deposits, is introduced into the ecosystem as fertilizer in the form of phosphate. Although some phosphorous is released to the active phase of the cycle by weathering processes, in either case it must be in a soluble salt form to be biologically effective. As the soluble salt travels with water, some is incorporated into protoplasm of fresh-water or terrestrial species; most of it escapes, eventually going to the sea. In the sea, some is deposited in shallow waters, but much is lost to the deeper waters. Some phosphorous, which is taken up by plankton (especially the zooplankton) becomes incorporated into fish flesh and is cycled through the biological activity of predators and decomposers. Part of this phosphorous is returned to terrestrial ecosystems through fishing activities of man and birds. Rich phosphorous deposits in the form of guano are formed by birds. Man produces fertilizer from fish or directly obtains phosphorous by eating marine products. In clay soil on land, phosphorous may become so tightly bound to clay particles that the roots of plants cannot absorb it, and then a fertilizer must be added for effective plant growth. Phosphorous introduced into the biotic phase of the cycle is passed along in the food web. Much of the phosphorous in an ecosystem results from the breakdown of dead organic material.

Attempts to study phosphorous cycling by means of radioactive phosphorous (P^{32}) have emphasized the diverse routes a nutrient may take through an ecosystem. Furthermore, the rate of uptake and retention varies interspecifically and intraspecifically.

Although no widespread deleterious effects are apparent when a great deal of phosphorous is removed from the biotic portion of the cycle and deposited in the relatively inert deep-sea sediments, eventually this slowdown in cycling could greatly affect man's ecosystem.

Various phases of the cycling of other nutrients have been investigated in addition to phosphorous. Ovington (1962) presented a balance sheet of important mineral nutrients in a woodland ecosystem for a 55-year period (Table 11-4), but because of the peculiar character of the individual ecosystem, he found it impossible to generalize on long-term trends. For more details, consult the earlier book of Hutchinson (1957), who summarized much of the information on cycling in lakes. Further, Ovington (1965) reviewed cycling in woodlands, and Polikarpov (1966) dealt with cycling in the marine environment.

Cycling of Radioactive Elements

Since the first atom blast initiated the Atomic Age, various types of devices have released radioactive materials into the atmosphere and underground, which enter biogeochemical cycles. The immediate influence of these substances was not known and the long-range effects still are unknown. Strontium-90, which is one of the better-known new substances, easily enters a cycle after being absorbed by plants. Herbivores feed on plants and carnivores feed on herbivores. A vivid illustration of the incorporation of strontium-90 and cesium-137 into a cycle and a relatively simple food web was documented by Palmer *et al.* (1963) who noted that in the Arctic, these substances are picked up by lichens, which are eaten by caribou, which in turn are eaten by carnivores, including man. Like calcium, strontium-90 is concentrated in those regions of the body, such as bone, where calcification occurs. The amount of strontium-90 accumulated in the Arctic at the time of this study was almost half the permissible dosage. Although the problem is not so acute in other regions of the world as yet, this example readily demonstrates what can happen in the future (see Chapter 12 for further discussion of this problem of man's altering of the environment). More detailed information on radioactive cycles is reported by Schultz and Klement (1963); Polikarpov (1966); Templeton, 1966; and Nelson and Evans, 1969.

SUMMARY

The analysis of ecosystems is a currently dominant subdivision of ecology. Research involves both descriptive (anatomical) studies on communities and the abiotic environmental component and investigation of the functional (physiological) attributes of ecosystems, such as energy flow and primary production.

A community is a dynamic system that is responsive to environmental changes. Because of the complexity of communities, and for other reasons, ecologists are in healthy dispute as to whether communities are discrete entities or form a continuum. The replacement of one community by another is known as ecological succession, and this replacement, which may lead to a climax community, in many cases proceeds on a predictable basis.

Table 11-4. Balance Sheet of Plantations of *Pinus sylvestris* for the 55 Years from Time of Planting.

	Na	*K*	*Ca* (*kg/half-acre*)	*Mg*	*P*	*N*
Trees						
Uptake by trees	132	1933	2272	431	413	4817
Present in living trees	36	150	272	64	41	453
Removed in harvested tree trunks	5	98	210	42	14	161
Left in crowns and roots of harvested trees	48	279	386	80	75	704
Litter fall from trees (leaves, branches, cones, etc.)	43	1406	1404	245	283	3499
Transfer from trees to soil	91	1685	1790	325	358	4203
Ground flora						
Uptake by ground flora	72	876	771	169	182	2058
Present in ground flora	4	22	19	5	3	40
Change in ground flora	+ 2	− 1	− 4	− 1	− 5	− 73
Litter fall from ground flora	70	877	775	170	187	2131
Litter						
Total return from trees and ground flora	161	2562	2565	495	545	6334
Present in litter layers	6	34	141	30	28	409
Change in litter layers	+ 6	+ 24	+ 115	+ 22	+ 19	+ 250
Released by decomposition	161	2538	2450	473	526	6084
Average annual uptake by trees and ground flora	3.7	51	55	11	11	125
Average annual change in ecosystem	+ 0.8	+ 3	+ 7	+ 2	+ 1	+ 11
Average annual removal in tree trunks	< 0.1	2	4	< 1	0.3	3
Average annual release by decomposition	2.9	46	44	9	10	111

Source: Ovington, 1962.

The mechanisms by which the radiant energy from the sun is incorporated into food by green plants and its subsequent passage through different trophic levels are fundamental to the understanding of ecological energetics. In addition to the cycling of energy through an ecosystem, various biogeochemical cycles occur, including naturally occurring elements (like nitrogen and oxygen) and man-made substances (like radioactive isotopes).

Comprehensive analyses of ecosystems have relevance to the basic understanding of the biosphere and to the painfully practical problem of how man can keep from ruining the environment.

References

Allee, W. C., A. E. Emerson, O. Park, T. Park, and K. P. Schmidt. 1949. *Principles of Animal Ecology.* Saunders, Philadelphia.

Becking, R. 1957. The Zurich-Montpellier school of phytosociology. *Botan. Rev.*, 23:411–488.

Buchsbaum, Ralph and Buchsbaum, Mildred. 1957. *Basic Ecology.* Boxwood Press, Pittsburgh, Pa.

Bunt, J. S. 1965. Measurements of photosynthesis and respiration in a marine diatom with the mass spectrometer and with carbon-14. *Nature,* 207:1373–1375.

Cole, L. C. 1946. A study of the Cryptozoa of an Illinois woodland. *Ecol. Monographs,* 16:49–86.

Connell, J. H. 1961. Effects of competition, predation by *Thais lapillus,* and other factors on natural populations of the barnacle *Balanus balanoides. Ecol. Monographs,* 31:61–104.

Curtis, J. T. 1959. *The Vegetation of Wisconsin.* University of Wisconsin Press, Madison.

Darnell, R. M. 1961. Trophic spectrum of an estuarine community, based on studies of Lake Pontchartrain, Louisiana. *Ecology,* 42:553–568.

Darnell, R. M. 1968. Animal nutrition in relation to secondary production. *Am. Zool.,* 8:83–93.

Dempster, J. P. 1960. A quantitative study of the predators on the eggs and larvae of the broom beetle, *Phytodecta olivacda* Forster, using the precipitin test. *J. Anim. Ecol.,* 29:149–167.

Dice, L. R. 1952. *Natural Communities.* University of Michigan Press, Ann Arbor.

Doty, M. S. (ed.). 1961. *Primary Productivity Measurement, Marine and Freshwater.* Proc. Conf. Primary Productivity Measurement, 10th Pacific Sci. Cong., Hawaii.

Elton, C. 1927. *Animal Ecology.* Macmillan, New York.

Engelmann, M. D. 1961. The role of soil arthropods in the energetics of an old field community. *Ecol. Monographs,* 31:221–238.

Engelmann, M. D. 1968. The role of soil arthropods in community energetics. *Am. Zool.,* 8:61–69.

Goldman, C. R. 1968. Aquatic primary production. *Am. Zool.,* 8:31–42.

Golley, F. B. 1968. Secondary productivity in terrestrial communities. *Am. Zool.,* 8:53–59.

Holme, N. A. 1950. Population dispersion in *Tellina tenuis,* da Costa. *J. Mar. Biol. Assoc. U.K.,* 29:267–280.

Holmes, R. W. 1961. A summary of productivity measurements in the southeastern Pacific Ocean. *In* M. S. Doty (ed.), *Primary Productivity Measurement, Marine and Freshwater,* pp. 18–57. Proc. Conf. Primary Productivity Measurement, 10th Pacific Sci. Cong., Hawaii.

Holmes, R. W., M. B. Schaefer, and B. M. Shimada. 1957. Primary production, chlorophyll and zooplankton volumes in the tropical eastern Pacific Ocean. *Inter-Am. Trop. Tuna Comm. Bull.,* 2:129–169.

Hutchinson, G. E. 1957. *A Treatise on Limnology,* Vol. I. Wiley, New York.

Kendeigh, S. C. 1944. Measurement of bird populations. *Ecol. Monographs,* 14:67–106.

Kormondy, E. J. 1969. *Concepts of Ecology.* Prentice-Hall, Englewood Cliffs, N.J.

Kozlovsky, D. G. 1968. A critical evaluation of the trophic level concept. I. Ecological efficiencies. *Ecology,* 49:48–60.

Lindeman, R. L. 1942. The trophic-dynamic aspect of ecology. *Ecology,* 23:399–418.

Manning, W. M., and Juday, R. E. 1941. The chlorophyll content and productivity of some lakes in northeastern Wisconsin. *Trans. Wisc. Acad. Sci. Arts Lett.,* 33:363–393.

McIntire, C. D., and H. K. Phinney. 1965. Laboratory studies of periphyton production and community metabolism in lotic environments. *Ecol. Monographs,* 35:237–258.

Miller, A. H. 1951. An analysis of the distribution of the birds of California. *Univ. Calif. Publ. Zool.,* 50:531–644.

Möbius, K. 1877. Die Auster und die Austernwirtschaft. Berlin [Transl., 1880. The oyster and oyster culture. *Rept. U.S. Fish Comm.,* 1880:683–751].

Moreau, R. E. 1944. Clutch-size: a comparative study, with special reference to American birds. *Ibis,* 86:286–347.

Nelson, Daniel J. and Francis C. Evans. 1967. Symposium on Radioecology. Proceedings of the Second National Symposium Held at Ann Arbor, Michigan, May 15–17, 1967.

Odum, E. P. 1959. *Fundamentals of Ecology,* 2nd ed. Saunders, Philadelphia.

Odum, E. P. 1968. Energy flow in ecosystems: a historical review. *Am. Zool.,* 8:11–18.

Odum, H. T. 1956. Primary production in flowing waters. *Limnol. Oceanog.,* 1:102–117.

Odum, H. T. 1957. Trophic structure and productivity of Silver Springs, Florida. *Ecol. Monographs,* 27:55–112.

Ovington, J. D. 1962. Quantitative ecology and the woodland ecosystem concept. *In* J. B. Cragg (ed.), *Advances in Ecological Research,* Vol. I, pp. 103–192. Academic Press, New York.

Ovington, J. D. 1965. Organic production, turnover and mineral cycling in woodlands. *Biol. Rev.,* 40:295–336.

Paine, R. T. 1966. Food web complexity and species diversity. *Am. Nat.,* 100:65–75.

Palmer, H. E., W. C. Hanson, B. I. Griffin, and W. C. Roesch. 1963. Cesium 137 in Alaskan Eskimos. *Science,* 142:64–65.

Pandian, T. J. 1967. Transformation of food in the fish *Megalops cyprinoides*. II. Influence of quantity of food. *Mar. Biol.,* 1:107–109.

Phillipson, J. 1966. *Ecological Energetics.* Inst. Biol. Studies *No. 1,* St. Martin, New York.

Polikarpov, G. G. 1966. *Radioecology of Aquatic Organisms; The Accumulation and Biological Effect of Radioactive Substances* [see Ch. 13]. [Transl. from Russian.] Reinhold, New York.

Pomeroy, L. W. 1961. Isotopic and other techniques for measuring benthic primary production. *In* M. S. Doty (ed.), *Primary Productivity Measurement, Marine and Freshwater,* pp. 97–102. Proc. Conf. Primary Productivity Measurement. 10th Pacific Sci. Cong., Hawaii.

Rice, T. R., and R. J. Smith. 1958. Filtering rates of the hard clam (*Venus mercenaria*) determined with radioactive phytoplankton. *Fish Bull. Fish Wildl. Serv.,* 58:73–82.

Ryther, J. H. 1959. Potential productivity of the sea. *Science,* 130:602–608.

Ryther, J. H., and C. S. Yentsch. 1957. The estimation of phytoplankton production in the ocean from chlorophyll and light data. *Limnol. Oceanog.,* 2:281–286.

Sanders, H. L. 1968. Marine benthic diversity: a comparative study. *Am. Nat.,* 102:243–282.

Schultz, V., and A. W. Klement, Jr. (eds.). 1963. *Radioecology.* Proc. 1st Natl. Symp. Radioecology. A.I.B.S., Washington; and Reinhold, New York.

Slobodkin, L. B. 1962. Energy in animal ecology. *In* J. B. Cragg (ed.), *Advances in Ecological Research,* pp. 69–101. Academic Press, New York.

Steemann-Nielsen, E. 1952. The use of radioactive carbon (C^{14}) for measuring organic production in the sea. *J. du Cons. Intern. Exptl. Mer,* 18:117–140.

Strickland, J. D. H., and T. R. Parsons. 1965. A manual of sea water analysis. *Fish. Res. Bd. Canada Bull. No. 125* (2d ed.).

Sverdrup, H. U., M. W. Johnson, R. H. Fleming. 1942. *The Oceans, Their Physics, Chemistry, and General Biology.* Prentice-Hall, New York.

Talling, J. F., and D. Driver. 1961. Some problems in the estimation of chlorophyll-*a* in phytoplankton. *In* M. S. Doty (ed.), *Primary Productivity Measurement, Marine and Freshwater,* pp. 142–146. Proc. Conf. Primary Productivity Measurement. 10th Pacific Sci. Cong., Hawaii.

Tansley, A. G. 1935. The use and abuse of vegetational concepts and terms. *Ecology,* 16:284–307.

Teal, J. M. 1957. Community metabolism in a temperate cold spring. *Ecol. Monographs,* 23:41–78.

Templeton, W. L., and A. Preston. 1966. Transport and distribution of radioactive elements in coastal and estuarine waters of the United Kingdom. Proceedings of the Symposium on the Disposal of Radioactive Wastes into Seas, Oceans and Surface Waters. IAEA Vienna (16–20 May, 1966).

Thomas, W. H. 1961. Physiological factors affecting the interpretation of phytoplankton production measurements. *In* M. S. Doty (ed.), *Primary Productivity Measurement, Marine and Freshwater,* pp. 147–162. Proc. Conf. Primary Productivity Measurement. 10th Pacific Sci. Cong., Hawaii.

Van Deventer, W. C. 1936. A winter bird community in western New York. *Ecology,* 17:491–499.

Westlake, D. F. 1965. Basic data for investigations of the productivity of aquatic macrophytes. *In* C. R. Goldman (ed.), *Primary Productivity in Aquatic Environments,* pp. 229–248. Mem. 1st Ital. Idrobiol., 18 Suppl., University of California Press, Berkeley.

Whittaker, R. H. 1956. Vegetation of the Great Smoky Mountains. *Ecol. Monographs,* 26:1–80.

Whittaker, R. H. 1961. Estimation of net primary production of forest and shrub communities. *Ecology,* 42:177–180.

Whittaker, R. H. 1965. Dominance and diversity in land plant communities. *Science,* 147:250–260.

Whittaker, R. H., and G. M. Woodwell. 1968. Dimension and production relations of trees and shrubs in the Brookhaven Forest, New York. *J. Ecol.,* 56:1–25.

Woodwell, G. M. 1967. Radiation and the patterns of nature. *Science,* 156:461–470.

Woodwell, G. M., and R. H. Whittaker. 1968. Primary production in terrestrial ecosystems. *Am. Zool.,* 8:19–30.

12

Man and His Biosphere

We cannot command nature except by obeying her.

—Francis Bacon

ALTERATION OF THE ENVIRONMENT

Although the question of the origin of the Earth has not yet been resolved, we do know that this planet is a part of the solar system and that it has been in orbit around the sun somewhere between 4.5–5 billion years. Since its formation, the Earth has been undergoing constant change. Mather (1964)* provides a very graphic picture of the history and dynamics of these changes:

> To get some idea of a geologist's view of the Earth and its history, an ideal instrument would be a time-lapse camera such as those used to photograph the progress of a flower from bud to full bloom. One can imagine a motion picture of the Earth, taken by a camera on a platform some thousands of miles out from the Earth's surface, with one picture of the same hemisphere taken every 5,000 years. After nearly a billion years, we would have a movie of truly epic proportions, telescoping a large part of the history of our planet into a $3\frac{1}{2}$ hour drama.
>
> In such a motion picture the Earth would appear to be alive, its exterior writhing in spasms. Great wrinkles—mountain ranges and canyons—would appear in land that a few moments before had been flat. Lands previously covered by shallow water would emerge and other lands would be flooded by spreading seas. Glaciers and running water would file the mountain ranges into jagged peaks, then down to low hills and finally back to flat valleys once more. Green jungles would change suddenly into stark deserts. Great gashes,

* From K. F. Mather, *The Earth Beneath Us,* 1964, p. 9, Random House, Inc.

such as the Rift Valley in Africa, might open up in a few seconds. Volcanos would fairly jump out of the surface and then be worn away in a minute or two. Vast ice sheets would expand over immense areas then retreat and expand and retreat again, carving the land and leaving behind new rivers, lakes and soils. This is the kind of rapidly changing world the geologist sees.

The idea that the Earth is a changing body was not generally accepted until the beginning of the nineteenth century. Until then, most scientists believed that the Earth was created through a series of catastrophic happenings, and from that time on had remained unchanged. Then, in 1797, the geologist James Hutton suggested that because man cannot readily see now how certain features of the Earth came about does not mean that it is necessary to call upon catastrophic occurrences to explain them. Rather, the very small, almost imperceptible changes on the surface of the Earth, such as the wearing away of a stream bank or changes in the shoreline, can eventually alter the surface of the Earth. This idea is now well accepted among scientists and has become the center of geological thought (Herrin, 1967). Yet, today this knowledge is largely ignored. Man is busy building dams and rivers, filling in marshes, cutting through new river channels, building new canals, polluting the environment with toxic substances—apparently disregarding any of the possible consequences.

Water Resources

Some consequences of such environmental manipulation are already quite obvious. One example is the shortage of available fresh water. Of the four minor planets, Earth is characterized as being the watery planet, since water occurs not only in oceans, lakes, and ponds, but also in the ground and in the atmosphere as vapor. Now, as never before, fresh water is in enormous demand in agriculture, industry, and domestic use. It has been estimated that if all the fresh water in the world were distributed equally over the Earth, there would be enough available to meet the world's needs up to about the year 2000 (Hickman, 1966). However, the fresh-water supply is not distributed equally over the Earth. Within the United States, the area southwest of the Rockies has a chronic water shortage while the region around the Great Lakes has an excess of water. Actually, North America and northern Europe are about the only water-rich areas in the world. Almost half of India, China, and North Africa are arid. Two-thirds of Australia and Asia Minor, and all of Arabia are also desertlike areas. Water shortages exist from time to time in many other areas where there is abundant rainfall.

Still, actual lack of water is not the main reason for current water shortages—rather, it is the gross misuse and mismanagement of available water. For too many years the water systems have been used as convenient waste disposal units for by-products from the breakdown of chemical, physical, and biological materials. These products that defile the environment are generally referred to as *pollutants*. A very striking example of such pollution is found among the Great Lakes, which represent one of the largest accessible supplies of fresh water in the world. Already one of these, Lake Erie, serving Detroit, Toledo, Cleveland, and Buffalo, is so polluted that it has been called a health hazard (Congressional Committee hearings on water pollution, Rochester, N.Y., July 29–30, 1966), and serious concern has been expressed over the state of purity of the others. Man has also affected the level of the water table by decimating forests, thereby creating runoff and drought problems. Where industry places heavy demands on the water supply by digging deep wells, the water table in the surrounding areas is apt to drop appreciably.

The causes of pollution are legion. It has been estimated that there is now six times as much pollution in rivers, streams, and lakes as there was 60 years ago (Terry, 1966). Greater concentrations of people, coupled with increased industrialization, contribute markedly to this pollution. When organic pollutants decompose, they make demands upon the dissolved oxygen supply in the water. These organic pollutants include compounds such as proteins, fats, soaps, carbohydrates, oil, tar, dyes, and synthetic detergents. In Lake Erie, for example, organic pollution from the Detroit area is so great that almost all dissolved oxygen has been depleted from 1400 sq miles of the central basin bottom waters. In addition, another 1600 sq miles are intermittently very low in dissolved oxygen (Carbine, 1962). The decline in the more desirable types of fish can be correlated closely with this depletion of dissolved oxygen in Lake Erie, which started in the 1950s. Instead of lake trout, blue pike, lake herring, sauger, and whitefish, there are perch, smelt, carp, and catfish. Many of the same pollutants that deplete the oxygen supply are also rich sources of nitrogen and phosphorous. These nutrients cause certain types of plant and animal life to flourish wildly. Thus Lake Erie and some other similarly polluted bodies of water support spectacular algal blooms, great populations of plankton, worms, and midges.

This whole process, wherein the waters become rich in dissolved nutrients and low in oxygen, sets in motion an aging process, termed *eutrophication*. In nature, a similar aging process occurs, but at a far slower pace. After a lake is formed, nutrients gradually flow in from drainage streams. This enrichment gives rise to the aquatic food web, ranging from algae and plankton to fish. As living matter continues to increase, organic deposits pile upon the bottom and the lake becomes smaller, shallower, and warmer. The type of animal life changes and rooted plants begin to fill the bottom. The lake then becomes a marsh, shoreline vegetation takes over, and the lake eventually disappears.

In most lakes near metropolitan areas where sewage is discharged into the water, this same process is occurring at a greatly accelerated pace (Fig. 12-1). In many of these large population centers, sewage systems are inadequate. Since the sewage systems for domestic wastes also serve as storm sewers, these systems cannot handle both rain and wastes during rain storms, and consequently large amounts of untreated sewage pours into the lakes. Modern sewage plants in operation, however, can

Figure 12-1. Water pollution on Lake Michigan. (Van Bucher/Photo Researchers.)

efficiently remove suspended solids and the oxygen-consuming substances. Unfortunately, even the best of these sewage treatments does not remove the nutrients, and there is no known inexpensive method for doing so. Mihursky (1967) has suggested that one solution is to release these nutrients into a controlled environment and use them to culture and harvest algae and zooplankton.

Some pollutants discharged into the water, either by accident or design, are extremely toxic and immediately affect living organisms. Oil spills have killed water birds, mammals, fish, and vegetation (Hunt, 1965). Some synthetic detergents contain alkylbenzene sulfonate, a substance which in concentrations above 2 ppm can adversely affect fish (Henderson *et al.,* 1959). Some aquatic plants cannot grow in waters that contain 2.5 ppm or more (Klein, 1957). Yet, concentrations of alkylbenzene sulfonate above 12 ppm have been reported in sewage effluents (Henderson *et al.,* 1959).

When the oil tanker *Torrey Canyon* was wrecked off the coast of Cornwall in 1967, over 2 million gallons of detergent were dumped into the sea to combat the oil pollution. In a recent report on the effects of this disaster, scientists at the Marine Biological Laboratory at Plymouth concluded that the detergents were more harmful to marine life than the oil. The detergents were toxic to the limpets and barnacles on the rocks along the beaches. These invertebrates not only act as scavengers, but also help to control the seaweed growth, and as a result, the beaches where there was mass mortality of these animals became green with heavy growths of seaweed. In areas where oil but not detergents reached them, there was no great mortality among the limpets and barnacles (Corner *et al.,* 1968).

Inorganic pollutants from industrial wastes such as acids, alkalies, salts, heavy metals, and soluble salts have adverse effects on living organisms (Hunt, 1965). Acids and alkalies greatly alter the pH of the water, which in turn affects the resident organisms. Sometimes fresh water is made brackish by drainage from salt mines, by effluents from water-softening plants, or where there is large-scale road salting and the salts are subsequently washed into storm sewer waters. Salt in these concentrations, of course, is fatal to fresh-water organisms.

One problem just beginning to be common throughout the industrialized world is that of thermal pollution. Although many industrial processes produce heat as a waste product, the demand of the steam electric-generating station utilities for water as a heat-transfer medium is greater (Fig. 12-2) than all other industrial processes combined (Mihursky, 1967). Projections are that by 1985, the power plants will use one-quarter of the total fresh-water runoff of the United States for cooling

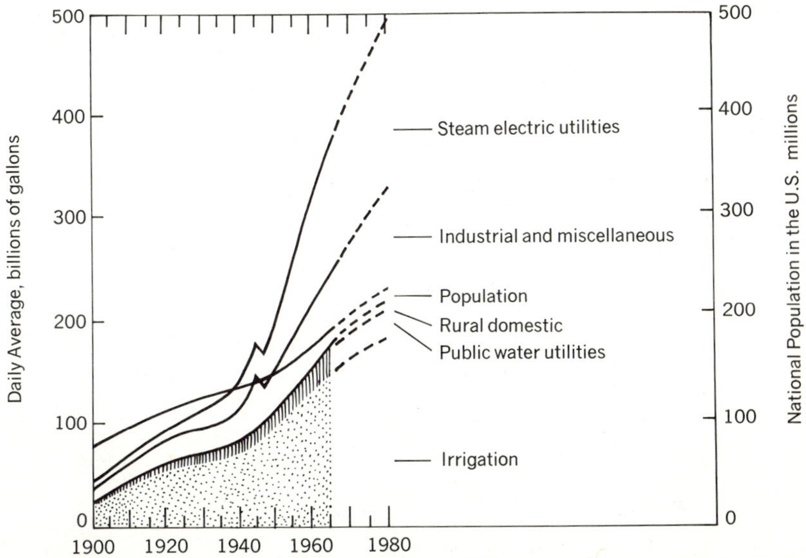

Figure 12-2. Water use in the United States. (After Mihursky, 1967.)

(Singer, 1968). If electrical demands continue to double every 6–10 years, there will be a 30- to 256-fold increase in electricity needs by the year 2010 (Mihursky, 1967). To meet this demand, nuclear installations are being built in addition to the conventional fossil-fuel plants. Since nuclear units have a larger heat rejection than the more conventional plants, the demand for water will be even greater. To satisfy this demand for water, many plants are being built in estuaries and near large bodies of fresh water. Uncontrolled dumping of waste heat into the estuaries and lakes obviously will have serious ecological consequences. It is highly unlikely that the demand for electricity will decrease in the future. Furthermore, at the present time and within the foreseeable future, there are no other methods of producing electricity without the resultant heat load (Cootner and Löf, 1965).

Two approaches have been suggested to solve the problems of thermal pollution. One involves constructive uses of the waste heat, which could be employed in climate control, desalting of water, or heating buildings (Singer, 1968). The other approach would be to redirect these temperature loads into controlled ecosystems (Mihursky, 1967). Waste heat could be used to stimulate the breakdown of sewage, to stimulate the growth of shellfish, or to introduce new species of desirable organisms into areas that were previously too cold to support them. The problems of thermal pollution are manageable now; they may not be ten years from now if proper adjustments are neglected.

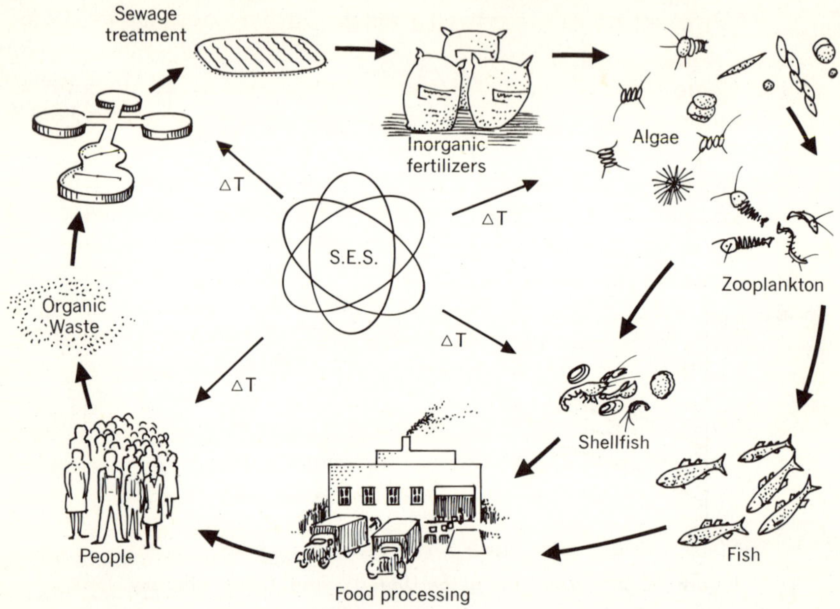

Figure 12-3. Possible recycling approach to converting waste materials into useful food stuff. S.E.S.: steam electric stations. (After Mihursky, 1967.)

Mihursky (1967) has presented a schematic diagram to illustrate the possibilities of using two troublesome water-pollution waste products: the waste heat from the electric power plants and the nutrients from organic wastes (Fig. 12-3). Essentially this would involve recycling, which could convert waste materials into useful food stuffs, a process that would, of course, require much research and truly multidisciplinary effort.

Pesticides

Organic pesticides have become extremely widespread in water, air, and the soil. Pesticides of one type or another probably have been used since the time of primitive man, but it has only been since World War II that the use of these compounds has become extensive. Furthermore, although early pesticides such as arsenic were poisonous to man and animal alike, those now in use are far more deadly and persistent than earlier ones. During the 1950s there were isolated scientific papers pointing out the dangers of these new insecticides, but it was not until Rachel Carson wrote *Silent Spring* in 1962 that they were brought to public attention.

Most pesticides can be classified as either chlorinated hydrocarbons or organic phosphorus insecticides. Although both types are deadly, chlorinated hydrocarbons have the added feature of persisting for very long periods of time, and they are characteristically non-biodegradable. The first chlorinated hydrocarbon to be used widely was DDT (dichlorodiphenyltrichloroethane). It was actually synthesized 65 years before its effectiveness as a pesticide was recognized. After World War II, use of DDT became very widespread, for it was assumed to be safe because it killed insect pests but did not immediately kill man. Destruction of the insect pests increased farm production enormously. Thus, synthetic pesticide production increased accordingly: in 1947, 124,259,000 lb were produced compared with 637,666,000 lb in 1960 (Egler, 1964). As a result of this widespread and often indiscriminate use of these insecticides, they have become an integral part of the biological, geological, chemical cycles of the earth.

Although DDT is not the deadliest pesticide available, it has been in use for a longer period of time than has any other modern insecticide. Moreover, it is present throughout the ecosystem. Measureable amounts of DDT residues (DDT and its metabolites, DDE and DDD) can be found in the air, soil, man, and in animals thousands of miles from where the DDT was applied originally (Wurster and Wingate, 1968). Generally, pesticides are applied to rid an area of particularly noxious insect pests, and often they are effective. Unfortunately, this shotgun method kills desirable living organisms just as effectively as the pests. For example, DDT in oil (when applied this way, it is particularly toxic) was sprayed over large areas of a Canadian forest to kill the spruce budworm *Choristoneura fumiferana*. The drainage system in the forest lying in the spray zone was contaminated by the insecticide; many young salmon were killed and the stocks reduced to a very low level. When DDT was sprayed in a concentration of $\frac{1}{2}$ lb per acre, the salmon underyearlings were reduced by 90–98 percent compared with unsprayed areas; larger fish were reduced by 70 percent. When the concentration of DDT cut to $\frac{1}{4}$ lb per acre, young fish were reduced by 50 percent and the older ones by about 20 percent. These harmful effects extended 30 or more miles below the spray zone. Four to six months after the spraying, with the onset of cold weather, there were significant numbers of delayed mortalities among the fish that had been in the contaminated water (Elson, 1967).

Scores of other examples illustrate that numbers of animals have been reduced owing to pesticide contamination. Population reductions have been reported for crayfish, shrimp, amphipods, isopods, annelids, fish, and blue crabs following single applications of DDT in the range of 0.1–0.3 lb per acre (reviewed by Woodwell *et al.*, 1967). When DDT was sprayed to control Dutch elm disease in New Hampshire, many migrant birds were killed. A dosage of 1.9 lb per acre was used, despite the controversy over whether or not DDT was actually effective in controlling Dutch elm disease (Wurster *et al.*, 1965).

The immediate effects of direct absorption of pesticides into the ecosystem are fairly dramatic; the accumulation of persistent pesticides in food chains is even more so. Woodell *et al.* (1967) analyzed DDT residues in animals in an east-coast estuary and found not only an increasing concentration of DDT residues as the size of the animal increased, but also greater concentrations of DDT residues in higher carnivores than in those at lower trophic levels (Table 12-1). Thus the total residues in plankton were 0.04 ppm, while at the terminal trophic level, they were 75 ppm in the ring-billed gull. There was twice as much residue in the needle

Table 12-1. DDT Residues (DDT + DDE + DDD) in Samples from Carman's River Estuary and Vicinity, Long Island, N.Y., (ppm wet-wt of whole organism) with Proportions of DDT, DDE, and DDD Expressed as Percentage of Total. Letters in Parentheses Designate Replicate Samples.

Sample	*DDT Residues (ppm)*	*Percent of Residue as*		
		DDT	*DDE*	*DDD*
Water	0.00005			
Plankton, mostly zooplankton	0.040	25	75	Trace
Cladophora gracilis	0.083	56	28	16
Shrimp†	0.16	16	58	26
Opsanus tau, oyster toadfish (immature)*	0.17	None	100	Trace
Menidia menidia, Atlantic silverside*	0.23	17	48	35
Crickets*	0.23	62	19	19
Nassarius obsoletus, mud snail*	0.26	18	39	43
Gasterosteus aculeatus, threespine stickleback*	0.26	24	51	25
Anguilla rostrata, American eel (immature)*	0.28	29	43	28
Flying insects, mostly Diptera*	0.30	16	44	40
Spartina patens, shoots	0.33	58	26	16
Mercenaria mercenaria, hard clam*	0.42	71	17	12
Cyprinodon variegatus, sheepshead minnow*	0.94	12	20	68
Anas rubripes, black duck	1.07	43	46	11
Fundulus heteroclitus, mummichog*	1.24	58	18	24
Paralichthys dentatus, summer flounder†	1.28	28	44	28
Esox niger, chain pickerel	1.33	34	26	40
Larus argentatus, herring gull, brain (d)	1.48	24	61	15
Strongylura marina, Atlantic needlefish	2.07	21	28	51
Spartina patens, roots	2.80	31	57	12

Table 12-1. (*Continued*)

Sample	*DDT Residues (ppm)*	*Percent of Residue as* DDT	DDE	DDD
Sterna hirundo, common tern (a)	3.15	17	67	16
Sterna hirundo, common tern (b)	3.42	21	58	21
Butorides viresceus, green heron (a) (immature, found dead)	3.51	20	57	23
Larus argentatus, herring gull (immature) (a)	3.52	18	73	9
Butorides virescens, green heron (b)	3.57	8	70	22
Larus argentatus, herring gull, brain‡ (c)	4.56	22	67	11
Sterna albifrons, least tern (a)	4.75	14	71	15
Sterna hirundo, common tern (c)	5.17	17	55	28
Larus argentatus, herring gull (immature) (b)	5.43	18	71	11
Larus argentatus, herring gull (immature) (c)	5.53	25	62	13
Sterna albifrons, least tern (b)	6.40	17	68	15
Sterna hirundo, common tern (five abandoned eggs)	7.13	23	50	27
Larus argentatus, herring gull (d)	7.53	19	70	11
Larus argentatus, herring gull‡ (c)	9.60	22	71	7
Pandion haliaetus, osprey (one abandoned egg)§	13.8	15	64	21
Larus argentatus, herring gull (f)	18.5	30	56	14
Mergus serrator, red-breasted merganser (1964)†	22.8	28	65	7
Phalacrocorax auritus, double-crested cormorant (immature)	26.4	12	75	13
Larus delawarensis, ring-billed gull (immature)	75.5	15	71	14

SOURCE: Woodwell *et al.* (1967).
* Composite sample of more than one individual.
† From Capirco Island, 20 miles (32 km) WSW of study area.
‡ Found moribund and emaciated, north shore of Long Island.
§ From Gardiner's Island, Long Island.

fish as in *Fundulus*, on which it feeds. It is also of interest that the water had only 0.00005 ppm of DDT residues, while birds at the top of the food chain had residues about a million times greater (Fig. 12-4). This process is termed *biological magnification*. In nature, accumulations of DDT apparently can reach a point where they are fatal, and there is evidence that a constant attrition in animal populations results from accumulated lethal concentrations of DDT (Woodwell *et al.*, 1967).

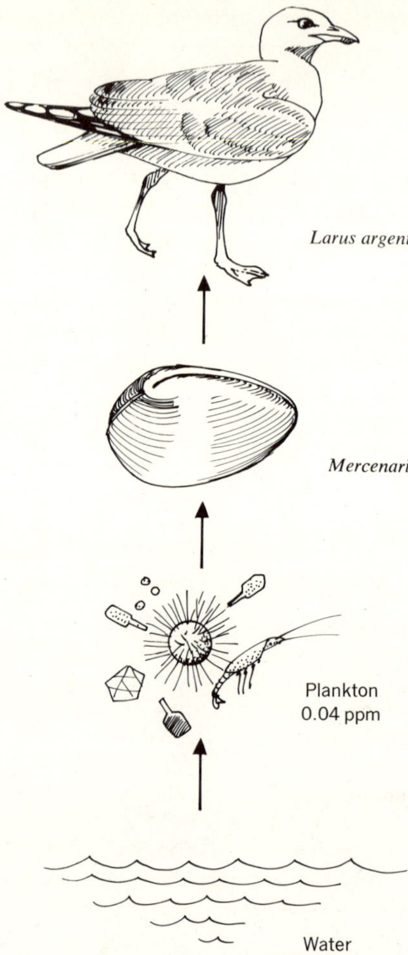

Figure 12-4. An example of biological magnification of DDT residues. (Based on data from Woodwell *et al.*, 1967.)

Although the effects of DDT have been more thoroughly studied than other chlorinated hydrocarbons, the others are equally dangerous. Even sublethal concentrations of the chlorinated hydrocarbons are not without effect. The temperature-acclimation mechanism in salmon can be upset by small doses of insecticide, resulting in extreme sensitivity to cold water (Ogilvie and Anderson, 1965). Furthermore, sublethal concentrations of chlorinated hydrocarbons frequently have been linked with reproductive failure in birds. In laboratory controlled experiments, Koeman *et al.*

(1967) evaluated the dangers of these compounds by measuring the concentration of dieldrin in the blood of embryos and chicks hatched from eggs treated with this compound before incubation. The insecticide did not affect the rate of hatching, but following hatching, mortality rates were high. The authors concluded that this result is due after hatching to absorption of the yolk material containing insecticide. Since, in nature, recently hatched birds often go without food for prolonged periods, yolk material is absorbed rapidly. Sufficient concentration of insecticide in the yolk, then, can be fatal.

Measurable amounts of chlorinated hydrocarbons have also been detected in bird eggs in nature in areas far removed from where the insecticide had been applied. One very striking example has been reported in the Bermuda petrel, which breed on a few tiny, uninhabited islets near Bermuda—never treated with DDT. In feeding habits, these birds are the terminal members of a pelagic food chain, and their feeding ground covers much of the North Atlantic. These birds undoubtedly concentrate insecticides present in the lower trophic levels, since residues of DDT were observed (Wurster and Wingate, 1968) in eggs, chicks, and adults (Table 12-2). Such examples provide graphic evidence that many parts of the marine environment far from coastal areas are contaminated with pesticides. Ocean currents and river drainages certainly contribute to the dis-

Table 12-2. Residues of DDT (ppm wet wt) in Eggs and Chicks of Bermuda Petrel. Collected in Bermuda in March 1967; Proportions of DDT, DDE, and DDD Expressed as Percentages of Total.

Sample	Residues (ppm)	Percentages DDT	DDE	DDD
A, egg[b]	11.02	37[a]	58[a]	5[a]
A, egg[b,d,e]	10.71	34[a]	62[a]	4[a]
B, addled egg[b]	3.61	15	65	20
C, chick in egg[c]	4.52	33	64	3
D, chick in egg[c]	6.08	33	62	5
D, chick brain[c,e]	0.57	30	54	16
E, chick, 1–2 days old	6.97	29[a]	66[a]	5[a]
Av	6.44	31	62	7

SOURCE: Wurster and Wingate (1968).
[a] Identity confirmed by thin-layer chromatography.
[b] Egg showed no sign of development.
[c] Fully developed chick died while hatching.
[d] Analysis 5 months later by Wisconsin Alumni Research Foundation, which also detected dieldrin at 0.02 ppm.
[e] Not included in averages.

tribution of pesticides. However, airborne dust containing concentrations of chlorinated hydrocarbons also dump comparable amounts over the open oceans far from the coastal areas where they were applied (Risebrough *et al.,* 1968).

Organic phosphates are very deadly poisons, but they do not persist and become concentrated as do the chlorinated hydrocarbons. As with other insecticides, the organic phosphates are not specific for insect pests only, but will kill all living organisms indiscriminately. There is also danger to the person using the spray, or the possibility of accidental contact with food that has been recently sprayed. Since these pesticides attack the nervous system, they are closely allied to certain nerve gases developed for use in war.

Overwhelming evidence indicates that the chlorinated hydrocarbons, particularly DDT, have become incorporated into the ecosystem. The problem is further compounded by the developing resistance of insects to pesticides. Yet these compounds are still used widely. Although pesticides are far from obsolete, when it is necessary to use them their use should be carefully governed.

There are, however, other approaches to insect control. One of the oldest is to use a biological control, that is, the introduction of a parasite or predator which specifically attacks an insect pest. A classic example of biological control is found in the cottony cushion scale (*Icerya*) and the lady bird beetle (*Rodolia*). Cottony cushion scale was accidentally introduced into California from Australia, where this insect was kept under control by the lady bird beetle. In California, however, the control beetle was missing and cottony cushion scale spread rapidly, causing widespread damage to citrus trees. The lady bird beetle was subsequently introduced into California, and the scale was rapidly controlled. Since the scales are the essential food supply for the beetles, the beetles have not been able to undergo any prolonged explosive growth and become pests.

Other methods also have been successful in eliminating insect pests. Screw-worm flies, which prey on livestock, for example, have been eliminated by sterilizing adult male flies with gamma radiation and then releasing them to breed fruitlessly with wild females (Knipling, 1960). This method has been used with success on numerous other insect pests. Cytoplasmic incompatibility has also proved successful in controlling certain insects, particularly mosquitoes. Geographically separated populations of mosquitoes are compatible as far as chromosomal genes are concerned, but not in cytoplasmic genes. When different populations are mixed, many dominant lethals are introduced, and within five or six generations, the mosquito population can be decimated (Randal, 1967). Craig and Hickey (1967) have found a strain of mosquitoes that are 90 percent male. Since this is an inherited characteristic transmitted from one generation to the next, females can be eliminated within about two months in the laboratory.

Another promising method involves juvenile hormones (Williams, 1967). All insects secrete juvenile hormones at certain periods of their life cycle, but at other stages the hormone must be absent or the insect would not develop normally. The use of such a hormone has certain inherent advantages: it will not harm other plants or animals, nor is it probable that they can become resistant to it. On the other hand, beneficial insects can be affected as well as pests, but this is not regarded as an insurmountable problem. Such studies indicate that it is possible to find methods of insect control other than shotgun insecticide treatment.

Radioactive Contamination

Most elements have various kinds of atoms, each of which has a slightly different structure, some being radioactive and some not. These elements are referred to as *isotopes*. Those known as radioactive isotopes are unstable and will disintegrate. In the process of disintegration, the nuclei comprising a radioactive element return to a stable, or a more stable, state by throwing off excess energy in the form of alpha, beta, or gamma radiation. As radiation passes through matter, it may drive electrons from their orbit in the atom, leaving the atom with a positive charge. The free electron becoming attached to another atom causes it to be negatively charged. This energy-transfer process is called *ionization*, and is believed to be the chief cause of injury to protoplasm. All life as we know it has evolved in the presence of low levels of ionizing radiation from natural sources, but within the past two decades, living organisms have been subjected to vastly increased amounts of ionizing radiation resulting from the disintegration of artificial radioactive isotopes.

All three ionizing radiations have important biological effects. The alpha and beta radiations are corpuscular; that is, they have a definite mass that consists of atomic or subatomic particles which travel only a short distance through air or water. When they are stopped, their energy is transferred to the object they strike, so that large amounts of local ionization are produced. Alpha and beta particles do not easily penetrate matter. The greatest biological effect results when they are ingested, absorbed, or in some way deposited internally. Gamma radiation, in contrast, travels great distances and penetrates biological material easily without ingestion (Fig. 12-5). Gamma-ray effects depend on the distance from the source, number and energy of the rays, and the density of the mass object.

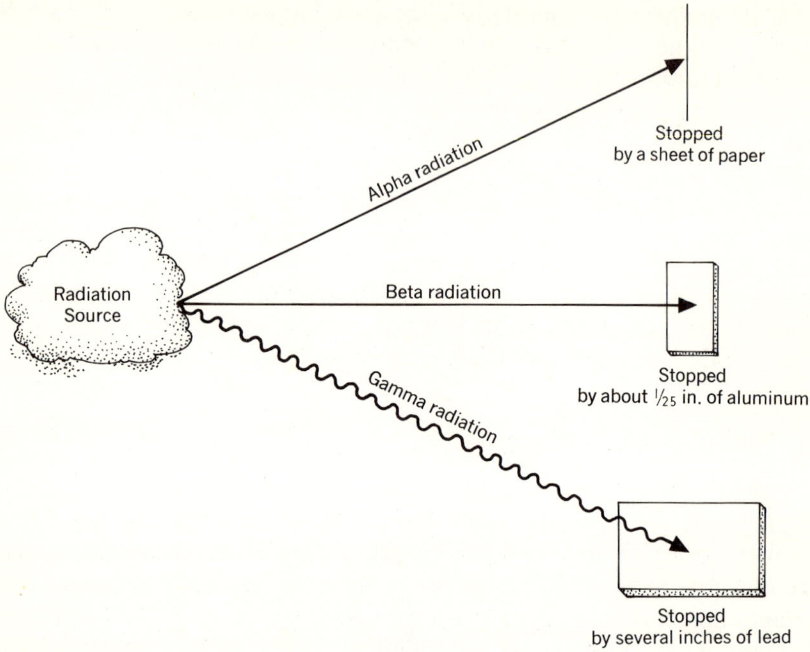

Figure 12-5. Types of ionizing radiations.

Each radioactive isotope has a characteristic rate of disintegration, indicated by its half-life. The half-life, or rate of decay, is constant and unaffected by environmental factors. For example, calcium-45, the isotope of calcium loses half its radioactivity every 160 days. Isotopes vary in rate of decay from a few seconds to many years. Odum (1959) defined three groups that are important ecologically: those which occur naturally and contribute to background radiation, such as uranium-235 and radium-226; those that are metabolically important elements, such as calcium-45, carbon-14, or zinc-65, commonly used as tracers in metabolic studies; and those produced by the fission of heavy elements, such as uranium and thorium. Many fission products can replace vital elements in protoplasm and thus become concentrated in living systems. Unlike chemical poison, the threshold for radiation damage is unknown. The smallest amounts can affect living systems. Therefore, toxic effects of radiation are often expressed as that amount which causes reproductive failure or as the amount necessary to kill 50 percent (LD_{50}) or more of a certain species. Determination of radiation damage is further complicated because some stages in the life cycle are more sensitive to radiation damage than are others. Genetic changes and pathological conditions

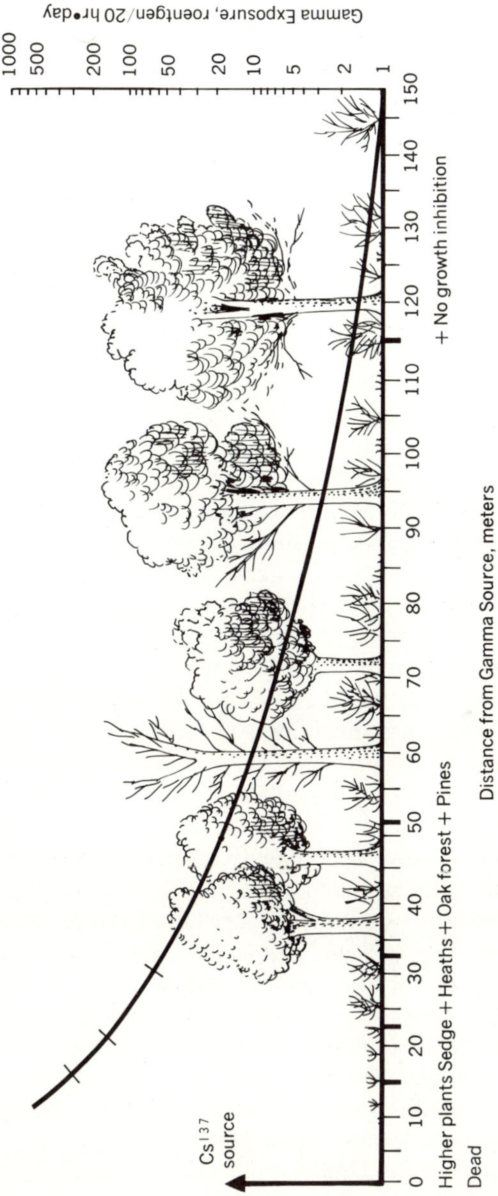

Figure 12-6. Pattern of radiation damage to an oak-pine forest after a six-month exposure.

frequently are not manifested in irradiated organisms for some time after exposure.

Ecologically, one of the greatest hazards of radioactive contamination is its differential effect among living animals. A lethal dosage for one group in an ecosystem may leave another group totally untouched, thus completely changing the structure of a community. A very graphic example of this has been shown by Woodwell (1967). An oak-pine forest was irradiated with cesium-137, one of the long-lived radioactive substances. Successional changes following radiation exposure were then plotted. The pattern of radiation damage to the oak-pine forest after six month's exposure is shown in Fig. 12-6. The most dramatic change was the marked reduction in species diversity along the radiation gradient (Fig. 12-7a). With exposure of only one-fifth (160 roentgens) maximum radiation dosage delivered, diversity was reduced by 50 percent in the herbaceous community. The plants killed were not replaced by equal numbers of radiation-resistant herbaceous plants. Instead, a radiation-resistant species of low-lying herb invaded up to 20 percent of the ground surface. Insect populations changed dramatically (Fig. 12-7b) with a sharp increase in populations of insects that feed on dead organic matter. All higher plants were relatively more sensitive than lower ones. There was also a gradient in relation to size, smaller forms of life being more resistant than larger forms. These successional changes showed an interesting parallel between effects of radiation and successional changes associated with burning over a similar area.

The amount of radiation that this forest was subjected to is many times greater than any present in our environment now. However, with increased industrial usage, present levels could be appreciably raised very quickly. Nuclear industries and nuclear weapon tests have contaminated the environment with the fission products of heavy elements. Some of these fission products, such as iodine-131 and phosphorous-32 are short-lived and gradually are excluded from natural cycles. Others, however, such as strontium-90 and cesium-137, are long-lived products. These isotopes have become incorporated into the cycles of the atmosphere, soil, and water, and persist for long periods of time.

Strontium-90 is considered the most potentially dangerous isotope to man; this isotope is a beta-emitter. It behaves like calcium and it incorporates quickly into biogeochemical cycles. In vertebrates it becomes concentrated in bone tissue, where it can damage both bone and marrow cells, and possibly can cause tumor production. In terrestrial animals, strontium-90 is acquired chiefly through the grazing-food chain, particularly in areas where the rainfall is high and concentrations of calcium and other soil minerals are low, as in the arctic regions, for example. The dom-

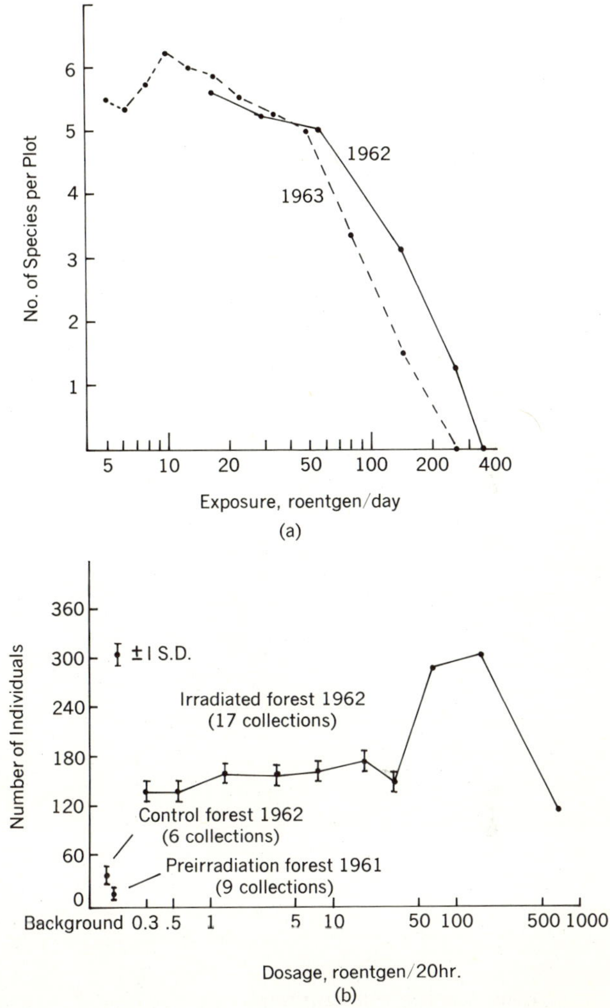

Figure 12-7. Changes in biota along the radiation gradient. (*a*) Reduction in species diversity. (*b*) Increase in populations of insects that feed on dead organic matter. (After Woodwell, 1967.)

inant plants in these areas are lichens, which absorb almost all radioactive isotopes that fall upon them. Radio-contamination moves up the food chain from lichen, to caribou and reindeer, to man. In other parts of the world, the route to man is chiefly via the grass-cow-milk pathway or through ingested plants and vegetables in which strontium-90 is concentrated (Fig. 12-8).

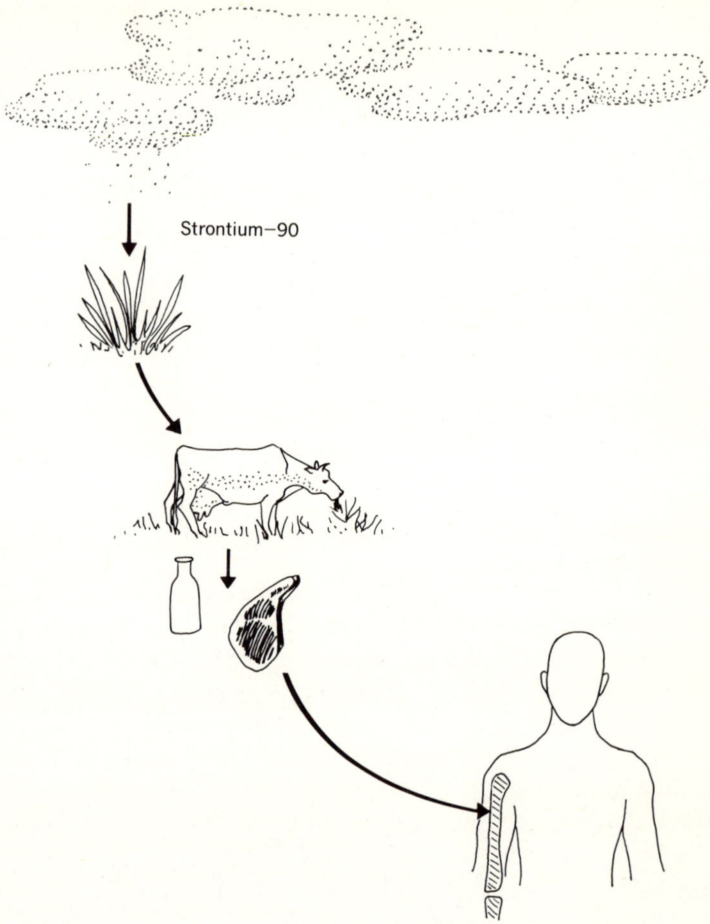

Figure 12-8. One route to man of radio-contamination with strontium-90.

In the marine environment, the greatest concentration of strontium-90, as well as other radioisotopes, is in the top 5 cm of the ocean. Instead of concentrating strontium-90 through a food chain, the marine animals absorb this and other isotopes primarily from the water (Fig. 12-9). Fish are the most sensitive to radiation damage, especially during the egg and larval stages. In laboratory experiments, increased abnormalities in larvae of marine fish were noted when the level of concentration of strontium-90 was at levels already present in the Irish Sea and in the Pacific Ocean (Polikarpov, 1966). Radioactive wastes of British nuclear plants are discharged into the Irish Sea, and the Pacific Ocean has become a nuclear proving ground. Algae are the most resistant of marine organisms. The

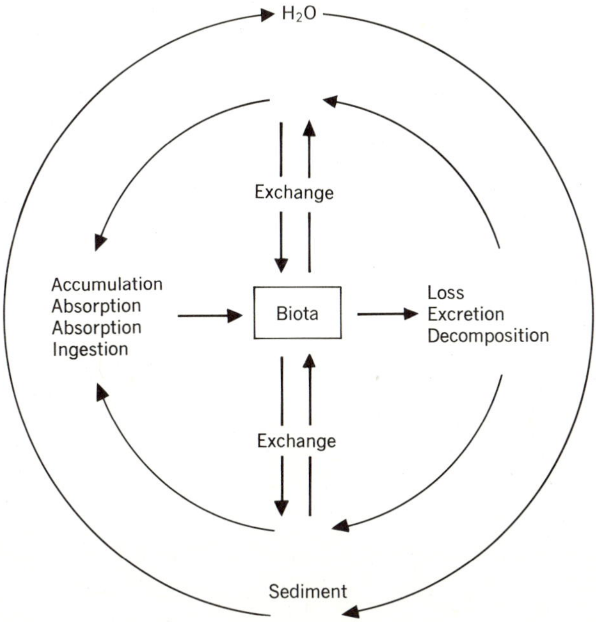

Figure 12-9. Cycling of radionuclides through the marine biota. (From Rice, 1965.)

strontium-90 accumulated during the life of these plants is returned to the environment through detritus when they die; thus it is repeatedly cycled through marine organisms.

Widespread nuclear weapons testing has stopped, but a substantial reservoir of both strontium-90 and cesium-137 remains in the stratosphere as the result of earlier testing. The relative time over which strontium-90 will be deposited can be seen in Fig. 12-10. Although dumping of radioactive wastes into the seas and oceans from nuclear powered vessels is not currently a regular practice, accidents do occur, as in the case of the *Thresher,* for example. Nuclear reactors that produce electricity are potentially a source of contamination as well. There are now about 25 nuclear power reactors operating in the U.S. but between 1970–1975 64 more (all now being built) will raise the nuclear hazards. These are powered by uranium fuel, which must be reprocessed periodically to maintain the chain-reaction level. Strontium-90 and cesium-137, which replace potassium and concentrates in soft tissues such as the gonads, are by-products of the reprocessing. These wastes are stored in tanks that should prevent contamination of the environment for at least 1000 years, although some tanks are already leaking (Cole, 1968).

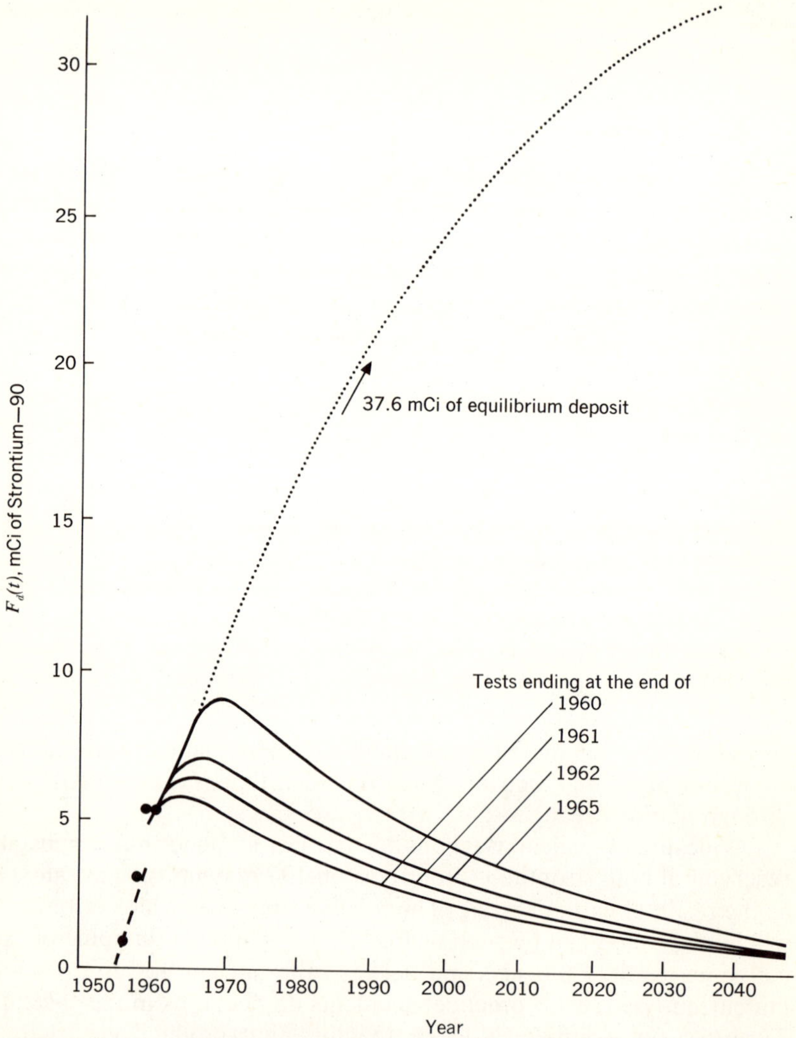

Figure 12-10. Future deposition of strontium-90, data for soil; dotted line denotes 6M Cur Sr-90/5 yr. (United Nations, 1962.)

Unquestionably, radioisotopes have contributed to the welfare of man. Isotopes used as tracers offer an invaluable tool in elucidating metabolic pathways and environmental processes. Radioisotopes are used extensively in medical diagnostic techniques and in treatment of various diseases. In coming years, they undoubtedly will be used more and more as energy sources for power plants and transportation vehicles. The dangers of radioactive contamination of the environment, however, must not be underestimated or overlooked, for they will continue to increase in the years ahead.

Air Pollution

The atmosphere surrounding the earth offers an enormous reservoir of about 5×10^{18} cu meters of air that dilutes gaseous and particulate wastes discharged into it. It is, however, not an unlimited reservoir, and although the natural processes of recycling have considerable capability in handling wastes discharged into the atmosphere, they cannot keep pace with the waste material now being discharged into the air (Fig. 12-11). The result is continuing accumulation of debris in the air.

Most pollutants discharged into the air are the products of man's advanced technological state: gases, particulates, heat, and aerosols. These materials, many of them a consequence of incomplete combustion of fossil fuels, are released into the atmosphere in enormous quantities. For example, one common air pollutant is sulfur dioxide. This gas is gradually broken down into sulfuric acid in cloud droplets or rain, corroding building stones, metals, and other materials. In 1960, 23.3×10^6 tons

Figure 12-11. Air pollution over New York City. (Aero Service.)

were released into the air; by 1980 it is predicted that this figure could reach 36.0×10^6 tons if no controls are instituted (Middleton, 1966). Another example is carbon dioxide. The oceans absorb considerable quantities of carbon dioxide from the atmosphere, and through biological activity it is precipitated as limestone. Today, however, more carbon dioxide is being discharged into the air than the oceans can absorb. It has been estimated that there has been a 15–25 percent increase in atmospheric carbon dioxide during the past 30–40 years (Sargent, 1967). There are two major effects of this increase in air pollutants: modification of the weather, and toxicity of waste materials to living organisms.

Weather modification is particularly noticeable near large concentrations of people. Hickman (1966) calculated that the amount of heat released daily into the atmosphere over urban areas is equal to 0.2 percent of that received by the sun, which is enough in itself to raise temperatures by 0.6°C. In the United States, this energy release is concentrated into three areas: the eastern seaboard, the Chicago-St. Louis central area, and the Los Angeles basin. Since the rocklike structure of the city's buildings and streets can conduct heat faster and store it longer than an equal volume of soil, it is not surprising to find temperatures several degrees higher than in nearby areas. It is not, however, only thermal pollution that can alter air temperatures. When large amounts of carbon dioxide are present in the atmosphere, the temperature of the earth's surface is raised. Carbon dioxide is more transparent to short-wave solar radiation coming to the earth than to long-wave heat radiation going from the earth outward. Thus the outgoing infrared radiation is trapped. This is termed the *greenhouse effect.*

Not all air pollutants raise air temperatures, however, and some may even cause significant cooling. Dusts and aerosols, for example, have caused a significant increase in atmospheric turbidity, both over cities and in nonurban areas (McCormick and Ludwig, 1967), which reflects incoming solar radiation so that solar energy reaching the earth's surface is reduced. Consequently, there is a significant cooling of air temperatures. Moreover, the turbidity also acts to stabilize the atmosphere, resulting in decreased rainfall. Hickman (1966) suggests that drought will be produced in proportion to the concentration of people.

Generally, pollutants are diluted and dispersed in the atmosphere by a combination of vertical temperature profile and wind. The air near the surface of the earth is warmer than that above it, and owing to reduced density, rises rapidly. The cooler air higher up mixes with rising air; this mass is warmed and rises higher yet. In combination with wind, the result is vigorous mixing and dispersion of pollutants. On occasion, however, a meteorological condition occurs which greatly retards dispersal, trapping the air pollutants over large areas; that is, air temperatures increase rather than decrease with height. Thus, air temperatures nearer the surface of the ground are cooler than those above it, and vertical mixing of the air is prevented. Since this temperature inversion inhibits mixing of air, pollutants are not dispersed above a certain level. In the absence of strong winds, the level is gradually lowered closer to the ground, where it remains until the weather pattern changes. These inversions have been associated with all major air-pollution disasters and alerts (Pack, 1964). Typically, this type of weather pattern occurs during "Indian summers," and seldom occurs during other seasons of the year.

Green plants are among the most susceptible of living organisms to toxic influences of air pollutants. In fact, green plants can compare with man-made instruments as sensors of air pollution. For example, both morphological and physiological abnormalities can be observed in lichens growing near industrial centers. In addition, rate and pattern of photosynthesis are altered markedly (Pearson and Skye, 1965). Sensitivity of green plants has serious ecological consequences, for these plants not only convert solar energy into food for use by other living organisms, but also influence the atmospheric carbon dioxide/oxygen balance. Air pollution, coupled with the loss of millions of acres of green plants that are paved under each year, is contributing significantly to the drastic reduction of the rate at which oxygen enters the atmosphere (Cole, 1968).

Effects of air pollution on human health are reflected in the morbidity and mortality curves for chronic respiratory diseases. Death rates are markedly increased during air-pollution disasters, but the subtle and chronic effects of exposure to low levels of air pollution must not be overlooked.

Recent efforts to reduce the amount of atmospheric pollution are concerned with two primary sources of air pollution: emissions from automobiles and sulfur from the incomplete combustion of coal and residual oil. Under federal pressure, new methods to improve combustion efficiency and suppress emissions of hydrocarbons are being developed by automobile manufacturers. Exhaust emitted from a 1960 car contained 3.5 percent of CO and 900 parts of hydrocarbons per million, whereas 1968 cars must meet standards permitting no more than 1.5 percent of CO and 275 parts of hydrocarbons per million (Abelson, 1968). Several methods have been suggested for reducing the amount of sulfur dioxide in the atmosphere. In New York City, for example, there is a limit on the sulfur content of the fuel burned. Another method being tried experimentally is that of recovering the sulfur in elemental form or as H_2SO_4 before it reaches the atmosphere. Since there is a large demand for sulfur and H_2SO_4, this type of control could convert a former pollutant into a useful product (Abelson, 1967).

Man-Made Changes on the Surface of the Earth

The building of dams, canals, and irrigation systems offers another example of the way that man can significantly alter the environment, often with undesirable and unanticipated side effects. The Welland Canal, for example, greatly improves shipping on the St. Lawrence Seaway, but at the same time it has allowed passage of alewives and sea lampreys into the upper Great Lakes, where they have become major problems. Alewives have undergone an explosive population increase since their introduction into the Great Lakes, and they are now the dominant species in Lake Michigan where they feed on eggs and larvae of more desirable fish. In recent years, thousands have washed ashore around Chicago, clogging the city's water filters as well as polluting the beaches. The lampreys have become a serious pest because they attach to healthy fish, killing or weakening them.

Not all changes *per se* are adverse, although it cannot be denied that such changes do have far-reaching ecological consequences. One of the best examples can be found in ecological changes occurring since the Suez Canal was opened in 1869 (Thorson, 1967). This canal connects the tropical Red Sea with the subtropical eastern Mediterranean. The water level at the Red Sea entrance was 26–40 cm higher than the water level at the entrance of the Mediterranean. Except during August, September, and October, the current flowed toward the Mediterranean. During these three months the Nile had its highest outflow, causing the current to run in the opposite direction; as a result, the water mass was stagnant. When the canal was first opened, the biological and chemical compositions of the Red Sea and the Mediterranean were very different. In the tropical Red Sea, the fauna was diverse, and the salinity was 42–43 ‰ at the entrance of the canal. In the subtropical Mediterranean, the diversity in fauna was limited, and the salinity where the canal entered the Mediterranean was 32–39 ‰ except at the highest Nile flood. During high flood, salinity reached 26 ‰. In addition, the area between the Mediterranean and the Red Sea, the Great Bitter Lake, had a salinity of 76–80 ‰. The Great Bitter Lake had a salt layer of 970 million tons, 13.4 meters thick. Thus, with such great salinity differences, it is not surprising that there was little faunal mixing. In fact, 60 years after the opening of the Canal, Steinitz (as quoted in Thorson, 1967) recorded only 15 valid species that had passed from the Red Sea to the Mediterranean and none had passed from the Mediterranean to the Red Sea.

Now, however, conditions have changed radically. At the Red Sea entrance, salinity is still 42–43 ‰. On the other hand, in the Great Bitter

Lake, the salt layer has dissolved, so that salinity is nearly half that present when the canal was first opened; it is now 43–44 ‰. In addition, the Aswan Dam has reduced the Nile flood so that salinity at the Mediterranean entrance is now about 40 ‰. The salinity barrier to the migration of animals, then, essentially has been removed. Furthermore, the canal has been deepened over the years so that the speed of the current has increased. During the past 40 years, eight times more species of animals have gone through the canal and have become established in the Mediterranean than in the first 60 years, and it is anticipated that in coming years there will be a steady and increasing number of Red Sea animals migrating into the Mediterranean. Eventually, the whole fauna and food webs in the Mediterranean will be completely altered. A number of migratory species of fish, shrimp, and pearl oysters already are being exploited commercially in the eastern Mediterranean.

Undoubtedly the proposed sea-level canal in Central America will also greatly alter the ecology of the areas involved. Here, water level differences between the Caribbean and Pacific sides will be great: the Pacific side will average about 6 ft higher, and could be as much as 18 ft higher, since the tides are out of phase. In addition, the waters of the Pacific are much colder than those of the Atlantic: salinity differs, and marine populations are dissimilar. Just what effect this will have on the faunal composition, on the ocean currents, or on the climate is unknown (Cole, 1968).

POPULATION CONTROL

Population regulation is a dire problem in our age. But it was dramatically brought to attention as long ago as 1798 by Thomas Malthus, who predicted that world population would increase faster than food supply:

> Through the animal and vegetable kingdoms, nature has scattered the seeds of life abroad with the most profuse and liberal hand. She has been comparatively sparing in the room and the nourishment necessary to rear them. The germs of existence contained in this spot of earth, with ample food and ample room to expand in, would fill millions of worlds in the course of a few thousand years. Necessity, that imperious all-pervading law of nature restrains them within the prescribed bounds. The race of plants, and the race of animals shrink under this great restrictive law. . . . Among plants and animals its effects are waste of seed, sickness, and premature death.

At the time Malthus made this projection, it seemed unnecessarily pessimistic to most people, but within the past two decades the absolute

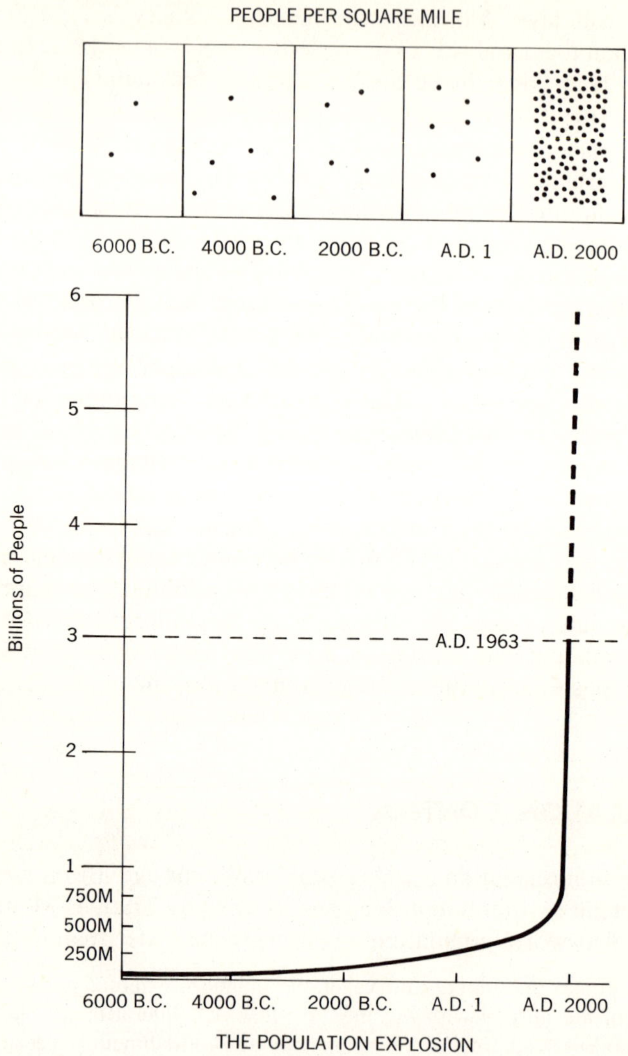

Figure 12-12. Increase in world population through the ages. (United Nations, 1962.)

and relative rates of increase have been so rapid that his projection has become a threatening possibility, unless the trend can be reversed. Figures of population increases can soon put the enormity of population gain into perspective. From the time of Christ until about the middle of the seventeenth century, the world population doubled. During the next 200 years it doubled again, redoubled in less than 100 years, until now it is doubling in some countries about every 20 years (Fig. 12-12). Demographers have moved the population projection figures upward about 20 percent since the middle of the 1950s. The United Nations estimate in 1954 for the population size in 1980 was between 3.3 and 4.0 billion persons; this figure, however, was reached by 1965.

In the earlier history of man, population size was limited primarily by disease and food supply. The last great outbreak of plague occurred about 1350, when the bubonic plague killed more than half the population in many parts of Europe, which is a greater fraction of the extant population than has died since then, even if deaths from the most devastating wars are included. Although many limited outbreaks of diseases have occurred since that time, plagues no longer have a significant influence on population size (Markert, 1966). Therefore disease probably will not limit human population densities in the future. Barring unforeseen disasters, only food supply will place the upper limit on population density.

The complex problems connected with regulation of population size were outlined recently in an analysis of the world food and population problem by the President's Science Advisory Committee (*The World Food Problem,* 1967). Perhaps one of the more significant points of this report is the higher rate of population increase in the less developed countries as compared to that of more developed ones. In the less developed countries, where two-thirds of the world's population resides, the time required to double the population size is from 18 to 27 years, whereas in most developed countries it takes from 55 to 88 years. By the year 2000, there will be four times as many people in these less developed areas as in the more developed ones (Table 12-3). Although most people surveyed in these less developed regions have expressed a desire for smaller families, the number of children desired is still greater than that in more developed countries (President's Science Advisory Committee, *The World Food Problem,* 1967). This difference is based on both social mores and high child mortality rates in underdeveloped countries. Although seemingly contradictory, child mortality rates must be reduced before the birth rate can be appreciably lowered.

It is not impossible, however, to reverse an ever-spiraling birth rate, and already in some countries birth rates have dropped sharply in just the past two decades, particularly in eastern Europe and Japan. In Hungary, there was a 40 percent decrease in the number of births from 1954 to 1962; in Rumania, a 36 percent reduction from 1955 to 1962; and in Japan, a 44 percent reduction between 1947 and 1965. The attitudes of national governments toward birth control have changed significantly during the past few years, and some countries, such as India, have conducted massive campaigns to promote family planning. Still, a great

Table 12-3. Projected Population of World Regions According to the United Nations "Medium" Variant, 1965 and 1985.

	Population (in millions)		*Percent Change*	*Each Area as a Percent of World Total*	
Area	*1965*	*1985*	*1965–1985*	*1965*	*1985*
World	3281	4746	45	100	100
More developed[a]	1032	1256	22	31	26
Less developed[b]	2249	3490	55	69	74
Africa	306	513	68	9	11
Asia	1828	2701	48	56	57
Latin America	245	436	78	7	9
Europe	440	492	12	13	10
North America	213	283	33	7	6
Oceania	17	25	44	1	1
U.S.S.R.	231	297	28	7	6

SOURCE: President's Science Advisory Committee, *The World Food Problem,* 1967.
[a] Includes Europe, the U.S.S.R., North America, Japan, Argentina, Chile, Falkland Islands, Paraguay, Uruguay, Australia, and New Zealand.
[b] Includes Asia (except Japan), Africa, Latin America (except Argentina, Chile, Falkland Islands, Paraguay, and Uruguay), Oceania (except Australia and New Zealand).

decrease in fecundity over the next 20 years may not solve the problem of overpopulation, for even a 30 percent decrease in fertility over the next 20 years would not prevent a population increase as high as 40 percent above the 1965 figures, or about 4.65 billion more people. In subsequent years, however, the effects of population control would be noticeable (Fig. 12-13). The high projection figure for the year 2000 (that is, without a decrease in fecundity) is 7.15 billion; the low projection figure is 6 billion. Thus, in the next 20 years, a reduction in population increase accompanied by increased food production would seem to be the key solution to the world population problem.

The less developed countries—where food supply is inadequate and where the largest population increases will occur over the next few years—have common characteristics other than too many people and too little food. On one hand, almost all these nations are in the tropics. Furthermore, although these countries actually occupy the same amount of land area as the more developed ones, they contain more than twice the number of people. In southeast Asia, middle south Asia, and mainland

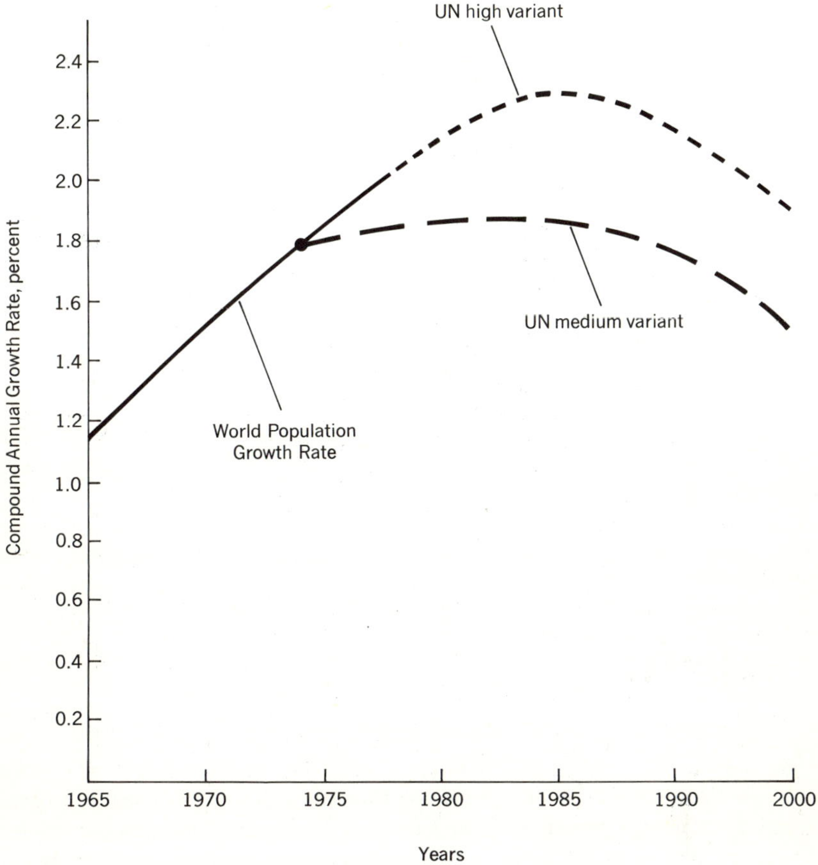

Figure 12-13. Effects of population control on population size up to the year 2000. (United Nations, 1962.)

China, there are about 1–3 acres per person. It might be pointed out, however, that the densities are not so high as those in Japan and western Europe and not all underdeveloped countries have these dense populations. In Africa and Latin America, for example, frequently there are 12–25 acres per person in areas where the land is desert, barren savannah, or tropical rain forest, all of which are quite unsuited for cultivation of crops.

In many underdeveloped countries, the physical resources for increased food production are undeniably present. Only in isolated situations, however, have these potentials been realized. In the report of the

President's Science Advisory Committee (*The World Food Problem,* 1967), it is pointed out that two main deficiencies must be overcome before significant improvements in food production can be expected in these areas: a general lack of knowledge about tropical soil and crop management systems, and a deficiency of economic and technological inputs, such as fertilizers. Too often the more developed nations have assumed that technology exported from their countries can solve every economic problem. As a result, when unmodified temperate-zone techniques have been transported to the tropical areas, notable failures have resulted. Many areas, for example, should not be clean-cleared and plowed. In steeply sloping areas, erosion will be a problem of great consequence, while in other areas the soil that is underlain with a doughy laterite will become irreversibly hardened when exposed to rain and sun. Large areas can be salvaged from forests if the existing vegetation is gradually thinned and replaced with food and industrial crops rather than being clean-cleared. Only when the proper techniques are devised can the potentials of a region be realized.

SUMMARY

Man has always modified the earth's environment, and until recently the earth has been amazingly resilient. Now, however, man's modifications have become so gross that the total environment is threatened. Even many physical factors in the environment that have remained unchanged for millions of years—such as temperature, oxygen, and carbon dioxide—are being modified through misuse of the environment.

Unfortunately, many abuses cannot be easily and quickly rectified. It has been estimated, for example, that if all use of DDT were suspended immediately, it would be at least 10 years before the biosphere would be relatively free of this chemical, and DDT is less toxic and no more persistent than some of the newer pesticides now in use. Persistent pesticides are only one of the long-term environmental problems that face us. Water pollution, radioactive contamination, an exploding world population, all need immediate attention.

Man's future survival on this planet may well depend on his ability to recognize that man himself is an integral part of the physical and biological environment that surrounds him. Failure to do so invites certain and irreversible disaster.

References

Abelson, P. H. 1967. Diminishing the role of sulfur oxides in air pollution. *Science,* 157:1265.

Abelson, P. H. 1968. Progress toward abatement of air pollution. *Science,* 160:257.

Carbine, W. F. 1962. Statement of W. F. Carbine. In *Joint Federal-State of Michigan Conference on Pollution of Navigable Waters of the Detroit River, Lake Erie, and Their Tributaries within the State of Michigan, March 27–28, 1962,* Vol. I, II. U.S. Public Health Serv., Washington, D.C.

Carson, R. 1962. *Silent Spring.* Houghton Mifflin, Boston.

Cole, L. C. 1968. Can the world be saved? *BioScience,* 18:679–684.

Cootner, P. H., and G. O. G. Löf. 1965. *Water Demands for Steam Electric Generation.* Resource for the Future, Inc., Washington, D.C.

Corner, E. D. S., A. J. Southward, and E. C. Southward. 1968. Toxicity of oil-spill removers ("detergents") to marine life: an assessment using the intertidal barnacle *Elminius modestus. J. Mar. Biol. Assoc., U.K.,* 48:29–47.

Craig, G. B., Jr., and W. A. Hickey. 1967. Genetics of *Aedes aegypti. In* J. W. Wright and R. Palmer (eds.), *Genetics of Insect Vectors of Disease,* pp. 67–131. Elsevier, New York.

Egler, F. E. 1964. Pesticides—in our ecosystem. *Am. Sci.,* 52:110–136.

Elson, P. F. 1967. Effects on wild young salmon of spraying DDT over New Brunswick forests. *J. Fish. Res. Bd. Canada,* 24:731–767.

Henderson, C., Q. H. Pickering, and J. M. Cohen. 1959. The toxicity of synthetic detergents and soaps to fish. *Sewage Ind. Wastes,* 31:295–306.

Herrin, E. 1967. The crust and continents. *In* H. Odishaw (ed.), *The Earth in Space,* pp. 68–76. Basic Books, New York.

Hickman, K. 1966. Oases for the future. *Science,* 154:612–617.

Hunt, G. S. 1965. The direct effects on some plants and animals of pollution in the Great Lakes. *BioScience,* 15:181–186.

Klein, L. 1957. *Aspects of River Pollution.* Academic Press, New York.

Knipling, E. F. 1960. Control of screw worm fly by atomic radiation. *A.A.A.S. Publ.,* 61:169–182.

Koeman, J. H., R. C. H. M. Oudejans, and E. A. Huisman. 1967. Danger of chlorinated hydrocarbon insecticides in birds' eggs. *Nature,* 215:1094–1096.

Malthus, T. R. 1798. *An Essay on the Principle of Population.* Johnson, London. [Reprinted in Everyman's Library, 1914.]

Markert, C. L. 1966. Biological limits on population growth. *BioScience,* 16:859–862.

Mather, K. F. 1964. *The Earth Beneath Us.* Random House, New York.

McCormick, R. A., and J. H. Ludwig, 1967. Climate modifications by atmospheric aerosols. *Science,* 156:1358–1359.

Middleton, J. R. 1966. Control of environment—economic and technological. In *Environmental Control,* pp. 53–71. U.S.D.A., Washington, D.C.

Mihursky, J. A. 1967. On possible constructive uses of thermal additions to estuaries. *BioScience,* 17:698–702.

Odum, E. P. 1959. *Fundamentals of Ecology,* 2d ed. Saunders, Philadelphia.

Ogilvie, D. M., and J. M. Anderson. 1965. Effect of DDT on temperature selection by young Atlantic salmon, *Salmo salar. J. Fish. Res. Bd. Canada,* 22:503–512.

Pack, D. H. 1964. Meteorology of air pollution. *Science,* 146:1119–1128.

Pearson, L., and E. Skye. 1965. Air pollution affects patterns of photosynthesis in *Parmelia sulcata* a corticolous lichen. *Science,* 148:1600–1602.

Polikarpov, G. G. 1966. *Radioecology of Aquatic Organisms.* Reinhold, New York.

President's Science Advisory Committee. 1967. *The World Food Problem,* Vol. II. U.S. Govt. Printing Office, Washington, D.C.

Randal, J. 1967. Insect control. *BioScience,* 17:551.

Rice, T. R. 1965. The role of plants and animals in the cycling of radionuclides in the marine environment. *Health Physics,* 11:953–964.

Risebrough, R. W., R. J. Huggett, J. J. Griffin, and E. D. Goldberg. 1968. Pesticides: transatlantic movements in the northeast Trades. *Science,* 159:1233–1236.

Sargent, F., II. 1967. Adaptive strategy for air pollution. *BioScience,* 17:691–696.

Singer, S. F. 1968. Waste heat management. *Science,* 159:1184.

Terry, L. L. 1966. Environmental health. Everybody's Business. *In* B. H. Jennings and J. E. Murphy (eds.), *Interactions of Man and His Environment,* pp. 1–7. Proc. Northwestern Univ. Conf., Jan. 28–29, 1965.

Thorson, G. 1967. Recent migration and larval transport via the Suez Canal (Preprint Abstr.). *Conference on Marine Invertebrate Larvae with Emphasis on Rearing.* Duke University Marine Lab., Beaufort, N.C., Sept. 2–10, 1967.

United Nations. 1962. *Report of the U.N. Science Committee on the Effects of Radiation.* General Assembly, 17th Session, Suppl. No. 16(A/5216), New York.

Williams, C. M. 1967. Third generation pesticides. *Sci. Am.,* 217:13–17.

Woodwell, G. M. 1967. Radiation and the patterns of nature. *Science,* 156:461–470.

Woodwell, G. M., C. F. Wurster, Jr., and P. A. Isaaccson. 1967. DDT residues in an east coast estuary: a case of biological concentration of a persistent insecticide. *Science,* 156:821–824.

Wurster, C. F., Jr., and B. Wingate. 1968. DDT residues and declining reproduction in the Bermuda petrel. *Science,* 159:979–981.

Wurster, C. F., Jr., D. H. Wurster, and W. N. Strickland. 1965. Bird mortality after spraying for Dutch elm disease with DDT. *Science,* 148:90–91.

Index